DATE DUE

JUL 17 1996			
GAYLORD			PRINTED IN U.S.A.

The Diffident Naturalist

Science and Its Conceptual Foundations
David L. Hull, Editor

THE DIFFIDENT NATURALIST
Robert Boyle and the Philosophy of Experiment

Rose-Mary Sargent

THE UNIVERSITY OF CHICAGO PRESS
Chicago & London

ROSE-MARY SARGENT earned her doctorate in philosophy at the University of Notre Dame. Now assistant professor in the Department of Philosophy at Merrimack College, she formerly taught at the University of New Mexico and has held postdoctoral fellowships at Northwestern University and the University of Minnesota.

The University of Chicago Press, Chicago 60637
The University of Chicago Press, Ltd., London
© 1995 by The University of Chicago
All rights reserved. Published 1995
Printed in the United States of America
04 03 02 01 00 99 98 97 96 95   1 2 3 4 5

ISBN (cloth): 0-226-73495-1
ISBN (paper): 0-226-73497-8

Library of Congress Cataloging-in-Publication Data

Sargent, Rose-Mary.
The diffident naturalist : Robert Boyle and the philosophy of experiment / Rose-Mary Sargent.
p.   cm.—(Science and its conceptual foundations)
Includes bibliographical references and index.
1. Boyle, Robert, 1627–1691.   2. Scientists—Great Britain—Biography.   I. Title.   II. Series.
Q143.B77S27   1995
530'.092—dc20   94-19205
[B]   CIP

To Ernan

Contents

Acknowledgments		ix
Introduction		1
	Overview	2
	Recent Trends	6
	Philosophical Issues	11
	Plan of the Study	14

PART I. LEARNING FROM THE PAST

1	The Philosophical Tradition	23
	A Philosophical Revolution	25
	A New Alternative	27
	The Pyramid and the Tree	31
	Boyle's Choice	35
2	The Legal Tradition	42
	English Common Law	44
	Experience and the Experimental Philosophy	50
	Boyle versus Hobbes	56
3	The Experimental Tradition	62
	The Physico-Mechanical Tradition	64
	The Alchemical Tradition	70
	The Medical Tradition	75

PART II. BEING A CHRISTIAN VIRTUOSO

4	Natural Theology	87
	The Book of Nature and Philosophical Worship	89
	A Free Inquiry	93
	The Corpuscular Philosophy and Physical Causality	98
	Causal Relations and the Essences of Bodies	103
5	Biblical Hermeneutics	109
	The Two Books	112
	The Interpretation of Scripture	115
	The Interpretation of Nature	122

Part III. Acting Experimentally

6   Observing   131
      Constructing the Factual Foundation   131
      Collecting Observations   138
      Assessing Credibility   145

7   Experimenting   159
      Creating an Artificial Environment   159
      The Contingencies of Experiment   165
      Making Experiments   170
      Experimental Strategies   176

8   Writing   181
      Composing Experimental Essays   183
      Exciting Curiosity   186
      Collaborating   189
      An Experimental History of Cold   193

Conclusion: The Experimental Process   205
      Boyle's Philosophy of Experiment   207
      The Significance of Boyle's Philosophy   212

Notes   217
Bibliography   315
Index   337

ACKNOWLEDGMENTS

At various stages in the production of this work I have benefited from the advice, encouragement, and support of numerous and diverse philosophers and historians. My earliest intellectual debt is owed to Dan Rochowiak, whose enthusiasm for history and philosophy of science was contagious. He guided me through an extensive and rigorous study of some of the best literature in the discipline during the course of which I first became acquainted with the work of Ernan McMullin, which in turn influenced my decision to continue my studies at the University of Notre Dame.

I would like to thank the Mellon Fellowships in the Humanities for funding my graduate education and my many teachers and fellow students at Notre Dame—especially Michael Crowe, Jim Cushing, Gary Gutting, Tim Shanahan, Craig Stillwell, and Mary Thomas—who patiently listened to my lengthy disquisitions on the virtues of Robert Boyle. In addition, I am indebted to the three official readers of my dissertation—Ed Manier, Phil Quinn, and Phil Sloan—for helping me to see the value of their various perspectives on history and philosophy, as well as to Steve Watson, an unofficial reader, who added a further dimension to my work by introducing me to historical and contemporary literature in hermeneutics.

Above all, Ernan McMullin deserves, and has, my deepest gratitude. I was honored when he agreed to direct my dissertation efforts, was simply astounded by the conscientiousness with which he fulfilled his duties, and am pleased that he has continued to take an active interest in my work. He first suggested that I revise my dissertation for publication. During the years that followed, as the scope expanded and the length of this work nearly tripled, he has read and critically commented upon my numerous reformulations. I do not think that I would have taken on such a complex project if it had not been for his constant encouragement and unwavering confidence in me. His wit and personal charm have made knowing and working with him a pleasure. In addition, his critical spirit, scholarly integrity, and synthetic vision have been a

ix

constant inspiration to me, although I am keenly aware of how far my efforts have fallen short of his example.

My self-induced academic wanderings after leaving Notre Dame have brought me into contact with many more influences. From 1987 to 1989 I benefited from discussions with my colleagues in philosophy at the University of New Mexico, especially Brom Anderson, Russell Goodman, Kevin Lavelle, and Fred Schueler. This experience was followed by a year as a postdoctoral fellow at Northwestern University. During that time I began work on this book in earnest, and I gained much from my association with Betty Jo Dobbs, Arthur Fine, and David Hull, all of whom encouraged my work and provided helpful advice. I have been equally fortunate in my current position at Merrimack College, where I have had the support of my colleagues in the philosophy department—Art Ledoux, Jerry Matross, Herb Meyer, and John Warren. I would also like to thank my colleagues in other departments, particularly Jack Amariglio, Tom Casey, Al DeCiccio, and Peter Ford, with whom I have taught interdisciplinary courses on early modern and postmodern thought. Peter Ford also deserves special thanks for reading large sections of the manuscript and giving me many helpful suggestions.

The final version of this book was completed during the 1993–94 year while I was a research associate in the program for Studies of Science and Technology at the University of Minnesota. I am grateful for the leisure that this appointment gave me to finish my work. I would like to thank Alan Shapiro for his sound advice on numerous occasions, and Ron Giere for reading and commenting on portions of the manuscript. I also learned much from informal conversations with John Beatty, Helen Longino, Jeff Ramsey, Roger Stuewer, and Ken Waters. Susan Abrams at the University of Chicago Press encouraged my efforts early on, and her advice during this final year has been invaluable. I thank also the anonymous referees of my manuscript for their many constructive suggestions, as well as David Hull for his insightful criticisms concerning the overall presentation of this book.

Many Boyle scholars, past and present, have influenced my work, and my debts to them should become clear throughout the following. I believe that Michael Hunter deserves special mention, however. I would like to thank him not only for the advice and criticism that he has given to me personally, but also for the way in which his contributions have elevated the state of Boyle studies generally. In 1991 he organized a symposium at Dorest to commemorate the three-hundredth anniversary of Boyle's death, in part because he was aware of how an exchange of ideas among researchers working on different aspects of Boyle's thought could help to shed light on its sophistication and complexity. Not only has Michael brought Boyle scholars together; he has also been instrumental in providing them with the tools that will further their research. In the 1980s he took on the task—which others before him had said was impossible—of compiling a catalogue of Boyle's unpublished papers and letters; it has since appeared as a guide for the microfilm collection of this material. In addition, he is presently working with Ted Davis on a new edition of Boyle's works that will finally correct the numerous infelicities present in the extant editions.

My research on Boyle's unpublished manuscripts was conducted at the Royal Society of London. I thank the president and board of the society for permission to use this material, as well as Sheila Edwards, Alan Clark, and Keith Moore for the valuable assistance they provided while I conducted research. I am also grateful to the National Science Foundation for providing me with a grant for my study at the Royal Society, as well as a later grant for the completion of the manuscript. I thank the referees of my NSF proposals for their helpful comments, and Ronald Overmann, director of the foundation's Program in History and Philosophy of Science, for his encouragement and advice.

Some of the ideas in this book first found expression in work that has appeared elsewhere. Two early papers, "Robert Boyle's Baconian Inheritance" and "Scientific Experiment and Legal Expertise," published in *Studies in History and Philosophy of Science*, provided the basis for expanded treatments of these subjects in chapters 1 and 2. In turn, portions of parts 1 and 3 of this book provided the basis for a chapter published in *Robert Boyle Reconsidered*, edited by Michael Hunter (Cambridge University Press, 1994), entitled "Learning from Experience: Boyle's Construction of an Experimental Philosophy." I have also presented papers based upon chapters from this book at a number of institutions including Boston University, the University of Colorado at Boulder, the University of Connecticut, Illinois Institute of Technology, the University of Minnesota, Northwestern University, Virginia Polytechnic Institute and State University, and Wesleyan University. I would like to thank all who commented upon these talks. In addition to those already mentioned, I especially thank Richard Burian, Bob Cohen, Allan Franklin, Steve Horst, Joe Pitt, Joseph Rouse, Sahotra Sarkar, Warren Schmaus, and Abner Shimony.

INTRODUCTION

ROBERT BOYLE WAS NOT a scientific genius. Strictly speaking, he was not a scientist at all. Rather, he was a natural philosopher who devoted his life to developing the details of a new way of knowing that he called the experimental philosophy. Experimental practices were not new, of course. Experiments had been performed in the "low sciences" of medicine and alchemy for many years, and by Boyle's time they had been introduced into the mechanical and mathematical disciplines by Galileo, Pierre Gassendi, and Marin Mersenne, among others. What was new was the way in which Boyle sought to give experimental practices a rational foundation—to construct a "comprehensive method" that would lead to knowledge in all areas of human concern.[1]

Numerous studies have been produced in this century concerning Boyle's work, but there remains a marked lack of consensus among historians and philosophers about who he was and what he did. In part, such conflicting interpretations could be viewed as a result of the failure to treat Boyle as a philosopher and to see his works in their entirety. His publications covered a wide range of topics, including chemical, mechanical, and physiological investigations; theoretical speculations concerning the usefulness and justification of the corpuscular philosophy; and defenses of natural theology and the Christian religion. At first sight, such eclecticism may appear to be the result of an unrestrained or unreflective curiosity. Yet when these individual researches are taken together, it becomes clear that they were designed by him to contribute to one complex and coherent philosophical project.

My purpose in this work is to provide a primarily philosophical explication of Boyle's experimental program by examining how he drew upon the work of numerous predecessors in its construction, how his religious ideals contributed to his notion of what it meant to live the life of a natural philosopher, and how he developed a sophisticated account of scientific discovery and justification by immersing himself in experimental practice. First, however, a brief, and necessarily selective, review of the secondary literature on Boyle will be helpful in order to avoid confusion concerning the scope of the present study. My intention is not to engage in theoretical debate about the merits of particular approaches to the study of past science but to lay out the terrain and locate my work within it so that the reader will not have false expectations concerning what is and what is not included here.

Overview

Aside from biographical accounts, there have been two dominant types of works on Boyle: "intellectualist" studies that have, for the most part, sought to provide rational reconstructions of Boyle's thought by referring to philosophical and scientific categories; and "contextualist" studies that have sought to explain Boyle's beliefs by an appeal to concepts developed in the field of the sociology of scientific knowledge.[2] Within the first genre, by far the most influential studies were those produced by Marie Boas Hall during the 1950s and 1960s. Hall used Boyle's published and unpublished works in order to provide comprehensive accounts of his mechanical and chemical theories.[3] Other scholars, such as Peter Alexander, James B. Conant, Norma E. Emerton, Thomas S. Kuhn, James G. Lennox, and Richard S. Westfall, also produced studies designed to exhibit the logical relations that held between Boyle's theoretical beliefs and the conceptual and empirical evidence that he brought forward in support of them.[4] Because these scholars primarily intended their works as examinations and explanations of Boyle's theoretical beliefs, they did little analysis of his methodology, although Alexander, Hall, and Westfall all noted that Boyle did not appear to follow a strictly empirical approach in his justification of knowledge claims about natural processes.[5]

One reason for questioning a straightforward empiricist interpretation of Boyle's methodology derives from the fact that he was an advocate of the corpuscular philosophy, which indicated, among other things, that he believed that physical processes could be explained by the deterministic causal action of the least parts of matter. Empiricists, on the other hand, normally maintain that human knowledge is limited to the discovery of descriptive accounts of observable regularities.[6] In the 1930s Philip P. Wiener offered a "pragmatic" interpretation of Boyle's epistemology in order to resolve the apparent tension between his methodology and his ontology.[7] Later works on Boyle's methodology tended to take the form of predecessor studies.[8] In the 1960s Larry Laudan, for example, noted the similarities between the corpuscular philosophies of Boyle and Descartes and argued that Boyle therefore followed a type of Cartesian "method of hypothesis."[9] In response to Laudan, G. A. J. Rogers argued for a stronger Baconian influence on Boyle's methodology, although he noted that a strict empiricist interpretation of either philosopher was not warranted.[10]

Beginning in the 1960s and continuing today, another group of scholars, including Gary B. Deason, Eugene M. Klaaren, Barbara J. Shapiro, and Henry

G. van Leeuwen, has studied Boyle's theological beliefs in an attempt to reex-
amine the content of his ontological ideas, which in turn could possibly be
used to resolve the apparent tension between his ontology and his epistemology.[11] Although theology was taken to be "external," and thus irrelevant, to
scientific concerns in the earlier positivist tradition, these works showed how,
in the absence of an explicitly stated metaphysical position, Boyle's ontology
could be constructed from an examination of his religious writings. Despite the
existence of alternative interpretations of Boyle's epistemology and the doubts
expressed about his empiricism, work in this area has led to the widespread
acceptance of an empiricist interpretation of Boyle's epistemology.[12] One of
the most influential studies was that produced in 1972 by J. E. McGuire.

McGuire cited passages in which Boyle spoke of the passivity of matter
and the unlimited power of God, to argue that for Boyle the universe was composed of a set of nonrelated particulars, which in turn indicated that what was
then the "received view" about Boyle's commitment to a mechanistic universe
was incorrect. According to McGuire, Boyle believed that there was no true
physical causality in the world. All natural phenomena are produced by the
immediate activity of a God who is able to change the order of things at any
time, and thus knowledge about the world must be limited to the discovery of
phenomenal laws about manifest occurrences.[13] While such an interpretation
of Boyle's ontology is consistent with passages in which he appeared to advocate an empiricist methodology, it did not completely resolve the tension, because it left unexplained those other passages where Boyle spoke of the predictive power of hypotheses and the "discovery of the true genuine causes"
operative in nature.[14] McGuire's work did show, however, that in order to understand Boyle's method, it would be necessary to ascertain his general ontological beliefs about the nature of the world as the object of study for which
he had devised his method.

In the 1970s scholars committed to a more contextualist approach to the
history of science began to express opposition toward intellectualist histories.
In particular, traditional historians of ideas were criticized for their reliance
upon an "internalist" understanding of past science. It was said, for example,
that their rational reconstructions were based almost solely upon textual analysis and that their decisions concerning the selection of relevant factors that
could be used for the explanation of past scientific episodes were based upon
twentieth-century philosophical categories that were historically inappropriate. Unlike earlier Mertonian studies dating from the 1940s and 1950s, in
which social categories were employed only for the explanation of "irrational"
elements in past science, contextualist studies were designed in order to produce a more "symmetrical" treatment by which past successes as well as failures
were to be explained via categories of interest as developed in the sociology
of scientific knowledge (SSK).[15]

David Bloor, a leading exponent of the "strong programme" in SSK, has maintained that because theoretical beliefs are underdetermined by empirical evidence, such evidence cannot adequately explain actors' beliefs, and the student of past science must therefore look to social interests in order to achieve a complete explanation for actors' decisions.[16] He believes, for example, that "it is necessary to look at the social context" in order to explain Boyle's preference "for an inert and passive rather than an active and self-moving matter."[17] Bloor's position has found support in the work of contextualist historians such as Christopher Hill, James R. Jacob, Margaret C. Jacob, and Charles Webster.[18] According to the Jacobs' influential 1980 study, for example, Boyle's matter theory was "designed to combat two threats, heresy and social insubordination," and his advocacy of corpuscularianism can thus be explained by reference to the function that it played in Restoration society. Building upon McGuire's analysis, they argue that because Boyle's corpuscular philosophy postulated the passivity of matter and the necessity of God's governance of the world, it could be used to support the conservation of the social order that followed from a dominant Anglican Church. While monistic natural philosophies, such as Hobbesian materialism, "tended to dissolve hierarchy, . . . hierarchical social order found support in the Christian dualism newly shored up by the corpuscular philosophy of the reforming Puritans."[19] As Bloor sees it, the corpuscular philosophy thus gained acceptance because in it "the world was made to prefigure the dependence of civil society on an involved, active and dominant Anglican church," and thus the view could be used "to bolster up the social and political policies" of Restoration England.[20]

In this type of contextualist study, the focus tended to be restricted to an analysis of how the sociopolitical interests of involved scientists influenced their acceptance of particular theoretical constructs concerning the constitution of the natural world. A newer trend, however, has emerged in the past two decades that is directed more toward the examination of scientific practice as distinct from the theoretical products of that practice. H. M. Collins, David Gooding, Bruno Latour, Michael Lynch, Andrew Pickering, and Steve Woolgar, for example, have attempted to understand how the evidence that is used for the justification of knowledge claims is itself constituted by social conditions surrounding the practical life of the laboratory.[21] In Boyle studies, this newer focus can be seen in Steven Shapin and Simon Schaffer's 1985 study, *Leviathan and the Air-Pump.* Shapin and Schaffer sought to investigate the means by which various boundaries, such as that between the natural and the social worlds, came to be established in the seventeenth century and to explore how experimental procedures at the same time came to be accepted as reliable indicators of factual claims concerning the natural world. Their analysis moved the discussion of Boyle's experimentalism to another level of sophistication, as they recorded in minute detail, for example, the numerous problems faced by Boyle and his contemporaries in their attempts to replicate intricate pneumatic experiments.

Although Shapin and Schaffer's focus follows the newer trend, their work is not as radical as that by Latour, Lynch, or Woolgar. As I will discuss more fully below, these writers have questioned the explanatory categories employed by traditional SSK theorists such as Bloor and Collins. Shapin and Schaffer, on the other hand, made extensive use of such sociological categories in their analysis of Boyle's experimental practice.[22] They showed quite well, for example, that the boundaries set up by Boyle and his contemporaries were not self-evident and that other methods and boundaries were equally acceptable, but they then went on to explain why Boyle's program proved successful by referring to the way in which the practices embedded within it for "the generation and justification of proper knowledge were part of the settlement and protection of a certain kind of social order."[23] They argued that the experimental program itself, by setting up boundaries between the factual and the theoretical and admitting the testimony of only those witnesses deemed credible by the closed community of experimenters, carried an important social message, by which the acceptance of experimental practice can be explained: "The general form of an answer to the question of Boyle's 'success' begins to emerge, and it takes a satisfyingly historical form. The experimental form of life achieved local success to the extent that the Restoration settlement was secured. Indeed, it was one of the important elements in that security."[24]

The numerous and often acrimonious debates that have ensued between intellectualist and contextualist historians of science are in large part the result of the fact that both parties are engaged in the same type of explanatory project. Both approaches seek to explain the reasons for actors' beliefs, and the major difference between them resides in the choice of which factors will be deemed explanatory. The intellectualist historian tends to find an appeal to the logical relationship that obtains between evidence and theory as sufficient to explain the acceptance of particular knowledge claims, whereas the contextualist finds such an appeal whiggish or naive and insists instead upon locating the social and cultural factors that caused actors to view the theoretical beliefs as adequately supported by the evidence.[25] In this sense, both approaches advocate a global understanding of science. Although contextualists have criticized intellectualist historians for their use of "cherished philosophical norms" through which they attempt to explain past scientific episodes by subsuming them under an idealized understanding of scientific practice, SSK theorists have merely replaced such norms with their own equally global explanatory categories, particularly with what Pickering has called the "distinctive sociological concept of interest."[26]

Recent Trends

In the 1990s a number of sociologists have become more vocal in their criticism of the explanatory project of traditional SSK studies. Lynch, for example, has advocated an ethnomethodological approach to the study of science, wherein his main design is to "describe the ensemble of actions that constitute a practice," and has rejected Bloor's contention that such an approach reduces to a type of "anticausal irrationalism."[27] In a different but related criticism, Latour has argued that the type of "social realism" advocated by Bloor and Collins is unwarranted because social concepts cannot be used as unproblematic explanatory resources. As Latour sees it, the global concept of interest embedded within traditional SSK explanatory accounts of science is actually a remnant of "modernism" that serves to perpetuate the positivist distinction between the natural and the social worlds.[28] For this reason, while he generally approves of the way in which Shapin and Schaffer displayed the historical origin of such a distinction in *Leviathan and the Air-Pump*, he remains critical of their work because they failed to go beyond the traditional explanatory project of SSK and thus tended to reinforce the very distinction that they had sought to deconstruct.[29]

While critics such as Latour, Lynch, and Woolgar have different motivations for their opposition to traditional SSK studies, one thing that they share is their perception of the increasingly conservative and dogmatic attitude expressed by their opponents. Contextualist historians in the SSK tradition tend to be satisfied with the present state of science studies and are dismissive toward those studies that are not firmly within their camp.[30] Philosophers and philosophically-minded historians are not likely to agree with the extreme forms of relativism championed by these radical critics of SSK, yet a similar reaction against the entrenchment of social contextualism can be seen in some newer works in the history and philosophy of science. In particular, contextualists are being criticized for their failure to provide any new insights into the intellectual dimension of past science. As Shapin noted early on, the contextualist project is designed to build upon intellectualist history: "The demonstrated connections between one set of ideas and another are the necessary starting points for historians who would put an additional set of contextualist questions to the materials."[31] Such a description of this project seems to betray an assumption that the history of ideas is in some sense complete. As the contextualists have repeatedly pointed out, however, the accuracy of the earlier intellectualist histories may be questioned because of the way in which logicist and positivist conceptions of science were used in their construction. A fresh examination of the intellectual content of past science therefore seems to be required, and such attempts should not be dismissed simply on the grounds that they are not in line with the current fashionable status of contextualist studies.

These criticisms should not be confused with earlier responses toward SSK that involved debates about which categories ought to be used for the explanation of past scientific episodes.[32] Today's critics are not calling for a return to the older forms of rational reconstruction. As Michael Hunter has described it, the goal of these newer historical studies is to provide more nuanced accounts of individual thinkers while resisting the temptation to reduce their thought to some set of putatively timeless philosophical, scientific, or sociological categories.[33] In reaction to the "derogatory attitudes" of some in the SSK camp toward those who wish to "study scientists in their own rights," Thomas Söderqvist has suggested that this new approach could be called "existential" because these studies attempt to display the meaning of science "for each individual in a particular culture in a particular era."[34] Such a project calls for a more localized approach than either of the two earlier ones.[35]

There is a shift of emphasis in these newer studies away from the explanatory and toward the descriptive. When explanations are provided, they are much more complex than those offered by previous studies, because the goal of the newer approach is to identify all of the factors relevant for understanding an individual's thought, whether social or epistemic, in the actor's own categories. Epistemological categories such as "empiricism" or "rationalism," for example, are rejected because they do not provide sufficient explanatory power, particularly for historical figures who lived and worked prior to the philosophical elaboration of such categories in the nineteenth century. The vague historical category of "influence" is also questioned by this newer approach. Predecessor studies may provide significant insights into the intellectual development of a particular thinker, but they become unduly speculative and of little explanatory value when the attempt is made to reduce a past thinker's work entirely to that of another based upon a small set of select similarities. Equally problematic is the practice employed by many contextualists who use the social status of a thinker to speculate about which sociopolitical interests exerted a direct causal influence on the intellectual decisions made by that individual.

Two significant book-length studies on seventeenth-century scientific figures manifest this newer approach. In his recent study of Bacon, Antonio Pérez-Ramos acknowledges that contextualists have aided our understanding of the social context of past science, but he also notes that because they have neglected the analysis of past ideas, their work remains "philosophically unilluminating."[36] He has sought to provide a more "general level of conceptual exploration and analysis." While broadly intellectualist, the study is more localized and historically sensitive than earlier rational reconstructions, particularly in the way in which Pérez-Ramos has attempted to correct common interpretations of the Baconian notions of *inductio, forma,* and *opus* by tracing their intellectual roots and examining the role that they played in Bacon's entire system.[37] Daniel Garber has displayed a similar purpose in his study of Descartes, in which he sought to achieve an integrative "understanding of Descartes' thought" by focusing upon his "own writings," because "much of Descartes' interest for us, as for his contemporaries, lies in his texts and his ideas."[38] Although Garber refers to his work as somewhat "old-fashioned" because of this focus, he also displays a new historical sensitivity, as can be seen in the way in which he rejects earlier interpretations that, while plausible, cannot be justified by reference to Descartes's texts.[39] Neither of these studies is explanatory in the older sense. That is, while both attempt to achieve a fuller, more nuanced understanding of their subjects, neither tries to explain why Bacon or Descartes was successful in the ideas that he developed. Indeed, both studies show the way in which the systems constructed by these two figures were not successful.[40]

The trend away from the two traditional approaches to the history of science can also be found in recent Boyle studies. Michael Hunter, William Newman, and Lawrence M. Principe, for example, have produced detailed accounts of Boyle's lifelong interest in alchemy. Although the alchemical interests of other historical figures, such as Newton, have been clearly documented, this aspect of Boyle's thought was neglected, in part because historians until recently tended to accept Hall's claim that Boyle had rejected all forms of mysticism and alchemy early in his career.[41] In a similar manner, John Henry has examined Boyle's work on occult qualities, an area that had also largely been ignored by earlier intellectualist historians because such work did not fit neatly into the "modern" conception of science.[42] On the other hand, unlike contextualist historians who have discussed Boyle's extreme toleration, modesty, and "scrupulosity" as rhetorical devices that could be used to advance the political and social aspirations of experimentalism, scholars such as John Harwood, Michael Hunter, and Malcolm Oster have sought to trace the personal idiosyncrasies and life experiences that shaped Boyle's character.[43] Indeed, as Hunter has effectively argued, these studies show a decidedly "dysfunctional" side of Boyle: some of his public actions were "potentially disruptive" to social stability and thus preclude a "simplistically functionalist reading."[44]

No widespread agreement about the content of Boyle's thought emerges in these newer studies. There is agreement, however, that a vast amount of work is left to be done before a full understanding of Boyle will be achieved, and that the earlier accounts of Boyle are flawed because the desire to produce an explanatory account led to the imposition of twentieth-century conceptual categories upon his thought. More detailed and comprehensive studies of Boyle are required, not only because of the extreme complexity of his thought but also because of the sheer quantity of his work—over forty published works together with almost fifty volumes of unpublished material.

My study of Boyle's philosophy of experiment is designed to adopt most closely this newer historical approach. It is primarily an exposition of his ideas, in which the explanatory content has been restricted to a discussion in his own terms of why he believed that his method was superior to alternatives. My focus is broad because it has been necessary to discuss all of the elements that Boyle used in the synthetic construction of his experimental philosophy, yet it is also narrow at some points because many interesting and significant aspects of his thought are only discussed in relation to this one project.[45] I do not mean to imply by my advocacy of this newer approach that there is no value to be found in earlier types of studies. The two traditional approaches are quite good at providing answers to the questions that they ask, and they have both contributed greatly to our understanding of the details surrounding Boyle's life and thought. Other types of historical questions, however, still need to be answered, and the earlier approaches necessarily preclude an understanding of the complexity of Boyle's thought.[46]

Although contextualists follow what Timothy Lenoir has described as a "historically localized approach," their method is not as localized as it could be because of their adherence to the sociological principle of collectivism.[47] This principle can be seen in *Leviathan and the Air-Pump*: Shapin and Schaffer frequently define Boyle's position by reference to quotations from his associates, such as Robert Hooke, Joseph Glanvil, and Thomas Sprat, and even from his adversaries, such as Thomas Hobbes.[48] This practice, as well as the use of the phrase "form of life," seems to betray an assumption that a community of thinkers shared a unified experimental method, which can be used to define Boyle's work, and effectively ignores the fact that relevant differences existed between these historical figures.[49] In this sense, Shapin and Schaffer tend to treat Boyle as an icon, or an idealized representative of the whole, and focus almost entirely upon those elements in his thought that can be generalized to his community.

A more specific problem arises from Shapin and Schaffer's attempt to show experimentalism as a form of life, because it led them to focus almost exclusively upon a limited set of Boyle's air-pump experiments, particularly those about which controversy arose, either from the failure of fellow experimentalists to replicate them or from the disagreements among his contemporaries about their proper interpretation. Shapin and Schaffer provide a detailed and interesting account of the problems surrounding these experiments, but they have little to say about many of the other air-pump experiments reported by Boyle and all but ignore his more numerous chemical and medical trials that played a significant role in his own methodological discussions concerning the problems with experimental practice. Shapin and Schaffer's strategy may be appropriate for determining how Boyle's work was perceived at the time and how social factors led members of his society to find a general form of experimentalism attractive, but one can question how well this strategy produced an accurate account of the details of Boyle's practice.

This discussion is not meant as a direct criticism of the work produced by Shapin and Schaffer, because such criticism would only be warranted if they had intended to provide a detailed explication of the intricacies involved in Boyle's methodological and epistemological position. I do not believe that that was their intention, although a number of their readers do appear to have taken their work as the final word on Boyle's philosophical project. Because contextualist accounts do not answer questions in this area but rely upon earlier intellectualist accounts to supply the explanans for their explanations, it is the latter type of accounts at which most of my criticism will be directed. To the extent, however, that *Leviathan and the Air-Pump* and other works in its genre are the latest and most influential promoters of certain "received views" concerning Boyle's epistemology, it may certainly appear to some readers that they are my main target.

My purpose in this study is to show how Boyle constructed a complex, critical, and dynamic approach to knowledge acquisition. He resisted strict analytic categories and disciplinary boundaries because he thought that such approaches would close the door to future lines of inquiry.[50] Instead he developed a unique experimental philosophy by way of an eclectic synthesis of the best elements from what have come to be known as various empiricist, rationalist, and pragmatist traditions. Thus his thought cannot be reduced to any one epistemological category. Because Boyle was not a strict empiricist, the adequacy of other accepted opinions, such as the widespread belief that he advocated the "passivity of matter," must be reconsidered. In a similar manner, it will be necessary to discuss at length the recent focus in contextualist studies upon Boyle's appeals to "matters of fact." Because such studies tend to see Boyle's factual category as coextensive with the positivist conception of a fact as a singular synthetic statement meant to refer to a particular event, they also tend to reinforce the empiricist interpretation of Boyle's epistemology.[51] Yet as Lorraine Daston has noted, the factual category was new to the seventeenth century, and there were many permutations of meaning among those who employed it.[52] In chapter 6 I will argue that for Boyle the factual was not a linguistic category used to refer to particular types of statements but an epistemic category designed to demarcate knowledge claims according to the degree of evidence provided in support of them.[53]

The emphasis upon linguistic analysis appears to be another common element in the two traditional approaches to Boyle studies. Intellectualist historians have tended to focus upon the semantic level of theory and reconstructions of the logical relationship between theories and the factual propositions said to support them. Although contextualists are more concerned with the deconstruction of texts, the proliferation of works on the rhetorical strategies and "literary technology" of Boyle have tended to perpetuate a concentration on the analysis of language.[54] In this study I focus on Boyle's texts but make no attempt either to reconstruct or to deconstruct his language. I look not at his so-called method statements but at his extended methodological discussions and how he illustrated his precepts by actual examples drawn from his extensive experience in the laboratory. Thus, while my concern is to explicate the epistemic facets of Boyle's experimentalism, I focus on how he developed it as a dynamic practice. In opposition to some recent literature in the SSK genre that creates a false dichotomy between knowledge and practice, I attempt to show how Boyle clearly recognized the epistemic dimension of the social elements embedded within his practice and how he developed a constructive and active notion of knowledge, significant for its sophistication and complexity, that has yet to be appreciated.[55]

Philosophical Issues

Although the text that follows is thus designed primarily as an exposition of Boyle's experimental philosophy, this does not mean that there is no normative content. In addition to seeing the need for a revised interpretation of Boyle's experimentalism, I also believe that a retrieval of his methodological and epistemological ideas can make a significant contribution to discussions in philosophy of science today. Such a claim may sound "Whiggish," but it is not any more so than the general conclusions that have been derived from previous rational and social reconstructions. Some presentism is unavoidable in historical studies. Indeed, the very act of selecting cases for study, together with decisions concerning which data are relevant for the examination of such cases, must necessarily involve some preconceived notions.[56] A problem arises not from the use of present-day interests in the study of the past but from the failure to make such interests explicit.

As David L. Hull has argued, a limited degree of presentism can be valuable if one wishes to learn something from history. At least since the time of Kuhn's *Structure of Scientific Revolutions*, philosophers of science have sought to use historical studies for insights into the processes involved in the development of science. While a number of these studies have focused on individual scientists, Ronald N. Giere has recently questioned the value of scientific biography for philosophical purposes because individual lives may not be representative of science as a whole.[57] To the extent that historical case studies are to be used as evidence for the efficacy of particular developmental models of science, Giere is correct. But this point need not negate the philosophical significance of a study on Boyle's thought. As I indicated earlier, I will be looking at Boyle not as a scientist but rather as a philosopher who developed methodological and epistemological precepts by a reflective analysis of his own extensive laboratory experience.

A discussion of how a retrieval of Boyle's ideas can make a contribution to today's philosophical concerns is different from such decidedly "Whiggish" practices as claiming that Boyle's concerns were the same as ours or that his experimental program was successful because it was in some sense the right way in which to initiate modern science. Indeed, as this study will show, many of Boyle's concerns were markedly dissimilar from those of today. His deeply-held theological commitments, for example, that led him to see the pursuit of natural philosophy as a religious duty have little in common with modern conceptions of science. Boyle's theological beliefs also provided him with a model of the world as a complex interrelated mechanism. This model in turn produced in him a cautious attitude toward the technological implementation of the knowledge claims of his day, unlike the confidence expressed by many of his contemporaries. In addition, a number of Boyle's methodological discussions are difficult to understand unless one realizes that many of the terms he used, such as *fact, experience,* and *probability,* did not carry the same connotations for him as they do for us. Yet despite all of these dissimilarities, his general conception of the active and future-oriented character of the experimental philosophy remains relevant and insightful, particularly in light of the numerous philosophical studies today that have been designed to display the active component in knowledge acquisition. As Thomas Nickles has recently argued, some presentism should be tolerated in historical studies because a "strong historicism" would, perhaps paradoxically, lead to the reinforcement of a "passive 'spectator' theory of knowledge."[58]

In opposition to traditional theory-dominated studies of science, works by philosophers and philosophically-minded historians, such as Nancy Cartwright, Allan Franklin, Peter Galison, and Ian Hacking, have concentrated on experimental activity.[59] As Hacking has argued, such studies are necessary because "reasoning is not a purely sedentary art. It includes a lot of doing, not just arguing or thinking."[60] The philosophical focus has shifted in these works away from an examination of how theories come to be justified and toward the study of how the data used in theory justification come to be constructed and accepted as evidential resources. In their examination of the activity of knowledge construction, these writers share a similar project with some of the sociologists discussed above, yet they tend to differ in the conclusions that they draw from their studies. To varying degrees, the writers of these philosophical studies reject the notion that the artifactual component of our knowledge is a cause for epistemic suspicion and instead argue that our successful manipulation and creation of phenomena provide grounds for the rational acceptance of numerous knowledge claims.[61]

The approach to science studies followed by these philosophers of experiment clearly has parallels with the localized approach followed by some historians. Other philosophers, such as Arthur Fine, Giere, and Hull, continue to seek a more general characterization of the nature of scientific practice, yet they also have rejected an appeal to timeless philosophical categories and instead have sought to provide more nuanced and context-sensitive accounts.[62] Fine, for example, has replaced global philosophical realism with his discussion of the "natural ontological attitude" that "seeks to ground scientific belief in reasonable practice."[63] Giere and Hull have both developed evolutionary accounts that, while quite different at the explanatory level, nevertheless stress the dynamic, open-ended, and fallible nature of scientific practice.[64] They all recognize the essential role played by social interaction, yet they also do not find this a sufficient reason to accept a radical relativist thesis concerning the knowledge claims of science. As Giere notes in his analysis of the geological debates of the 1960s, plate tectonics became an accepted model because the data were "so strong, and the connections between the data and the rival models ... so obvious, that most nonepistemic values" were "simply overwhelmed."[65]

A marked pluralism is present in today's historical, philosophical, and sociological studies of scientific practice, and I would agree with Franklin that this lack of consensus represents strength, not weakness, because the "absence of an accepted framework . . . can only encourage further study." Recently Joseph Rouse has described the work of Fine, Hacking, and others as "postmodern" in the sense that it joins "trust in local scientific practice with suspicion toward any global interpretation of science." In a similar manner, Pickering has described the trend in sociological analysis "to question . . . taken-for-granted distinctions" as a "trademark of 'postmodern' thought."[66] *Postmodernism* is a decidedly awkward term, and it remains somewhat ambiguous because it often means different things in different disciplines. As a general label for a specific style of reasoning in philosophy of science, it may also lead to some rather paradoxical associations. For example, because Boyle resisted "modern" ideals, such as a search for deductive certitude or a metascientific grounding of science, and sought instead to develop criteria for rational acceptance based upon his awareness of the constructive and fallible nature of our knowledge claims, he could be classified as postmodern.

The paradoxical nature of such a conclusion results, of course, from the temporal connotation of the term *postmodern*. Stephen Toulmin, in his recent analysis of the stages of modernity, however, has shown that there are clear similarities between the "premodern" ideals embedded within the humanism of the sixteenth century and the postmodern thought of today. According to Toulmin, prior to 1600 "theoretical inquiries were balanced against discussions of concrete, practical issues," whereas after 1600 "most philosophers" became "committed to questions of abstract, universal theory, to the exclusion of such concrete issues."[67] In contrast to the "modern" ideal of logical rigor, Toulmin discusses how the humanists' demand for reasonableness fostered "modesty" and "toleration" and the development of disciplines such as "casuistry" that became outmoded from the seventeenth century onward.[68] Toulmin notes that today Western culture is entering a new phase, which could be considered either as a "third phase in Modernity" or as a "new and distinctive 'post-modern' phase," that will ultimately oblige "us to reappropriate values from Renaissance humanism that were lost in the heyday of Modernity." Particularly in the scientific realm, we must "abandon the assumption that physics is the 'master' science, which gives an authoritative model of rational method to all science and philosophy," and instead acknowledge the shift in emphasis away "from abstract laws of universal application to particular decipherments of the complex structures and detailed processes embodied in concrete aspects of nature."[69]

A postmodern reading of Boyle does not require that we attribute to him a precognition of today's preoccupations, but rather demands that we place him more accurately in his historical context. In particular we should guard against the tendency to conflate his views with those of Newton and Descartes. Boyle never saw the ascendency of Newtonianism, dying as he did just four years after the publication of the *Principia,* and he severely criticized the Cartesians as well as the alchemists and atomists for what he perceived to be their premature systematization of theoretical speculations. Unlike many of his contemporaries, he retained a humanistic concern for practical studies, as Hunter's recent work on Boyle's lifelong interest in casuistry clearly shows.[70] In this sense, Boyle was not a "modern," and it is perhaps because of this that he has not been seriously studied as a philosopher in this century. Boyle has appeared as a poor philosopher precisely because it is difficult, if not impossible, to characterize his thought in modern categories. With the newer conceptual tools developed in recent work in the philosophy of science, however, the way is now clear for a renewed interest in and appreciation for the sophistication and complexity of his thought.

Recent works in history, philosophy, and sociology of science influence the present study in the sense that they provide a new way in which to understand Boyle's texts. In addition, this study was partly motivated by the contribution that I believe Boyle's works can make to normative issues in the philosophy and social studies of science. My main concern, however, is to provide an explication of Boyle's thought. Although normative issues will be pointed out as they occur, an explicit discussion of the ways in which Boyle's philosophy of experiment may be relevant for today will be reserved until the conclusion.

Plan of the Study

Boyle used elements from a number of different areas of learning to construct and justify his experimental philosophy. His argument structure was quite complex. In each chapter below, I examine individual portions of his argument. None of these chapters can stand alone, however, because each represents only a part of the totality of Boyle's thought. To understand his complete argument against the use of mathematics as an overriding model of inquiry, for example, it is necessary to see how his opposition originated with considerations drawn from the philosophical, legal, chemical, and medical traditions and was reinforced by the difficulties that he encountered with making exact mathematical determinations in his own experimental trials.[71] A great amount of detail is necessary to convey the full import of his individual arguments, but as with Boyle's own works, the danger is that here the larger argument may tend to be overshadowed by this detail. The following synopsis is provided as a general overview of how the arguments that Boyle constructed from different areas of learning converged to form the final complex argument in support of his experimental program.

For ease of exposition, this study is divided into three parts. Part I, "Learning from the Past," is designed to show how Boyle brought together various elements from the work of his predecessors and contemporaries and set up the general epistemological framework within which he would work. I have chosen the vague term *tradition* to refer to these areas of learning in order to stress the fact that Boyle explicitly used them as epistemic resources and that they need not be thought of as direct causal factors in the development of his philosophy. In addition, these traditions are discussed from Boyle's perspective, which does not always agree with historical accounts today. But it is Boyle's perspective, not the correctness of his interpretation, that is important for understanding how he was influenced by these traditions.

Chapter 1, "The Philosophical Tradition," begins where Boyle began, with his study of and reaction to three general philosophical alternatives: Aristotelianism, Cartesianism, and Baconianism. Boyle never rejected any of these philosophies in their entirety. At the methodological level, however, he favored Bacon's notion of an "inverted order of demonstration." According to Boyle, the causal processes operative in nature could not be known by rational analysis alone or by a hasty extrapolation from a few instances of sense experience. Rather, he agreed with his fellow countryman that one must first compile vast natural histories that could serve as a full experiential foundation for the discovery and justification of causal knowledge. Boyle also agreed with Bacon's notion that such knowledge could be useful, not only for utilitarian purposes but also because the ability of an experimenter to produce beneficial effects could serve as a "sign" of the truth of the knowledge upon which the experimental productions had been based.

Chapter 2, "The Legal Tradition," examines the way in which Boyle used analogies with the common law of England to illustrate his understanding of the methodological precepts that he had learned from Bacon. His notion of experience, for example, was similar to the broad historical notion employed in the legal sphere that went beyond a simple appeal to sense perception. Boyle also used the decision-making procedures of trial courts to illustrate his notion of experimental reasoning. He believed that a conclusion was rationally justified only when all of the relevant and procurable evidence produced a "concurrence of probabilities" in its favor and there was no evidence to the contrary. Boyle's defense of experience and experimental reasoning by his use of legal analogies is illustrated at the end of this chapter by an examination of how his response to Hobbes's criticism of experiment was similar to Chief Justice Matthew Hale's response to Hobbes's criticism of the common law.

Chapter 3, "The Experimental Tradition," examines how Boyle expanded upon and modified Bacon's precepts, particularly those having to do with the constructive nature of knowledge, by the insights that he gained from his study of, and reaction to, predecessors and contemporaries in three general areas of inquiry: physico-mechanics, alchemy, and medicine. He read widely in the works of Galileo, Mersenne, Blaise Pascal, Paracelsus, Jan Baptista van Helmont, Galen, and William Harvey, and he generally believed that all these thinkers had made positive contributions to natural philosophy. Further improvements could only be made if naturalists would break down the disciplinary boundaries that separated these areas of inquiry. In order to establish a true corpuscular philosophy of the matter and motion of bodies, for example, discoveries of the mathematical regularities of motion made by mechanists ought to be joined to the discoveries about the material properties of bodies made by chemists. In addition, Boyle found numerous practical examples of how Bacon's inverted order of demonstration worked in the inference patterns employed in chemistry and medicine. The chapter ends with a discussion of how Boyle used Harvey's theory of the circulation of the blood as an example of the way in which a theory about an unobservable causal process could be discovered and justified by inferences drawn from a variety of effects produced by experimenters following different lines of inquiry.

Part II, "Being a Christian Virtuoso," is designed to show how Boyle's Christianity provided him with an ontological foundation for his advocacy of the epistemological framework discussed in part I. These chapters are not meant to provide a strictly monocausal account of the effect of his theology upon his natural philosophy. Although Boyle began his theological studies slightly earlier than his natural investigations, his mature methodological and ontological views were worked out in unison. His studies in natural theology and biblical hermeneutics combined to provide him with a conception of the natural world as the object of study for which his experimental method of inquiry was uniquely suited. At the same time his experimental investigations into nature produced in him a greater admiration and reverence for the work of God the creator.

Chapter 4, "Natural Theology," sets out Boyle's view of God's free act of creation that resulted in a great cosmic mechanism governed in a deterministic manner by the laws of motion impressed by God upon the primordial constitution of things. Boyle believed that natural phenomena were physically produced by the powers that bodies possessed by virtue either of the internal configuration of their parts or their relation to other bodies in the universe. The primary goal of his experimental method was to discover these causal powers, which he termed the forms and qualities of bodies. Although he believed in the deterministic character of natural processes, Boyle also believed that God's wisdom in framing the universe was infinite. Because of the complex structure of the object studied and the fallibility of finite human reason, causal processes could not be known a priori but had to be discovered experimentally.

Chapter 5, "Biblical Hermeneutics," discusses the relationship that Boyle believed held between natural and revealed religion. Natural theology was useful for establishing belief in God, but revealed religion was necessary for establishing particular truths of the Christian religion. From Boyle's Christian standpoint, God had written two books for man's instruction—the book of nature and the book of the Word—and both had to be studied in order for one to understand God's creation fully. Because both are complex, yet coherent, texts, a comprehensive and dynamic method of interpretation would be required for one to learn from them. This chapter examines Boyle's similar approaches to his study of the two books. His biblical hermeneutics employed the same type of experimental reasoning that he advocated for natural philosophy, and his hermeneutic principles can therefore shed light on his methodological dictates. For the study of either book, Boyle maintained that factual information first had to be compiled and interpretations of the data from individual areas had to be compared and reconciled before the attempt was made to construct a coherent account about the meaning of that book. He warned his readers, however, that interpreters had to remain open to the possibility that new information might lead to future revisions of their original interpretations.

Part III, "Acting Experimentally," is designed to show how the dynamic and comprehensive approach advocated by Boyle in his programmatic statements was put into practice and how his practice in turn led him to produce a more sophisticated account of the epistemology of experiment. Boyle "acted" experimentally, in contrast to those of his contemporaries who simply performed experiments, and his activity included three components: observing, experimenting, and writing. These elements were all combined in the complex process of experimental reasoning by which the factual foundation was to be established, yet as activities they remained distinct. Thus I discuss them separately.

Chapter 6, "Observing," begins with a discussion of what Boyle meant by *matters of fact* and how observing and experimenting could contribute to the discovery of facts. The activity of observing is required because, according to Boyle, some processes cannot be investigated experimentally, and some experiments cannot be designed without information about the ways in which nature acts in areas different from those in which the experimenter lives and works. In this chapter I discuss Boyle's vast compilations of observations from his numerous foreign correspondents, along with his criteria for assessing the credibility of these witnesses. In addition, the reliability of the reports themselves had to be assessed. Boyle noted that corroboration of the occurrence of a particular phenomenon from independent witnesses was one test of reliability, but it would also be necessary to reconcile reported observations with other facts and theories about nature. Because of this involvement of theoretical knowledge in the process of observing, Boyle considered the factual category to be dynamic. New information or the refinement of theory could lead to a revision of those items previously admitted into the factual foundation.

Chapter 7, "Experimenting," begins with Boyle's discussion of how the creation of an artificial environment could improve the information compiled from ordinary observations. While experiment is a more expedient, and often more rational, way to make observations, Boyle was acutely aware of its limitations. Experimental trials, and the apparatuses used in their performance, require theoretical knowledge. He discussed how the lack of a theory or the use of a faulty theory could lead to "contingent" outcomes, in which the experimenter was either not able to repeat a previous result or not able to obtain the result that inferences from a particular theory would indicate. This chapter examines Boyle's practice of making experiments and his advice on how experimenters could alleviate some of the difficulties encountered in the laboratory. In line with his epistemological precepts concerning the necessity of a full informational basis, Boyle suggested two methodological strategies: repetition and variation. Naturalists must not rely on single experimental results but must repeat their trials and vary the circumstances of their production, in order to ensure that the results could be used as reliable pieces of evidence. Once again, however, the factual category remained dynamic for Boyle. Experimental facts could be overturned in the future in light of new information obtained from improved technical apparatuses or developments in theoretical understanding.

Chapter 8, "Writing," discusses how the naturalist must first collate and then communicate the results of experimental activity. This chapter is entitled "Writing" in order to distance it from current studies of scientific rhetoric that are mainly concerned with exhibiting the argument forms used to persuade readers of the truth of the knowledge claims made by the writer. Opposed to Cartesian or Hobbesian individualism, Boyle maintained that the experimental philosophy was to be a public, social enterprise. The experimental essay was his preferred mode of presentation because its loose structure allowed for the inclusion of detailed accounts of his trials along with tentative discussions concerning their theoretical significance. Such circumstantial reports could generate interest in the new philosophy and thus win converts to his program. Readers who wished to try experiments for themselves would be aided by his detailed instructions, while those of a more theoretical bent could use his accounts to judge the accuracy of his observations and the adequacy of his hypothetical speculations. Boyle recognized that his epistemological and methodological precepts could only be fulfilled by such a collaborative effort. This chapter looks at Boyle's programmatic statements, as well as the ways in which he used his essays to exhibit the actual collaboration that took place in his own laboratory. Writing is the final stage of experimental activity, but it is also the point at which the learning process began anew for Boyle. The chapter ends with a discussion of Boyle's *History of Cold*, his most extensive, yet inconclusive, study of a natural property, to illustrate all the points made about experimental activity.

In the conclusion, this study of Boyle ends with a summary of the main methodological and epistemological points of his experimental philosophy drawn from the preceding chapters, as well as an analysis of how his insights into the experimental process can contribute to current philosophical and sociological discussions about scientific practice.

The textual sources used for this study are the 1772 Thomas Birch edition of Boyle's published works and the collection of Boyle's unpublished drafts, letters, and notebooks housed at the Royal Society of London.[72] Because neither of these sources is fully reliable, however, I have sought to identify those general themes that recur throughout the full spectrum of Boyle's published and unpublished writing and have not relied upon isolated passages from these texts for the explication of his philosophy. In addition, I have provided substantial direct quotation from these works so that the reader will be able to see Boyle's words in the context in which they were written and thus gain an appreciation of his idiom. Boyle chose his words carefully, but his terms often have connotations that, not surprisingly, are different from those of today. To avoid misunderstandings that could arise from paraphrases, it has been crucial to find passages in which he explicitly stated the meaning of his terms and then to use his meanings when interpreting other passages in which the terms occur.[73]

Boyle believed that a slow but steady progress could be made by following his comprehensive approach. Yet he was acutely aware of the paucity of data in the informational basis and the lack of theoretical understanding possessed by his generation, and thus he often spoke of his "diffidence" concerning the truth of the facts and speculations that he presented to his readers. He planned to write a work entitled "The Diffident Naturalist," a draft outline of which is preserved among his unpublished papers.[74] I have chosen to use his title because I believe it best captures his general attitude toward the scientific process.[75] His diffidence was not a sign that he believed in the impossibility of knowledge acquisition but an expression of his desire neither to accept nor to reject any knowledge claim upon the basis of authority or faith in tradition. He had a keen sense of the defeasibility of experimental attempts to understand nature, but this awareness did not mean that all such attempts would actually be defeated. A flexible, critical attitude toward knowledge is a prerequisite for successfully achieving it, and Boyle sought to inculcate this critical attitude in his readers. As he saw it, experimental science was always on the verge of making new discoveries, which might themselves be overturned the next day.

The experimentalist learns but can never rest satisfied with what has been learned. As we learn from the past, we also correct the errors of the past. Boyle designed an experimental philosophy in which every defeat would itself be a step forward in the never-ending process of discovery. He clearly recognized the historical dimension of his studies. Boyle's talent for making detailed and exact observations, combined with his seemingly inexhaustible curiosity, led him to believe that he could best serve the interest of natural philosophy by playing the role of the "under-builder" who would lay the groundwork upon which others could build. His diffident attitude toward his own achievements was balanced by a confidence that, if faithfully followed, the experimental method would lead to rich rewards in the future. Indeed, he fully expected (and hoped) that his own scientific work would be eclipsed by future generations. As he wrote in an appendix to one of his last histories,

> I presume, that our enlightened posterity will arrive at such attainments, that the discoveries and performances, upon which the present age most values itself, will appear so easy, or so inconsiderable to them, that they will be tempted to wonder, that things to them so obvious, should lye so long concealed to us, or be so much prized by us; whom they will, perhaps, look upon with some kind of disdainful pity, unless they have either the equity to consider, as well the smallness of our helps, as that of our attainments; or the generous gratitude to remember the difficulties this age surmounted, in breaking the ice, and smoothing the way for them, and thereby contributing to those advantages, that have enabled them so much to surpass us.[76]

Part One

LEARNING FROM THE PAST

One

THE PHILOSOPHICAL TRADITION

IN 1635, EIGHT-YEAR-OLD ROBERT BOYLE arrived at Eton to commence his formal education.[1] The provost, Henry Wotton, admired the boy's keen observational skills and found his "pretie conceptions" entertaining.[2] Indeed, young Boyle's eagerness for learning and prodigious memory made him excel in his studies to such an extent that he was often exempted from regular school hours and privately tutored by the headmaster, John Harrison.[3] For a time, Boyle took pleasure in reading "the state adventures of *Amadis de Gaule,* and other fabulous and wandering stories," but in order to "curb the roving wildness of his wandering thoughts" produced by such reading, he felt the need to apply himself to the study of mathematics. He found that "the extractions of the square and cube roots, and especially those more laborious operations of algebra," were the "most effectual way" of disciplining his mind, since they "both accustom and necessitate the mind to attention, by so entirely exacting the whole man."[4]

By his fourth year at Eton, Boyle had become "addicted to more solid parts of knowledge." He no longer desired to pursue the "study of bare words" but thought it much nobler "to learn to do things, that may deserve a room in history, than only to learn, how congruously to write such actions in the gownmen's language."[5] In Boyle's recollections there is a tension concerning the content of his new line of study. In his "Account of Philaretus" he wrote that these studies were historical, while in another surviving manuscript he wrote that among the "more serious parts of learning," he was now "inclined to the study of natural philosophy" and had begun to read works on the "Aristotelian doctrine . . . whose principles I found generally acquiesced in by the Universities and schools and by numbers of celebrated writers."[6] About this same time, in 1639, Boyle's formal education came to an end. The earl of Cork removed his son from Eton and placed him under the tutelage of Isaac Marcombes, with whom the young man would shortly embark upon a projected three-year tour of the Continent.

Whether or not Boyle had been exposed to natural philosophy prior to his travels, we do know that he first read Diogenes Laertius and Epicurus while 23 on the Continent. It was at this time that he was also first "strongly tempted to doubt" the "solidity" of Aristotelianism, because he "met with many things" both in reading and in observation that were "capable to make me distrust the Doctrine."[7] In time he formulated three general objections to it. First, he "found Aristotle's Principles much more strongly asserted than proved by his admirers and by not inconsiderable arguments opposed not only by the Chymists in generall and great store of Moderne Physitians, but [by many] acute and famed Philosop[hers] . . . and the sect of the new Cope[rnicans] . . . , Ba[con, Ga]ssendus, Descartes and his sect to name no more."[8] Second, among the wonders of the Continent, he had "observed many things" that "were wholly unintelligible from Aristotle's theory which being grounded but upon a few obvious and not thoroughly examined appearances of things is much too narrow and slight to reach either all or the more abstruse effects of Nature."[9] In 1642, for example, upon his arrival in Florence, he became interested in the new telescopic observations that had been made in favor of Copernicanism, and he taught himself Italian in order to read "the new paradoxes of the great star-gazer *Galileo*, whose ingenious books, perhaps because they could not be so otherwise, were confuted by a decree from *Rome*."[10]

In addition to its lack of proof and its explanatory failures, Boyle objected to Aristotelianism because he found it to be "altogether barren as to useful productions and that with the assistance of it I could do no more than I could have done when I was a stranger to it."[11] The content of these objections was probably influenced to some extent by the general anti-Scholastic sentiment that he would have met with among the younger members of the aristocracy on the Continent.[12] The specific formulation of his objections, however, was most likely influenced by the situation that he encountered in London upon his return home in 1644.

Boyle had been stranded on the Continent for an additional two years because of the depletion of his father's fortunes during the Irish Rebellion. At the age of seventeen, upon news of his father's death, Boyle borrowed on the credit of his tutor and traveled to London with a vague plan of joining his older brothers who were fighting in the Irish wars. A fortunate "accident," however, brought him instead to the home of his sister Katherine, Lady Ranelagh, through whom he would meet Samuel Hartlib, John Milton, and others actively involved in a plan for the reformation of learning and education modeled after the precepts Bacon had advocated for the advancement of knowledge.[13] This stay in London had a significant impact on the direction of Boyle's future thought, as can be seen in an early letter to his old tutor, Marcombes. He wrote that among the "humane studies I apply myself to, are natural philosophy, the mechanics, and husbandry, according to the principles of our new philosophical college, that values no knowledge, but as it hath a tendency to use."[14]

In the same letter to Marcombes, Boyle wrote at length about the political revolutions taking place in the British Isles and about the hardships they were causing, indicating that he "was once a prisoner here upon some groundless suspicions." Despite the fact that he "quickly got off with advantage," from that day on he felt "forced to observe a very great caution, and exact evenness in my carriage," and sought to avoid political controversy by remaining neutral, and silent, on most political issues.[15] But in the world of ideas Boyle was less reticent. While he often remained neutral toward the specific theories of nature found in the competing philosophical systems of his day, he became a vocal proponent of the general philosophical revolution that advocated the abandonment of the schools' traditional learning.

A Philosophical Revolution

The rhetoric of revolution, of the rejection of stagnant Aristotelianism in favor of a more fruitful approach to the study of nature, is common in the writing of seventeenth-century natural philosophers. But this revolt cannot be discussed without qualification. The Aristotelian corpus was not rejected in its entirety; only those parts of it that were seen as pernicious to the advancement of learning were called into question. Indeed, in his earliest writing on the subject, Boyle noted that although many of the inventions of recent years had been made by "some luckyness or sagacity" and not with the help of Aristotle's principles, he still "much reverenced the rare abilities of that Great Philosopher who, what he would have done had his vast reason had experiments to work upon, we may guess by his excellent treatises concerning animals."[16]

Almost twenty years later, in his preface to *The Origin of Forms and Qualities,* Boyle was still concerned with giving Aristotle his due:

> I here declare once for all, that, where in the following tract, or any other of my writings, I do indefinitely depreciate *Aristotle's* doctrine, I would be understood to speak of his physicks, or rather of the speculative part of them (for his historical writings concerning animals I much esteem) nor do I say that even these may not have their use among scholars, and even in universities, if they be retained and studied with due cautions and limitations.[17]

Aristotle's followers had not exercised caution, however. Instead, they had constructed an elaborate system of natural philosophy laden with extravagant embellishments. Boyle noted that because of the "oftentimes dark and ambiguous" nature of Aristotle's expressions, many worthless disputes had been generated among the school philosophers.[18] If they had truly desired to advance learning, they should have "imployed as much dexterity to expound the mysteries of nature, as the riddles of the schoolmen, and laid out their wit and industry to surmount the obscurity of her works, instead of that of *Aristotle's*."[19]

Aristotelian theories had come into disrepute as nature came to be more closely examined by philosophers such as Bacon, Galileo, Descartes, and Mersenne. To this work Boyle would add his own contributions. In his *New Experiments Physico-Mechanical* (1660) he used his air-pump trials to argue against the notion of nature's abhorrence of a vacuum, and in 1661 he used chemical experiments to argue against the Aristotelian theory of four primary elements in his *Sceptical Chymist.* In 1666 he combined philosophical considerations with his experimental trials to argue against the theory of substantial forms in his *Origin of Forms and Qualities,* and in his *Free Inquiry into the Vulgarly Received Notion of Nature* of 1686, he added theological considerations to argue against the notion of nature in general as some type of independent entity capable of purposive action. In this last work, he again stressed that it was not the opinions of Aristotle to which he was opposed, but "those opinions, that are by the generality of scholars, taken for the *Aristotelian* and Peripatetick doctrines, by which, if he be misrepresented, the blame ought to light upon his commentators and followers."[20]

Boyle's overriding objection in all of these works was to the Scholastic "way of philosophizing."[21] In his *Origin of Forms and Qualities,* for example, he argued that the barrenness of the doctrines of the "scholastical philosophers" was largely a result of the mode of reasoning they used in support of such theories. Their language was "obscure" and "perplexed," and their discourses upon natural subjects "consist so much more of logical and metaphysical notions and niceties than of physical observations and reasonings, that it is very difficult for any reader but of an ordinary capacity to understand what they mean, and no less difficult for any intelligent and unprejudiced reader to acquiesce in what they teach."[22] The arguments of the Scholastics are "metaphysical or logical," Boyle said, rather than "grounded upon the principles and phaenomena of nature," and therefore "respect rather words than things."[23] Whenever a "solid argument" based upon the new learning was brought against them, the Aristotelians would "endeavour to elude it by some pitiful distinction or other; which is usually so groundless, so unintelligible, or so nugatory, or so impertinent to the subject, or at least so insufficient for the purpose it is alledged for, that to vouchsafe it a solicitous confutation might question a writer's judgment with intelligent readers." Boyle did not have the "leisure to wrangle about terms," nor did he have the "inclination."[24] Indeed, this practice was not merely barren but also prejudicial to a real understanding of nature. The "adorers" of Aristotle's physics had done "little more than wrangle, without clearing up . . . any mystery of nature, or producing any useful or noble experiments."[25]

The "grand mistake" of such logicians was that they attempted to draw physical conclusions concerning the existence of real entities from their metaphysical speculations concerning names and definitions.[26] This way of reasoning had led them to their faulty theories. The Aristotelian way of ascribing "but nominally understood qualities" to the explication of manifest phenomena was a "so general and easy way of resolving difficulties" that it allowed naturalists "to be very careless and lazy."[27] As Boyle argued, when "we mistake names for things," then "we are prone to conclude that the faculties and qualifications" of a body to which we have assigned a name signifying a form "are due to this form . . . as if this form were some distinct and operative substance that were put into the body as a boy into a pageant. . . . Whereas indeed what we call the form . . . seems oftentimes to be rather a metaphysical conception in our mind than a physical agent that performs all things in the body it is ascribed to."[28]

Because the Scholastic logic focused on names, and the diversity of names "doth not always infer a diversity of physical entities," Scholasticism remained in the realm of metaphysical speculation and failed to increase our knowledge of the physical world.[29] Boyle was more interested in learning about what things were "in themselves, not what, logician or metaphysician will call them in the terms of his art; it being much fitter in my judgment to alter words, that they may better fit the nature of things, than to affix a wrong nature to things that they may be accommodated to forms or words that were probably devised, when the things themselves were not known or well understood, if at all thought on."[30] In order to reform language so that it would better reflect the nature of things, an alternative to the Scholastic way of philosophizing was required.

A New Alternative

Boyle found his alternative in the history of philosophy. According to him, most of the philosophers who preceded Aristotle were corpuscularians who "exercised themselves chiefly either in making particular experiments and observations, as *Democritus* did in his manifold dissections of animals; or else applied mathematicks to the explicating of a particular phaenomenon of nature."[31] Despite the fact that in his writings there are "considerable footsteps" of this earlier way, "particularly in his books of animals," Aristotle was responsible for "having attempted to deduce the phaenomena from the four first qualities, the four elements, and some few other barren hypotheses . . . and having upon these slight and narrow principles reduced physics into a kind of system, which the judicious modesty of the Corpuscularians had made them backward to do."[32] The Scholastic followers of Aristotle ignored the "footsteps" and "contented themselves with hotly disputing, in general, certain unnecessary, or at least unimportant questions." By doing so they had "remained utter strangers to the particular productions of that nature, about which they so much wrangled, and were not able to give a man so much true and useful information about particular bodies, as even the meanest mechanicks . . . can do."[33]

It was the seventeenth-century "revival of the corpuscularian philosophy" that had led to Aristotle's doctrines' being "by more than a few derided as precarious, unintelligible, and useless."[34] According to Boyle, this revival consisted of two features: the attempt by Bacon "to restore the more modest and useful way practiced by the antients, of enquiring into particular bodies, without hastening to make systems"; and the Cartesian program, which advocated "the application of geometrical theorems, for the explication of physical problems."[35] The "experimental and mathematical way of enquiring into nature" was once again gaining ascendancy because of the work of these "restorers of natural philosophy," and their direct influence upon Boyle's early objections to Aristotelianism is undeniable.[36]

Descartes and Bacon had both decried the barrenness of Scholasticism. Descartes likened the disciples of Aristotle to the "ivy which never seeks to climb higher than the trees which support it." He complained that "the obscurity of the distinctions and principles they use makes it possible for them to speak about everything as confidently as if they knew it." Since there was "no point" in their philosophy that "is not disputed," however, it seemed clear to Descartes that the discovery of truth could not be achieved "by means of the disputations practiced in the schools."[37] Bacon also maintained that "not much can be known in nature by the way which is now in use."[38] As he said, the "subtilty of nature and operations will not be enchained in those [syllogistic] bonds." The disputations of the schools could not reach to the truth of things because "arguments consist of propositions, and propositions of words, and words are but the current tokens or marks of popular notions of things."[39]

For both philosophers, Scholasticism was also in large part responsible for the rise of skepticism. Although philosophy had been "cultivated for many centuries by the most excellent minds," Descartes noted that there was as yet nothing in it that was not "doubtful."[40] In a similar manner, Bacon argued that because of Scholasticism's endless disputations, it was "not without cause, that so many excellent philosophers became skeptics and Academics, and denied any certainty of knowledge or comprehension."[41] The Scholastics themselves promoted skepticism. Bacon noted that because they had "usurped a kind of dictatorship in the sciences and taken upon them to lay down the law with such confidence," they were forced, when faced with phenomena not explicable by their principles, to resort to "a device for exempting ignorance from ignominy." They would "rather lay the blame upon the common condition of men and nature than upon themselves."[42]

As Boyle had followed his two predecessors in their opposition to the barrenness of the learning of the schools, so too did he appreciate the deleterious effect that such nascent skepticism had upon the progress of philosophy. He noted with approval that "so judicious a friend to philosophy and mankind as Sir *Francis Bacon*" had "in several places" argued that "men's opinions of the impossibility of doing great matters" are "one of the chief obstacles to the advancement of real and useful learning."[43] The author of *The Sceptical Chymist* was not himself a skeptic. In that work, Boyle wrote that "though sometimes I have had occasion to discourse like a Sceptick, yet I am far from being one of that sect." It is "prejudicial to philosophy" to "propose doubts to persuade men, that all things are doubtful and will ever remain so." Rather, much as Descartes had done earlier, Boyle sought to "propose doubts not only with design, but with hope, of being at length freed from them by the attainment of undoubted truth; which I seek, that I may find it."[44]

To avoid the barrenness of the school learning, Boyle advocated a return to the earlier corpuscular philosophy. The occult forms and qualities of the Aristotelians would be replaced by hypotheses concerning the motions of the least parts of matter that make up bodies. In addition, he maintained that such hypotheses should be founded upon a firm informational basis about natural phenomena and thus urged philosophers to "often make and vary experiments; by which means nature comes to be much more diligently and industriously studied, and innumerable particulars are discovered and observed which in the lazy Aristotelian way of philosophizing would not be heeded."[45] Because his experimental approach is often linked with the "empirical" Baconian program while his hypothetical approach is more akin to Cartesian "rationalism," Boyle's advocacy of these two aspects of the new philosophy has sometimes been taken to betray an inconsistency, or at least an unresolved tension, in his thought.[46] A better way to proceed would be to question the too-simple dichotomy that has been constructed by the rigid application of nineteenth-century philosophical categories to discussions about these historical methods. Recent work by historians and philosophers, for example, has shown that Descartes and Bacon were not as extreme as the later traditions that bear their names.[47]

The important issue here is not whether Boyle was a Baconian or a Cartesian but how he perceived the traditions begun by his two predecessors and how he used their precepts in the formulation of his own version of the experimental philosophy. It is beyond the scope of this chapter to present a complete exposition of the full complexity of the thought of either Descartes or Bacon. The most that can be done here is to set out clearly the methodological and epistemological precepts to which Boyle referred when he discussed the way in which our knowledge of nature could be improved by the use of experimental procedures. As such, the discussion of Descartes and Bacon will be far from definitive. Instead, it will be based upon Boyle's perceptions of their precepts and will merely locate those passages in his predecessors' works that would be compatible with such perceptions. In addition, the following brief discussion of the similarities and differences between Descartes, Bacon, and Boyle is not meant to provide an explanatory account for why Boyle developed the epistemological position that he did. Rather, I intend to provide a preliminary description of the general content of his epistemology that will be filled in with more specific details in following chapters.

Both Bacon and Descartes made use of experiments and hypotheses to varying degrees, so Boyle could have been influenced by either in the methodological choices that he made. Bacon was not a strict empiricist. Rather, he called for a "true and lawful marriage between the empirical and the rational faculty."[48] As he stated in *The New Organon*, pure sense perception is an unreliable basis upon which to build a science, since "what in observation is loose and vague, is in information deceptive and treacherous."[49] Experiments were required as a rational constraint upon the superficial appearances given by our senses. They were also required as a way to go beyond those discoveries that lie "near to the senses" and "reach the remoter and more hidden parts of nature." In Bacon's well-known phrase, "The secrets of nature reveal themselves more readily under the vexations of art than when they go their own way."[50]

Despite Descartes's emphasis upon explanation derived from first principles, in part 6 of his *Discourse on the Method* he noted that "the further we advance in our knowledge, the more necessary" observations and experiments become.[51] The need for experiments was a product both of the generality of his principles and the complexity of nature. As he put it, "The power of nature is so ample and so vast, and these principles so simple and so general, that I notice hardly any particular effect of which I do not know at once that it can be deduced from the principles in many different ways; and my greatest difficulty is usually to discover in which of these ways it depends on them."[52] In his "Author's Letter" in *The Principles of Philosophy*, he expressed the same sentiment as Bacon concerning the need for apposite observations and experiments. Because there was often more than one way to explain the facts of common observation, more intricate knowledge of the effects of nature would be required as a way to determine which explanation was correct. The "majority of truths remaining to be discovered depend on various particular observations [*expériences*] which we never happen on by chance but which must be sought out with care and expense by very intelligent people."[53] Descartes did not publish vast compilations of observations and experiments in the same manner as Bacon had done, but one can see in his published works numerous references to the experiments that he personally performed. In addition, he was aware of, and approved of, Bacon's more empirical-sounding precepts. For example, in a letter to Mersenne in 1632, he wrote: "It would be very useful if some such person were to write the history of celestial phenomena in accordance with the Baconian method and to describe the present appearances of the heavens without any explanations or hypotheses. . . . Such a work would be more generally useful than might seem possible at first sight and it would relieve me of a great deal of trouble."[54]

Both of Boyle's predecessors also approved of the use of hypotheses. Bacon's preliminary "interpretations of nature" were hypothetical constructs formulated in order to guide scientific investigation.[55] Science was to rest upon an experimental foundation, but hypotheses were necessary for experimental design. "Since truth will sooner come out from error than from confusion," Bacon thought "it expedient that the understanding should have permission, after the three Tables of First Presentation (such as I have exhibited) have been made and weighed, to make an essay of the Interpretation of Nature in the affirmative way."[56] These interpretations, as instanced by his "First Vintage Concerning the Form of Heat," were supplied not as truths but as "powerful aids for the use of the understanding."[57] They were hypothetical superstructures to be used in the design of further experiments that would in turn help to confirm, refute, or refine the original conjectures.

Descartes's advocacy of the use of hypotheses was also in part a product of his recognition of their heuristic value. His well-known "hypothetical" story in part 5 of his *Discourse,* for example, was presented as a way to make intelligible how it would have been possible for God to create a universe simply by establishing laws to govern the motions of matter.[58] He used a similar strategy in his *Regulae,* when in rule 12 he argued that among the human faculties, the understanding alone is capable of knowing truth. He based his argument upon "certain assumptions" concerning the passivity of sense perception, which, he admitted, "not everyone will accept," but maintained that the use of such assumptions would do little harm to an argument "provided they help us to pick out the kind of apprehension of any given thing that may be true and to distinguish it from the kind that may be false."[59]

There is no tension at the methodological level between the joint use of experiment and hypothesis. Because both Descartes and Bacon advocated the use of these two approaches, Boyle had a significant methodological precedent to follow. A dichotomy does exist, however, at the epistemological level, where the question concerns the ultimate justification of the hypotheses constructed with the aid of experiment. It was at this level that Boyle would choose to follow the precepts of his fellow countryman.

The Pyramid and the Tree

The epistemological divergence between Bacon and Descartes can be seen most perspicuously in the metaphors that they employed to describe knowledge. In his *De augmentis* Bacon presented the pyramid as his model for the scientific enterprise:

> Knowledges are as pyramids, whereof history and experience are the basis. And so of Natural Philosophy the basis is Natural History; the stage next the basis is Physic; the stage next the vertical point is Metaphysic.

> As for the cone and vertical point (. . . namely, the summary law of nature) it may fairly be doubted whether man's inquiry can attain to it.[60]

Descartes chose the model of a tree to describe scientific pursuit in his *Principles:*

> Thus the whole of philosophy is like a tree. The roots are metaphysics, the trunk is physics, and the branches emerging from the trunk are all the other sciences, which may be reduced to three principal ones, namely, medicine, mechanics and morals.[61]

In both models, physics is located in the middle. It is not surprising, then, that similarities would exist between the individual methodological dictates and the actual scientific practices of a Baconian and a Cartesian. Dissimilarities between the two arise in discussions about the type of warrant that is appropriate for the justification of the knowledge claims produced by such methods.

Descartes rejected the traditional goal of explaining nature by the qualities and forms of bodies in favor of a more quantitative explanation relying on the categories of the mechanical philosophy, yet he retained the classical epistemological goal of a science of strict demonstration and sought to overcome skeptical criticisms by a metaphysical justification of man's ability to achieve certain knowledge in this area. In part 6 of his *Discourse,* where he spoke of the need for observations, experiments, and hypotheses, he also indicated that he still held to the ideal of logical rigor and maintained that he considered himself able to "deduce" the propositions that he referred to as hypotheses "from the primary truths" of his metaphysical foundation.[62] In his *Principles* he described the way in which he had proceeded in the discovery and justification of his scientific doctrines:

> First of all, I considered in general all the clear and distinct notions which our understanding can contain with regard to material things. And I found no others except for the notions we have of shapes, sizes and motions, and the rules in accordance with which these three things can be modified by each other—rules which are the principles of geometry and mechanics. And I judged as a result that all the knowledge which men have of the natural world must necessarily be derived from these notions; for all the other notions we have of things that can be perceived by the senses are confused and obscure, and so cannot serve to give us knowledge of any thing outside ourselves, but may even stand in the way of such knowledge.[63]

Because he identified the fallibility of human senses as the source of error, the a priori path was the natural one for Descartes to take to meet the skeptical challenge. While he advocated experimentation, it functioned more as illustration than as justification for his theoretical claims. As he wrote to Mersenne, "As for the more particular experiments, it is impossible not to make many that are superfluous, and even false, if one does not know the truth of things before making them."[64] Experiments could neither confirm nor refute a theory. If there was only one theory compatible with his first principles, then a conflicting experimental result would call into question the validity of the latter. In another letter to Mersenne, for example, he maintained that "even if [Beeckman] can make a thousand experiments to find [the pendulum acceleration] more exactly, I do not have to take the trouble to do these myself, if they cannot be explained by reason."[65]

While Descartes advocated the use of experiments and performed many himself, truth remained a matter to be determined by logic and metaphysical reasoning. As he stated in his first conclusion of rule 12, "There are no paths to certain knowledge of the truth accessible to men save manifest intuition and necessary deduction."[66] The true confirmation of a theory would be established by the logical connections that obtained between self-evident principles within a grand system, not by the theory's fit with empirical data. But if experiments cannot be made before one knows the truth and the experiments that are made will not be able to refute or even confirm a theory, then the role of experiment in the Cartesian system seems to be more illustrative than evidential.

It is at this point that the radical difference between Bacon and Descartes becomes apparent. Bacon retained more of the qualitative explanatory goal of the school philosophy than Descartes had, but he shunned metaphysical foundations and thus had no first principles from which to derive theoretical constructs. Rather, he wrote, it was "to experience [that] we must come."[67] In addition, he was critical of using the received Aristotelian logic of his day, because it "is not nearly subtle enough to deal with nature."[68] In his *New Organon* he presented a new physical logic that

> aims to teach and instruct the understanding, not that it may with the slender tendrils of the mind snatch at and lay hold of abstract notions (as the common logic does), but that it may in very truth dissect nature, and discover the virtues and actions of bodies, with their laws as determined in matter; so that this science flows not merely from the nature of the mind, but also from the nature of things.[69]

The mind had to be liberated from the confining chains of syllogistic logic. The best way to accomplish this end was to supply it with information from the "nature of things." "For the world is not to be narrowed till it will go into the understanding (which has been done hitherto), but the understanding to be expanded and opened till it can take in the image of the world, as it is in fact."[70] Any method that failed to produce such a liberating influence was to be rejected.[71] In particular, Bacon explicitly opposed the approach that Descartes was soon to develop: "That method of discovery and proof ac-

cording to which most general principles are first established, and then intermediate axioms are tried and proved by them, is the parent of error and the
curse of all science."[72]

Bacon's pyramid inverted the order of inquiry and proof. It is the slow
accumulation of particulars that will ultimately provide the foundation for
knowledge. To begin, natural histories must be constructed, but they also have
to be "drawn up on a better plan" than had been done previously.[73] In response
to the skeptics, Bacon maintained that they "ought to have charged the deceit
upon *the weakness of the intellectual powers, and upon the manner of collecting and concluding
upon the reports of the senses.*"[74] It was not so much that the senses were fallible
sources of information but that previous philosophers had relied too much
upon superficial appearances. By concluding too readily that such appearances
confirmed their elaborate systems, they had "done more to establish and perpetuate error than to open the way to truth."[75]

Bacon insisted that one could not simply take the information of the senses
at face value. The information is filtered through the mind, which, because it
is possessed by "idols," perceives things as if through an "enchanted glass."[76]
The old logic did nothing to hinder the influence of such idols. Bacon's new
logic was designed as a way to "devise and supply helps" for the understanding,
by which philosophers could "fortify themselves as far as may be against their
[the idols'] assaults."[77] The mind and the senses must work in unison. Experience will be the foundation, but it must be accumulated in a rational "regular
order," because "experience, when it wanders in its own track," is a "mere grop-
ing in the dark and confounds men rather than instructs them."[78] Because of
the ever-present effects of the mind's idols, natural histories must be collected
by many workers and, most important, must include the use of experiments "fit
and apposite, wherein the sense decides touching the experiment only, and the
experiment touching the point in nature and the thing itself."[79] Just as with
direct sense observation, so too must these experiments be done in a rational
manner in order to be of the "most use for the information of the understanding."[80]

> For the mechanic, not troubling himself with the investigation of truth,
> confines his attention to those things which bear upon his particular
> work, and will not either raise his mind or stretch out his hand for any-
> thing else. But then only will there be good ground of hope for the fur-
> ther advance of knowledge when there shall be received and gathered
> together into natural history a variety of experiments which are of no
> use in themselves but simply serve to discover causes and axioms, which
> I call *Experimenta lucifera*, experiments of *light*, to distinguish them from
> those which I call *fructifera*, experiments of *fruit*.[81]

Bacon's desire to produce an "active philosophy" entailed for him the notion that science is and must be constructed.[82] But unlike the Cartesian construction, where science grows as a tree from the reason of man into a logically interconnected system dependent upon first principles, the Baconian construction is based upon a careful and constrained experience of the world: "The best demonstration by far is experience."[83] Experience is the foundation of knowledge, and it is also that which will prove its worth:

> The use of History Mechanical is of all others the most radical and fundamental towards natural philosophy; such natural philosophy as shall not vanish in the fume of subtile, sublime, or delectable speculation, but such as shall be operative to the endowment and benefit of man's life; for it will not only minister and suggest for the present many ingenious practices . . . but further it will give a more true and real illumination concerning causes and axioms than is hitherto attained.[84]

Boyle's Choice

Because only a slight difference exists between the actual scientific practices inspired by Descartes and by Bacon, there has been controversy over whether Boyle was a Baconian or a Cartesian.[85] At the epistemological level, however, where one seeks a justification for the foundation of science and for the subsequent theoretical results based upon such a foundation, Boyle showed a clear preference for the Baconian way of experience. It supplied, he believed, the best means for discovering truths about nature.

While Boyle thought that Descartes was "the most ingenious" of the moderns, it would be improper to conclude that he therefore believed him to be the best of the moderns.[86] For Boyle, ingenious philosophies were those proposed by "speculative divisers" of hypotheses that had "hitherto more delighted than benefitted mankind." Boyle was not entirely opposed to the use of hypotheses, but he wanted them to be "built upon a more competent number of particulars," which "may free them from all imputation of barrenness."[87] For a theory "to be solid and useful," it must be built upon an "experimental history." He complained that the explications of "the most ingenious *Descartes*" were not as useful as they could have been because they "so depend upon his peculiar notions (of a *materia subtilis, globuli secundi elementi* and the like) and these, as it became so great a person, he has so interwoven with the rest of his hypotheses, that they can seldom be made use of without adopting his whole philosophy."[88]

Boyle reflected Bacon's sentiments against the construction of elaborate philosophical systems when he wrote that "it has long seemed to me none of the least impediments of the real advancement of true natural philosophy, that men have been so forward to write complete systems of it." The study of nature had just begun in earnest, but "these systems have been often found to persuade unwary readers, that all the parts of natural philosophy have been already sufficiently explicated; and, that consequently it were needless for them to put themselves to trouble and charges in making further inquiries into nature."[89] He did not go quite so far as his English predecessor, however, who had said that "those foolish and apish images of worlds which the fancies of men have created in philosophical systems must be utterly scattered to the winds."[90] Indeed, Boyle was not "so rigid as to be unwilling, that, from time to time, some very knowing writer should publish a system."[91]

First, if the systems were composed "by very intelligent persons," they could serve as an "inventory of what hath been already discovered" and thus prevent "the needless labour of seeking after known things." Second, if these works were presented not as complete systems but "as general principles (almost like the hypotheses of astronomers) to assist men to explicate the already known phaenomena of nature," then they would be useful, because the writers, "to make good their new opinions, must either bring new experiments and observations, or else must consider those, that are known already, after a new manner, and thereby make us take notice of something in them unheeded before." Since Bacon's time, systems published, particularly by Gassendi and Descartes, had proved useful, because "the curiosity of readers" had been "excited to make trials of several things," which seemed to be "consequences" of the new doctrines, in order to "either establish or overthrow them."[92]

The value of these systems was constrained to their heuristic usefulness, however. When Boyle defended the constructions of Descartes, for example, he did so by an appeal to Bacon's precept that it is "sometimes conducive to the discovery of truth, to permit the understanding to make an hypothesis," since, "by examining how far the phaenomena are, or are not, capable of being solved by that hypothesis, the understanding may, even by its own errors, be instructed."[93] These hypotheses were not to be accepted dogmatically as "established theory," but only as temporary "superstructures," and "though they may be preferred before any others, as being the . . . best in their kind as we yet have, yet are they not entirely to be acquiesced in, as absolutely perfect, or uncapable of improving alterations."[94]

It is true that the Cartesian system had a great influence on the development of Boyle's scientific views, but Boyle never accepted any claim without first putting it to experimental test. Descartes's theories provided Boyle with ideas for the design of experiment, but he solidly condemned the manner in which Descartes had sought to support them.

> And as for the way of arguing, so often imployed . . . and so much relied
> on by many modern philosophers, namely, that they cannot clearly con-

ceive such or such a thing proposed, and therefore think it fit to be rejected; . . . there is so much difference among men, as to their faculty of
framing distinct notions of things, and through men's partiality or laziness, many a particular person is so much more apt, than these men seem
to be aware of, to think, or at least pretend, that he cannot conceive,
what he has no mind to assent to, that a man had need be wary, how he
rejects opinions, that are impugned only by this way of ratiocination.[95]

Boyle's condemnation of this way of reasoning was motivated by the same concerns that had led him to reject the Aristotelian way of philosophizing. Again,
his views reflected the influence of Bacon, who had argued against the construction of "sciences as one would."[96] Bacon had urged philosophers to build
"in the human understanding a true model of the world, such as it is in fact,
not such as a man's own reason would have it to be."[97] Boyle agreed. In the
passage immediately preceding his criticism of the Cartesian way of reasoning,
he stated: "What we are to inquire after, is, how things have been, or are really
produced; not whether or no the manner of their production be such, as may
the most easily be understood by us . . . that way may often be fittest or likeliest
for nature to work by, which is not easiest for us to understand."[98]

   Boyle's rejection of Cartesian epistemology can also be seen in his discussions concerning mathematics. He did not go to the extreme that Bacon had
in rejecting its usefulness as a necessary element of natural philosophy, but he
did refuse to accept mathematical demonstration as a sufficient basis for the
new learning. Descartes's "way of arguing," modeled after mathematics, produced certainty, but the "mathematical way" of certainty did not ensure the
truth of the propositions thus proved. In mathematical demonstrations, some
things are always taken for granted. According to Boyle, even "in all his elements of geometry," Euclid

   contents himself to demonstrate his assertions in a mathematical way,
   and does not, that I remember, answer or take notice of any one objection; and the geometricians of our days think they may safely receive his
   propositions upon the demonstrations annexed to them, without knowing or troubling themselves about the subtleties employed by the sceptic
   Sextus Empiricus, or others of that sect, in their writings against the mathematicians, and all assertors of assured knowledge.[99]

The truth of mathematical derivations depends upon the truth of the propositions from which they begin, but most of these are mere definitions, the formulations of which are often decided upon arbitrarily. "Assertors of assured
knowledge" are those who are more concerned with the subtleties of their own
demonstrations than with discovering truths about the world. As early as 1647,
in a letter to Hartlib, Boyle had expressed appreciation for the problems associated with the mathematical way of inquiring into nature. He discussed how it

would "be perhaps but one remove from impossible, precisely to declare" among the "dissenting opinions of the *Ptolemeans,* the *Tychonians,* the *Copernicans* . . . and the other novelists" which ones had been "perfectly demonstrated." What one side considered as an "undeniable demonstration" the other would "either absolutely reject as a paralogism, or at least call in question as no more than a bare probability."[100]

More will be said about Boyle's views on mathematics in following chapters; at this stage it is sufficient to note that his opposition to the "mathematical way" was an opposition not to the "application of geometrical theorems" to natural phenomena but to the attempt by some to make geometry the overriding model of investigation. From his earliest published work on natural philosophy, his *New Experiments Physico-Mechanical,* Boyle followed Bacon's inverted order of demonstration. Only through the meticulous compilation of carefully made observations and experiments would there be a "way opened, whereby sagacious wits will be assisted to make such farther discoveries in some points of natural philosophy, as are yet scarce dreamed of." In the second edition of this work in 1662, he again stressed that his aim had been not to establish principles but rather "to devise experiments to enrich the history of nature with observations faithfully made and delivered," so that "men may in time be furnished with a sufficient stock of experiments to ground hypotheses and theories on."[101]

Boyle's repeated references to Bacon indicate a true allegiance to the latter's precepts, not an empty rhetorical homage to a famous fellow countryman.[102] In the first work in which he explicitly discussed the methodological issues of experimentation, *Certain Physiological Essays* (1661), he said that "provided experimental learning be really promoted," he would be content to be an "under-builder" and "dig in the quarries for materials towards so useful a structure as a solid body of natural philosophy."[103] For forty years he devoted his life to the compilation of catalogs on nature's effects, and for the last thirty years he published an almost staggering number of reports, notes, considerations, and speculations touching a large variety of natural phenomena. His essays and histories were always written for the future. He uncovered problems with established doctrines and suggested ways in which new experiments could lead to significant revisions. He suggested new hypotheses in the hope that by such speculation he would arouse the curiosity of others to investigate nature for themselves. He gave detailed accounts of his experiments so that those just beginning the study of nature would be able to learn the proper methods and thus continue his work. Most important, he produced vast histories so that future generations would have the material on which to build. In his *Experimenta et observationes physicae,* a collection of reports published in 1691, the year of his death, he indicated that he was still devoted to the way of the "illustrious Lord Verulam," who had "so early and so well, both urged the necessity of natural history, and promoted diverse parts of it by precepts and specimens."[104]

Boyle was steadfast in his adherence to Bacon's inversion of the order of demonstration. Before one could speculate about the causes operative in nature, one first had to determine which effects were produced. It would be useless, for example, to speculate about the cause for nature's abhorrence of a vacuum if it could be shown that in some instances a vacuum can actually exist. From the start of his public career, however, Boyle met with strong opposition to this program, not only from Aristotelians but also from systematic theorists working within the new philosophy, such as Hobbes and Spinoza. In response to Boyle's *New Experiments Physico-Mechanical,* Hobbes, in his *Dialogus physicus,* asked how Boyle could have aroused "the expectations of advancing physics, when you have not yet established the doctrine of universal and abstract motion (which was easy and mathematical)."[105] In a similar manner, Spinoza, in private correspondence with Henry Oldenburg, criticized Boyle's essay on the redintegration of nitre for Boyle's attempt to unlock the secrets of the corpuscular structure of matter by experimentation instead of by "reason and calculation."[106] These critics were opposed not to the performance of experiments but to the idea that natural philosophy could be constructed upon the basis of experimental proof. Boyle's refusal to start from rational and mathematical first principles, even though he styled himself as an advocate of the mechanical philosophy, clearly confused his two antagonists. Before these criticisms had actually been leveled against him, Boyle had already written a general defense of his approach in his "Proëmial Essay" in *Certain Physiological Essays.*

In this preface to his first collection of essays, Boyle noted that his readers would likely want an explanation of why "in the reasons and explications I offer of natural effects, I have not for the most part made an immediate recourse to the magnitude, figure, and motion of atoms, or of the least particles of bodies." He agreed with the new corpuscularians that atomistic explanations "are highest in the scale or series of causes" and that "the nearer the discovered causes are to those, that are highest in the scale or series of causes, the more is the intellect both gratified and instructed." But he did not agree with "some eminent Atomists" who maintained that "no speculations in natural philosophy could be rational, wherein any other causes of things are assigned than atoms and their properties."[107]

The ultimate explanations of nature would involve reference to the motion of the least parts of matter, and mathematics would be necessary for the determination of such motions, but more than abstract reasoning would be required for this task. Because "there are oftentimes so many subordinate causes between particular effects and the most general causes of things," philosophers ought "not to despise all those accounts of particular effects, which are not immediately deduced from those primitive affections of either atoms, or the insensible parts of matter." Boyle insisted that it was necessary to determine first the "in-termediate causes" of things, which, much like Bacon's intermediate axioms, would allow philosophers, "without ascending to the top in the series of causes," to "perform things of great moment and such, as without the diligent examination of particular bodies, would, I fear, never have been found out *a priori*, even by the most profound contemplators."[108]

In this last passage it is clear not only that Boyle was following Bacon in his gradual ascent to the truth by way of the discovery of intermediate causes but also that he appreciated the full import of the Baconian link between knowledge and power—those "twin objects" of *The New Organon*.[109] In his *Use-fulness of Experimental Philosophy* Boyle noted that the "famous distinction, intro-duced by the Lord Verulam, whereby experiments are sorted into luciferous and fructiferous" could be of "commendable use," but only "if rightly under-stood."[110] It was a mistake to think that while experiments could be sorted by their dominant end, there was no commerce between the two types. According to Bacon, "truth . . . and utility are here the very same thing." The practical benefits derived from experiment may give us power over nature, but they are of even "greater value as pledges of truth."[111] As Pérez-Ramos has recently ar-gued in great detail, most historians have ignored the presence, and the sig-nificance, of this "constructive" notion of knowledge in Bacon's philosophy. Experiments are not performed merely as tests for the hypotheses created by reason, nor are they designed primarily for utilitarian purposes. Rather, experi-ments may provide direct evidence for the discovery of the causal properties possessed by natural bodies.[112]

As we will see in following chapters, Boyle did understand and appreciate the significance of this active and constructive notion of knowledge. There was for him no strict demarcation between knowledge and practice. Our ability to use the power that experimental knowledge gives us to set up circumstances in which nature will produce the effects that we desire provides us with a sure sign that we have discovered the true causes operative in nature. As Boyle de-scribed it, "Man's power over the creatures consists in his knowledge of them; whatever does increase his knowledge, does proportionately increase his power." Luciferous experiments "whose more obvious use is to detect to us the nature or causes of things" may be "exceeding[ly] fructiferous." Although few experiments had as yet yielded such power, "by discovering or illustrating the nature or causes of things" they "may yet be very useful to man's interests." On the other hand, fructiferous experiments could be of use not merely to "advan-tage our interests" but also "to promote our knowledge." "Certainly there are few fructiferous experiments" Boyle asserted, "which may not readily become luciferous to the attentive considerer of them. For by being able to produce unusual effects, they either hint to us the causes of them, or at least acquaint us with some of the properties or qualities of the things concurring to the production of such effects."[113]

As Bacon had called for a marriage of the rational and the empirical, so Boyle sought a "happy marriage" of the practical and the speculative parts of learning.[114] Not only had the order of inquiry and demonstration been inverted, but the very meaning of knowledge had been altered. No longer was the sign of knowledge to be the deductive certainty of the logical or mathematical systems of classical philosophy. Knowledge was now that which has "a tendency to use."[115] Of the three distinct philosophical alternatives of his day—Aristotelianism, Cartesianism, and Baconianism—Boyle clearly found the last to be the most promising for the advancement of knowledge as newly defined. In part, Boyle's adherence to Baconianism was a result of his almost total immersion in experimental practice that began in 1649. Before looking at the tradition of experiment within which he worked, however, it will be helpful first to examine in more depth how Boyle understood the Baconian program. In particular, it is necessary to understand what Boyle meant by his advocacy of experience as the foundation for the new learning and how such experience could provide an adequate means for discovering and justifying knowledge claims about the natural world.

Two

THE LEGAL TRADITION

BOYLE INSISTED THAT EXPERIENCE had to constrain the arbitrariness of individual reason and form the basis for a new way of doing science. Because of this, he is often characterized as having advocated an empiricist epistemology. Traditionally, empiricists have maintained that all knowledge is to be grounded in sense perception and that science is to be restricted to the knowledge of observable phenomena. As a consequence, the idea that we can ever determine the truth of theories that explain regularities by referring to unobservable entities and processes is rejected.[1] Given this definition, however, a problem of historical interpretation immediately arises: it is not clear that Boyle lived up to this empirical ideal in his actual practice. He believed that through experimentation one could learn about the configuration of the least parts of matter that are causally responsible for the manifest effects observed in nature.

Scholars have produced a number of studies in an attempt to account for Boyle's frequent appeals to experience while also addressing his actual theoretical leaps. Some have suggested, for example, that he was simply inconsistent, that he failed to see that his empirical way could not produce the results he claimed for it.[2] Others have attempted to make him a consistent empiricist by arguing that his corpuscular explanations were, and always would remain, hypothetical.[3] But while Boyle was cautious in his acceptance of corpuscular explanations, many passages in his work indicate that he did not believe that they would remain hypothetical forever. The experimental philosophy, as a method of discovery, was clearly meant to disclose truths about the world's hidden processes.[4]

The idea that there are conflicting tendencies in Boyle's thought has in large part arisen from the widespread practice of placing an empiricist gloss on his appeals to experience.[5] But the experimental philosophy of both Bacon and Boyle was presented by them as a real epistemological alternative that would overcome the failings of classical rationalism *and* empiricism. It was to be a new method of discovery that employed a new argument form for the justification of such discoveries. It was new, however, only from the perspective of traditional philosophy. The epistemological alternative had already been developed in some detail by the legal profession in England, where it was argued that the foundation of knowledge rested upon an experience of the workings of the law that had been gathered over many years. In the legal sphere, experience was a broad notion that included the use of background knowledge, skill, and expertise in the process of rational decision making. The purpose of this chapter is to show how the experimentalists' use of legal analogies when speaking of experience provides a good reason for rejecting an empiricist interpretation of their method.

The legal overtones in the works of Bacon and Boyle—their references to trials, witnesses, and testimony—are hard to ignore, but the significance of these locutions for the justification of experimental science has yet to be fully appreciated. Some previous studies have focused upon the notion of "law" in the two realms.[6] These analyses have proved unsatisfactory, however, because the analogy does not appear to be used in this substantive sense. Boyle, for example, maintained that "law" in the scientific sense should be understood metaphorically. Laws established by states for the right behavior of citizens were something that an individual ought to obey, but it was improper to say that matter "obeyed" the laws of nature.[7] These latter laws were strictly mechanical accounts of the regular workings of nature, and their formulation owed much more to the mathematical than the legal tradition.[8] While the analogy does not hold at the substantive level, however, it does at the methodological level. Experimentation, for both Bacon and Boyle, was said to be the method by which nature is put on trial and made to reveal its hidden workings.[9] Since the analogy was used in such contexts to clarify the methodological dictates of the experimental philosophy, it is reasonable to suppose that an understanding of the procedures employed by the legal profession in seventeenth-century England can provide an insight into the methodology of experiment.

The following analysis of the legal tradition is not offered as an explanatory account for the source of Boyle's ideas concerning experimental proof. Other scholars, such as Michael Hunter and Barbara J. Shapiro, have produced detailed accounts of the developmental factors that led Boyle and other seventeenth-century figures to accept an experientially-grounded natural philosophy.[10] Nor is this analysis an attempt to explain experimentalism's political function in English society. Contextualist historians such as Julian Martin, Shapin and Schaffer, and Charles Whitney have argued that legal references show the incursion of politics into the new experimental science. Although the historical accuracy of such studies has been called into question, a more important problem arises from the fact that these studies tend to dismiss the epistemic factors embedded within cultural practices.[11] My aim here is not so much to explain the seventeenth-century acceptance of experimental procedures as to investigate how Boyle used a socially accepted practice to illustrate his method. That is, instead of speculating about the political or cultural implications of the experimental method because of its connection to a social practice, I will look at how the epistemic dimension of a social practice could provide a valuable resource for the introduction of an innovative idea in science. In particular, I will examine the methodological and epistemological aspects of the English lawyers' arguments in defense of the practice of "common law" and then discuss how Boyle used similar arguments as part of his justification of experience as the foundation of the new experimental science.

English Common Law

The law of the land that governed property rights and criminal trials for theft, murder, and treason was the "common law." The law itself was extremely complex, and only those educated at one of the four Inns of Court, such as Bacon, would be fully learned in it. The procedures employed, such as trial by jury, though, were familiar to most Englishmen.[12] As its name suggests, common law was not that which resulted from the command of an individual sovereign or governing body alone; it was derived from the custom of the country and was thus said to embody a system of laws and procedures unique to England and its history.

The tradition of common law differed in a number of ways from that of Roman law, as practiced on the Continent. Most important, the traditions were based upon radically different foundations. Roman law took the Code of Justinian as its model and consisted of a dialectically constructed and codified body of legal doctrine based upon rational first principles and the citation of authorities. Roman law also concerned itself with such academic exercises as the question of the validity of Lazarus's will after he had been brought back to life by the hand of Christ.[13] Common law, on the other hand, was not a uniform system capable of codification. As precedent law, it largely consisted of chronological collections of past cases in the Yearbooks and Register of Writs. Cases were recorded in order to yield a "continuity of experience" in the law, but the compilations were not systematic. Thus, learning the law was difficult and time-consuming.[14]

At the beginning of the seventeenth century a number of prominent lawyers called for moderate reforms that would introduce more order into the English legal system and thus simplify the learning process, but they resisted the temptation to impose strict codification. Bacon, for example, in his *Maxims of the Law* rejected the attempt to set down rules "into a certain method and order" and the practice of arguing "upon general grounds," both of which he believed to be faults of the Roman law tradition.[15] He did want to reduce the laws "to more brevity and certainty," however, by compiling digests in which general maxims could be "gathered and extracted out of the harmony and congruity of cases."[16] The lack of system in the common law, while inconvenient, was generally viewed not as a weakness but as part of the law's strength. It meant that the law would be flexible enough to allow for expansion to meet new needs and modification to accommodate unusual cases.[17] A question about a particular law's interpretation, for example, was to be resolved by a thorough examination of how the law had been applied in past cases, arranged "in such order and method as they should take light from one another."[18] These precedents, while not strictly binding, provided a useful guide for interpretation, and the common opinion based upon a large number of cases carried more authority than the opinion in any individual case.[19]

Sometimes, however, laws were needed for which there were no precedents, as when the just distribution of property became an issue with the unification of Scotland and England under James I.[20] In this case the sovereign, the Parliament, and lawyers were to work together to determine the proper legal procedures. As Bacon noted, in a situation like this there was a need for some type of "transcendent reason," and the lawmakers were "more bound to resort to the infallible and original grounds of nature and common reason." He immediately qualified this remark, however, when he urged the lawmakers "to fix our consideration upon the individual business in hand, without wandering or discourses," that is, to stay as close to the particular instance as possible without rising to too high a level of abstraction. In his "Report of the First Day's Conference," addressing the question of the naturalization of Scots that had arisen from the union, Bacon wrote that the participants had attempted to use all of the resources available to them "to lay open such inducements as did lead us to our opinion, and those were fetched out of the law of reason, out of the law of nations, and out of the civil law," because "in a doubtful case in laws, inducement may many times help the judgment as well as proofs."[21] After much debate, the body of lawmakers decided that Scots should be considered as naturalized citizens of England, but the House of Commons refused to pass the act that would make this opinion law.

To get around the impasse created by the Commons's refusal to act, a test case was proposed to resolve the issue. In "The Case of the Post-Nati of Scotland," Bacon argued that the plaintiff, Robert Calvin, an infant born after James's accession, had a legal right to inherit property under the laws of England.[22] He defended this claim by arguing that a positive decision on naturalization could be reached from a careful consideration of the consequences of the other laws of union that had been passed by Parliament. Bacon's use of the flexibility of interpretation, which allowed for the application of existing law to new cases, was successful. The court decided in favor of the plaintiff, and a precedent was established for future cases involving the legal status of Scots.

Flexibility was not the lawyers' sole concern, however. They also maintained that their practice was able to yield a greater degree of certainty than the Roman one—a certainty based upon a vast amount of experience. Early in the seventeenth century Sir Edward Coke asserted that Roman law was less certain than common law because it depended upon a "number of interpretations and glosses" that gave rise to "so many diversities of opinions, as they do rather increase than resolve doubts and uncertainties."[23] In contrast, common law was based upon "the resolutions of Judges in Courts of Justice . . . reported in our books, or extant in judicial records or in both, and therefore, being collected together, shall (as we perceive) produce certainty."[24]

Instead of the "analytical" approach used in Roman law, the common lawyers advocated a "historical" approach, wherein individual reason would be constrained by the experience embodied in actual judicial decisions collected over hundreds of years. The danger inherent in the use of pure reason was that it was more liable to lead to arbitrariness and uncertainty because it was overly speculative and not grounded in the actual problems associated with legal decisions. In order to be a competent judge of legal affairs, one had to acquire "legal reason"—an instinctive ability to reason on the law, which could not be taught but only resulted from a deep and prolonged exposure to the working of the law.[25] As Coke described it, the common law is "not to be decided by natural reason but by the artificial reason and judgment of law, which law is an act which requires long study and experience, before that a man can attain to the cognizance of it."[26] Once equipped with this artificial perfection of reason, a lawyer would have the experience necessary for the discernment of similarities and differences in past cases and for the analogical application of these precedents to new cases. In the construction and application of a law, experience had to guide the use of reason.[27] But clearly, this notion of experience refers to the acquisition of background knowledge, not to today's notion that is commonly associated with sense perception or mere observation. Also, when the lawyers claimed an experiential foundation for the justification of their decisions, they appealed not to a mere accumulation of facts but to a sophisticated process of interpreting the facts. Reason was not (and could not be) excluded, but it had to be restrained.

Bacon and Coke had called for moderate legal reforms, but the revolutionary period of 1640–60 brought with it calls for more radical reforms that included revision of the legal procedures themselves. Levellers, for example, wanted a system more like the analytic approach of the Roman code, whereby all the laws would be reduced to one book that could be read and understood by any literate citizen.[28] Hobbes also advocated the "analytical" approach in his *Dialogue between a Philosopher and a Student of the Common Laws of England*. But unlike the Levellers, who wished to abolish a professional elite in the cause of equality, Hobbes wanted to restore the absolute authority to the sovereign, whose power he believed had been usurped by the privilege that the lawyers claimed for themselves in lawmaking.[29]

The lawyer of the *Dialogue,* a somewhat feeble representative of Coke, began with a rather clear statement of the meaning of "legal reason," which was to be understood as "an artificial perfection of reason, gotten by long study, observation, and experience, and not of every man's natural reason, and not the product of one man's reason."[30] The strength of the common law came from the fact that "by so many successions of ages it hath been fined and refined by an infinite number of grave and learned men." The philosopher of the dialogue responded that "there is no reason in earthly creatures but human reason" and denied that professors of law had any power over the formulation of laws, since by definition the law was "the command of the sovereign."[31] According to the philosopher, laws must be grounded in natural reason, whose marks are consistency and certainty. Since the common law did not possess such virtues, it was clearly irrational and so should be abolished in favor of the analytic method practiced on the Continent.[32]

In "Reflections by the Lord Chief Justice Hale on Mr. Hobbes His Dialogue of the Law," Matthew Hale took up the defense of the procedures of common law at the point where Hobbes's lawyer had given in.[33] He replied to Hobbes that while all humans possess a faculty of natural reason, this faculty had to be exercised on a particular subject before one could be considered qualified to make judgments in that field. It was a combination of a reasonable faculty "habituated to it [the subject] by use and exercise" that "denominates a man a mathematician, a philosopher, a politician, a physician, a lawyer." The faculty of reason can be "variously applied and directed," and different subjects demand different skills that must be learned from practice.[34] Rather than speculating on the first principles of moral philosophy, common lawyers sought practical solutions to actual problems.[35] Practical experience must restrain natural reason more so than it does in the case of those "speculators that take upon them to correct all the governments in the world and to govern them by certain notions and fancies of their own."[36]

Hale did not claim that the use of such experience would yield infallible judgments but maintained that it was preferable to adopt laws that had proved successful over time rather than to follow the dictates of individual reason: "The unknown, arbitrary, uncertain judgment of the uncertain reason of particular persons, hath been the prime reason, that the wiser sort of the world have in all ages agreed upon some certain laws and rules and methods of administration of common justice, and these to be as particular and certain as could be well thought of."[37] Because it avoids the idiosyncrasies of individual judgments, "legal reason" is "preferable before that arbitrary and uncertain rule which men miscall the law of reason." It is more reasonable to adopt those laws "made by a hundred or two hundred persons of age, wisdom, experience and interest before a law excogitated by myself" and to prefer "a law by which a kingdom has been happily governed four or five hundred years than to adventure the happiness and peace of a kingdom upon some new theory of my own." Experience is not merely the foundation of the law; it also serves as proof of its effectiveness: "Long experience makes more discoveries touching conveniencies or inconveniencies of laws than is possible for the wisest council of men at first to foresee."[38]

Hobbes's analytic and logical criticisms of the common law were misplaced because, as Hale explained, the legal sphere differs in important ways from the sphere of mathematics.

> Of all kind of subjects . . . there is none of so great a difficulty for the faculty of reason to guide itself and come to any steadiness as that of laws. . . . And therefore it is not possible for men to come to the same certainty, evidence and demonstration touching them as may be expected in mathematical sciences and they that please themselves with a persuasion that they can with as much evidence and congruity make out an unerring system of laws and politics equally applicable to all states and occasions . . . deceive themselves with notions which prove ineffectual, when they come to particular application.[39]

As Bacon had noted in the previous generation, "Civil knowledge is conversant about a subject, which of all others is most immersed in matter, and with most difficulty reduced to axioms."[40] Hale agreed with this assessment and argued that not only is a mathematical demonstration not possible, but it is not appropriate in the legal realm. Experience must form the foundation of the law, because only from experience could one come to the knowledge necessary for rational legal decisions: "Men of observation and experience in human affairs and conversation between man and man make many times good judges, yet for the most part those men that have great reason and learning . . . are most commonly the worst judges that can be because they are transported from the ordinary measures of right and wrong by their over fine speculations, theories, and distinctions." The common law was an institution that had been successful for many years, and as Hale said, it is "a foolish and unreasonable thing" to "expect a mathematical demonstration to evince the reasonableness of an institution or the self evidence thereof."[41] Unlike a mathematician who could consider the relations between abstract definitions, a lawyer needed experience of how similar cases in the past had been resolved and how these precedents should apply to present cases.

Matters of law were to be tried and ultimately decided in the courts by a method of proof, sometimes referred to as "moral demonstration." It is important at this point to note that while a mathematical model of proof was rejected in part because it was not possible to attain, the lawyers did not reluctantly settle for moral demonstration as an inferior mode of proof. Rather, they believed that for the subject, it was superior to the analytic methods of mathematics, logic, and dialectic. The lawyers did not reject the use of logic and dialectic entirely, though. Such tools were necessary for the development of natural reason, but experience of the law had to be cultivated for the development of legal reason.[42] Experience and moral demonstration played equally significant roles in the determination of the matters of fact that took place in the courts, and here again, there was a marked difference between Roman law and common law.

Roman law required complete proof (*probatio plena*) before a verdict could be reached. Proof did not consist in the balance of persuasion. Rather, strict mechanical rules of evidence were followed. The canonical method of taking evidence consisted of questioning witnesses in private and then introducing their written depositions as evidence to a closed court. Based upon this evidence, the judicial bench would determine the facts of the case and pass legal rulings on its own findings.[43] Different pieces of evidence carried predetermined numerical values (from 0 to 1), and by a simple arithmetical calculus, complete proof was achieved when these values totaled 1. If the evidence provided only a *probatio semi-plena*, then other measures, such as torture, were employed to obtain certainty.[44]

In common law, on the other hand, trials were a public affair, and witnesses testified in an open court. A jury of twelve men would deliver its verdict on the facts of the case, and the bench would pass judgment on the verdict.[45] Duties in the trial court were sharply defined: while the judges were to decide "matters of law," it was the responsibility of the jurors to decide "matters of fact."[46] The jury was free to assess the value of individual pieces of evidence, and the common law courts did not require complete proof in order for them to reach a verdict. The method of proof, originally derived from the Anglo-Saxon law of the ninth century, was that of "compurgation." Jurors were chosen on the basis of the knowledge that they had of the case before the court, and they frequently came from the same neighborhood as the accused. Compurgators (witnesses) would take an oath, and the truthfulness of their testimony would be judged by the jury. While by the seventeenth century the procedure had changed somewhat—jurors' prior knowledge was not always possible or desirable—in practice the jury frequently retained its self-informing role.[47] If the evidence was incomplete, for example, the jury could proceed upon its own knowledge. The final decision was to be based upon the probable merits of the cases put forth by the accused and the accuser, and the case would be decided in favor of the one whose account appeared to the jury most likely to be true. The "proof" of the case was said to consist in the "finding of a body of reasonable men," according to the probabilities of the case.[48]

The jury was charged with assessing "the credibility of witnesses and the force and efficacy of their testimonies." All testimony given under oath was not to be treated as though it was of equal value. Indeed, as Hale wrote, if jurors had "just cause to disbelieve what a witness swears, they are not bound to give their verdict according to the evidence, or testimony of that witness; and they may sometimes give credit to one witness, though opposed by more than one."[49] The members of the jury had to reach decisions about matters of fact "remote from our sense," that is, about past occurrences they had not directly witnessed, with the help of their own past experience concerning the likelihood of the witnesses' credibility and of the event's having occurred.[50] A juror had to "infer and conclude from the testimony . . . by the act and force of his understanding."[51] In opposition to the notion of a mathematical demonstration and the Roman law's notion of strict numerical probability, the common law offered a model of demonstration in which experience was fundamental for the reasonable resolution of cases. This notion of an experiential foundation relied heavily upon the idea of an expert.

The expert was one who had developed his reason by experience in a specific area and was thus the most qualified to judge in that area.[52] The jury would have expert status in the judgment of the veridical nature of the witnesses' testimony, by virtue of past experience of the reputations of the accused and the witnesses, and thus of the likelihood of the matter of fact's having occurred. The judge (a lawyer) was the expert, who, by reason of his long experience of the workings of the law, derived in part from his mastery of the "common erudition" learned at the Inns, was able to deliver the best judgment on matters related to the interpretation of the facts and the resolution of the case. In the process of rational adjudication, the use of background knowledge was necessary both for the establishment of the facts and for the application of past cases that would determine the relevance and interpretation of the facts. This same broad notion of experience, and the type of demonstration grounded upon it, played an integral role in the attempts by English experimentalists to justify the knowledge-producing character of their enterprise. This can be seen most clearly in the case of Boyle, who, expanding upon the methodological precepts of Bacon, made frequent use of the lawyers' arguments in the interest of advancing the cause of experimental philosophy.

Experience and the Experimental Philosophy

Bacon thought that the sense "by itself is a thing infirm and erring."[53] When he advocated experience as the "solid foundation" for natural philosophy, he clearly did not have the type of experience in mind that is today equated with simple sense perception. Rather, *experience* referred to the skill and expertise that could be gained from a close and prolonged acquaintance with the subject being studied. The compilation of background knowledge would enable a philosopher to be able to judge the credibility of the testimony presented to the senses and to decide what could be concluded from such testimony. It was not the experience of the "sooty empirics" who had "contributed to supply the understanding with very bad materials for philosophy and the sciences," because of their reliance upon "observations, careless, irregular, and led by chance; tradition, vain and fed on rumor; practice slavishly bent upon its work; [and] experiment, blind, stupid, vague, and prematurely broken off."[54]

In order for advances to be made in natural philosophy, the intellect must as well "be supplied with fit matter to work upon, as with safeguards to guide its working." Natural histories would provide this informational basis, but they had to be "gathered on a new principle."[55] As he had called for reform in the compilation of legal histories, so too did Bacon appreciate the need for a better plan of gathering the information for natural histories. They could not be compiled from simple sense perceptions, because "by far the greatest hindrance and aberration of the human understanding proceeds from the dullness, incompetency, and deceptions of the senses; in that things which strike the sense outweigh things which do not immediately strike it, though they be more important." Because previous philosophers had relied upon the superficial appearances of things, "all the more subtle changes of form in the parts of coarser substances" had remained "unobserved."[56] Until these things are "searched out and brought to light, nothing great can be achieved in nature," for "the truer kind of interpretation of nature" will only be achieved when experiments "fit and apposite" have been made to reveal its hidden workings.[57] "To the immediate and proper perception of the sense, therefore, I do not give much weight; but I contrive that the office of the sense shall be only to judge of the experiment, and that the experiment itself shall judge of the thing."[58]

These histories were not meant as mere empirical catalogs of facts about nature, nor were they to be confined to those things which would be immediately useful. Rather, "the object of the natural history" was "to give light to the discovery of causes."[59] To find such causes, philosophers should not prematurely attempt to "resolve nature into abstractions" but should "dissect her into parts; as did the school of Democritus, which went further into nature than the rest."[60] While one must start from a careful consideration of the particulars of nature, the causes underlying the occurrence of these particulars could not be discovered by simple enumerative induction, which is a "puerile" thing. Hidden causes have to be inferred from the manifest effects observed in nature. But all speculation about such causes could be held as "no way established" until they had been "submitted to new trials and a fresh judgment has been thereupon pronounced."[61] As in a court of law, procedures had to be followed in the right order. First the philosopher had to determine the facts of the case, including those "remote from sense," and the testimony presented would have to be assessed in part by one's previous experience about what nature was capable of doing. Only when such evidence had been collected and verified could the philosopher then go on to judge what could be concluded upon the basis of such testimony.

Boyle understood the experiential foundation of the new philosophy in much the same way as Bacon had. He criticized other philosophers not only for their general reliance upon logic and demonstration but also for their improper use of the information of the senses. They had "presumed to give us general axioms upon insufficient inductions, and without thoroughly penetrating the differing natures of the things included in those comprehending axioms."[62] Because they had contented themselves with "the superficial account given us of things by their obvious appearances and qualities," they were little better than the Baconian "spider in a palace," who,

> taking notice only of those objects, that obtrude themselves upon her senses, lives ignorant of all the other rooms in the house, save that wherein she lurks; and discerning nothing either of the architecture of the stately building, or of the proportion of the parts of it in relation to each other, and to the entire structure, makes it her whole business, by intrapping of flies, to continue an useless life; or exercise herself to spin cobwebs, which, though consisting of very subtile threads, are unserviceable for any other than her own trifling uses.[63]

Before speculating about the most general "axioms metaphysical, or universal," one should first discover those "axioms collected or emergent; by which I mean such as result from comparing together many particulars."[64] The "judgments of reason" upon which such axioms are based are only "fit to be relied on, according as the informations they are grounded on are more or less certain and full."[65] Although almost half a century had passed since Bacon's "Great Instauration," Boyle still felt the need to caution his contemporaries about

> how incompleat the history of nature we yet have, is, and how difficult it is to build an accurate hypothesis upon an incompleat history of the phaenomena it is to be fitted to; especially considering, that . . . many things may be discovered in after-times by industry or chance, which are not now so much as dreamed of, and which may yet overthrow doctrines speciously enough accommodated to the observations, that have been hitherto made.[66]

It was because of the incompleteness of the informational basis that Boyle was content to be the "under-builder." As Bacon had noted earlier, however, the compilation of complete natural histories was not the end product of the new experimental philosophy, but merely the preliminary stage of inquiry designed to give the understanding fit matter to work upon.

Boyle did not discount the use of reason at this early stage of investigation. But he did argue that reason must be "improved by meditation, conferences, observation and experiments" that "need not destroy a dictate of reason, but only give it a limitation and restrain it."[67] Natural histories have to be compiled in a rational manner, and reason is required in order to ask the right questions of nature and interpret the answers received. The "experimental philosopher is not a mere empiric . . . who too often makes experiments, without making reflection on them, as having it more in his aim to produce effects, than to discover truths."[68] To gain experience of the world, one must "make reflections on the information of the senses" and not simply receive sense impressions passively.[69] Experience provides the matter, the information, upon which reason is to be employed, but "the understanding remains still the judge."[70] The combination of the information gained from sense perception and the learned expertise built up over years of practice provides the evidence that the philosopher must use in the construction of theories about nature: "It cannot but be a satisfaction to a wary man to consult sense about those things, that fall under the cognisance of it, and to examine by experiences, whether men have not been mistaken in their hypotheses and reasonings."[71] If reason is not restrained by such experience, then it is quite likely that the Baconian idols, those preconceived notions that all investigators possess, will prejudice the outcome of the study. For Boyle, reason could be compared to "an able judge, who comes to hear and decide causes in a strange country."[72]

> For the general notions he brings with him, and the dictates of justice and equity can give him but a very short and imperfect knowledge of many things, that are requisite to frame a right judgment about the cases, that are first brought before him; and before he has heard the witnesses, he may be very apt to fall into prejudicate opinions of things. . . . But when an authentic and sufficient testimony has cleared things to him, he then pronounces, according to the light of reason, he is master of; to which the witnesses did but give information, though that subsequent information may have obliged him to lay aside some prejudicate opinions he had entertained before he received it.[73]

The judge decides according to reason, but only after his reason has been improved by sufficient testimony. Not only are premature judgments likely to be faulty, but they can also prejudice one's ability to reach a right judgment in the future if they become set, like idols, in the mind of the one who is to judge. In natural philosophy, the inquirer whose reason has been habituated to learning and made expert by a prolonged exposure to the workings of nature will know best how to judge, and conclude from, the testimony presented. Such an expert will know better than to leap to premature judgments before all of the evidence is in. Natural philosophy must be "built upon two foundations, reason and experience." Both of these elements are required, and it is misleading to think of them as conflicting. Rather, they are complementary aspects of a complex method designed to allow the sensory and rational faculties to be used to their best advantage, primarily by providing constraints upon each other. Reason "is not degraded" by the requirement of experience. Rather, "by her own dictates," reason is obliged "to take in all the assistance she can from experience . . . and by the fuller accounts of things she receives from those informations, to rectify, if need be, her former and less mature judgments."[74]

The experimental philosophy was designed as a way to discover "emergent axioms," those truths that, while they rarely admit of exceptions, may not be "unlimitedly true."[75] That is, the truths obtained from it will not possess the metaphysical necessity of Cartesian first principles but will be limited to a knowledge of how things are actually produced in the world as it is presently constituted. But the experimental philosophy was also designed as a way to provide rigorous testing procedures by which such limited truths could be proved. As Boyle's method of discovery was modeled after that used in the legal realm, so too was his method of justification. He set up a standard of rational assent, quite similar to that employed by the lawyers, in which demonstration was to be based upon a balance of probabilities.

> When we are to judge, which of two disagreeing opinions is most rational, i.e. to be judged most agreeable to right reason, we ought to give sentence, not for that, which the faculty, furnished only with such and such notions, whether vulgar, or borrowed from this or that sect of philosophers, would prefer, but that, which is preferred by the faculty, furnished, either with all the evidence requisite or advantageous to make it give a right judgment in the case lying before it, or, when that cannot be had, with the best and fullest information, that it can procure.[76]

A "judgment of reason" is that "which takes in the most information procurable, that is pertinent to the things under consideration."[77] A suspension of judgment is required prior to the collection and verification of evidence. Sometimes conflicting evidence will require a suspension of judgment until the conflict has been resolved, either by a reassessment of the evidence or by a theory that can explain the apparent conflict. But there are times when there is no reason to doubt. If all of the evidence points to one side of the issue and there is no evidence on the other side that would militate against it, then the reasonable course would be to follow the standard of assent and affirm the conclusion.

Moral demonstration can yield undoubted assent. Of course, one could be mistaken, but a general skepticism about knowledge acquisition does not provide a specific reason to doubt. While judgments remain fallible, they are rational to accept, and it would be unreasonable to continue to doubt in the face of overwhelming evidence. Moral demonstration is demonstrative. Its strength, as a mode of proof, is clearly exhibited in a passage in which Boyle described moral demonstration as that which is

> made up of particulars, that are each of them but probable; of which . . . the practice of our courts of justice here in *England*, afford us a manifest instance in the case of murder, and some other criminal causes. For, though the testimony of a single witness shall not suffice to prove the accused party guilty of murder; yet the testimony of two witnesses, though but of equal credit, that is, a second testimony added to the first, though of itself never a whit more credible than the former, shall ordi-narily suffice to prove a man guilty; because it is thought reasonable to suppose, that, though each testimony single be but probable, yet a con-currence of such probabilities, (which ought in reason to be attributed to the truth of what they jointly tend to prove) may well amount to a moral certainty, *i.e.*, such a certainty, as may warrant the judge to proceed to the sentence of death against the indicted party.[78]

A conclusion built upon a concurrence of probabilities "cannot be but allowed, supposing the truth of the most received rules of prudence and principles of practical philosophy."[79]

The probabilism expressed here is not one that reflects any in-principle hypothetical nature of knowledge, however. Nor is it a product of likelihoods. Boyle contrasted moral demonstration with the kind of reasoning found in Pas-cal's wager argument, which is often cited today as the beginning of our mod-ern notion of quantitative probability. The wager argument produces a conclu-sion that is "of less cogency" than a moral demonstration that can "determine our resolves," but acting upon it can still be prudent.[80] Wagering, which con-cerns the practical aspects of reaching a decision under uncertain circum-stances, involves the figuring of likelihoods. One can act rationally "if all things considered," the outcome betted upon "appears more likely to be true, than not to be true."[81] Moral demonstration, on the other hand, is achieved only in the absence of specific reasons to doubt. It seems clear, then, that on the occa-sions when Boyle spoke of the probable truths reached by moral demonstra-tion, the term *probable* had the qualitative meaning "worthy of approbation" and not today's quantitative meaning expressed by degrees of likelihood.[82] Moral demonstration is compelling. If there is a concurrence of probabilities where all of the evidence favors a conclusion, then in order to be rational, one must assent to its truth.

As the standard for the rational acceptability of knowledge claims, concur-rence was used by Boyle in all of his scientific work. At the level of hypotheses, for example, he maintained that in order for them to "peaceably obtain discern-ing men's approbation," they must be grounded in experience.[83] A hypothesis is so grounded when one can show that it agrees "with all other phaenomena of nature as well as those it is framed to explicate."[84] Further, because the informational basis is as yet incomplete, hypotheses must be continually tested in the court of experience. It is the experimentalist's duty to perform such tests in order to ensure that no phenomena are inconsistent with the proposed hypothesis. As Boyle explained, an "excellent hypothesis" is one that enables "a skilful Naturalist to foretell future Phenomena, by their Congruity or Incongruity to it; and especially the Events of such Experiments as are aptly devised to Examine it; as Things that ought or ought not to be Consequent to it."[85]

In a manner reminiscent of Coke's and Hale's arguments about the effectiveness of the common law being a sign of its goodness, Boyle believed that the effectiveness of a hypothesis, its ability to account for new cases, was a mark of its excellence.[86] A hypothesis that accounts for all of the known phenomena is *probable*. A positive evaluation is called probable, rather than "true," simply because the evidence upon which it is grounded is incomplete. Boyle cited the fact that the recent invention of the telescope and "other philosophical instruments" had increased the amount of information in the natural histories upon which theories were to be built and there was reason to believe that these histories would continue to grow in the future.[87] A hypothesis would either remain probable or become improbable as new information came to light. More will be said about the actual applications of the concurrence notion to experimental practice in future chapters. Before concluding this chapter, however, it will be helpful to look at the sometimes heated debates that occurred between Boyle and Hobbes to illustrate the general epistemological notions that Boyle derived from the Baconian and legal traditions.

Boyle versus Hobbes

In his *Dialogus physicus* Hobbes criticized Boyle's manner of proof in much the same way as he had criticized the arguments of the lawyers. He argued that experiments could not establish either facts or theories about the world. One had first to find, by an "easy and mathematical" way, the "doctrine of universal and abstract motion."[88] Without such a doctrine, "however much work, method or cost be expended on finding the invisible causes of natural things, it would be in vain."[89] Again, as in his dialogue on the law, Hobbes stressed the use of natural reason by which definitions and axioms were to be settled upon and judged by the criteria of consistency and certainty. There are "two things" required to "make a legitimate hypothesis": the "first is that it be conceivable," the "other, that by conceding it, the necessity of the phenomena may be inferred."[90]

In the second edition of *New Experiments Physico-Mechanical* Boyle responded to Hobbes's methodological criticism.[91] He argued that in addition to Hobbes's two criteria for the acceptance of a hypothesis, he would "add a third, namely that it be not inconsistent with any other truth or phaenomena of nature."[92] In addition to the conceivability and internal consistency of a hypothesis, then, it also had to be externally consistent with all other hypotheses and facts about nature. By increasing the amount of information for which hypotheses must account, Boyle's third criterion provided a much stronger standard for their acceptance than that proposed by those who favored the way of mathematical abstraction. Few hypotheses will pass this strong standard, but when we find one that does, then we have reason to believe that we have made contact with the world and that our theory is more than a mere mental construct. Boyle defended this criterion in a number of his published works, where his arguments again closely followed those put forward by the lawyers in defense of their practice. In particular, he made the same type of argument against Hobbes's use of mathematical analysis in natural philosophy as Hale had made against its use in the law.

Boyle argued that, despite its fallibility, a demonstration produced by experimental science is superior to one produced by the mathematical way of reasoning. There is a complexity and subtlety in the physical world that pure mathematics is not able to capture. Geometry may produce an axiomatic system that assures the certainty of its conclusions by virtue of its logical structure, but Boyle was suspicious of such knowledge claims because they are "built upon suppositions and postulates . . . about which men are liable to slip into mistakes."[93] The arbitrary nature of the postulates put forth by "mathematical writers" meant for him that "the certainty and accurateness, which is attributed to what they deliver, must be restrained to what they teach concerning those purely-mathematical disciplines, arithmetic and geometry, where the affections of quantity are abstractedly considered."[94]

The high level of abstraction and generality employed by mathematicians made their proofs insufficient for the task of natural philosophy. Mathematics is a necessary component of natural philosophy, but proofs in the sciences cannot always follow the mathematical model of demonstration, because "there are many truths" that "by the nature of the things are not capable of mathematical or metaphysical demonstrations, and yet, *being really truths*, have a just title to our assent; it must be acknowledged, that a rational assent may be founded upon proofs, that reach not to rigid [i.e., mathematical] demonstrations, it being sufficient, that they are strong enough to deserve a wise man's acquiescence in them."[95] As we have seen, the experimental philosophy was designed as a way to discover and justify these types of truths. The third criterion was required because it was the only way in which one could be certain of having collected and accounted for all of the relevant information. As Boyle said, the purpose of a hypothesis is

> to render an intelligible account of the causes of the effects, or phaeno-
> mena proposed, without crossing the laws of nature, or other phaeno-
> mena; the more numerous, and the more various the particles are,
> whereof some are explicable by the assigned hypothesis, and some are
> agreeable to it, or, at least, are not dissonant from it, the more valuable
> is the hypothesis, and *the more likely to be true.* For it is much more difficult,
> to find an hypothesis, that is not true, which will suit with many phaeno-
> mena, especially, if they be of various kinds, than but with a few.[96]

Boyle's goal in natural philosophy was different from that of Hobbes. While Hobbes began with the most universal, abstract principles of motion, Boyle sought first to find the intermediate causes and axioms of nature, and the two men's methods differed accordingly. Given these differences, a general argument about the proper goal of science might have been appropriate, but it is not at all clear why they should have argued at the methodological level. Hobbes had publicly challenged Boyle's work, but even then, many of Boyle's associates, including Christian Huygens, had urged him not to waste his time responding to such "frivolous" objections.[97] As Boyle said, however, it was not so much the content of Hobbes's objections but his "fame and confident way of writing" that "might prejudice experimental philosophy in the minds of those who are yet strangers to it." Boyle noted that although Hobbes in an earlier work, *Mathematicae hordiernae,* had approved of experimental investiga-tions, in his *Dialogus* he had nevertheless "(by an attempt, for aught I know, unexampled) endeavoured to disparage unobvious experiments themselves, and to discourage others from making them. Which if he could by his dialogue effect, I dare be bold to say, he would far more prejudice philosophy by this one tract, than he . . . can promote it by all his other writings."[98]

In his *Dialogus* Hobbes had made it look as if Boyle's experiments had been arbitrarily performed and interpreted in order to prove the existence of a vac-uum. In response, Hobbes maintained, somewhat paradoxically, that Boyle's experimental results actually confirmed his own plenist doctrine: "Your colle-giates have as yet in nothing advanced the knowledge of natural causes, but that one of them hath found out an engine, in which there may be such a motion of the air excited . . . that the hypotheses of Mr. *Hobbes,* before probable enough, may be thence made more probable."[99] Although Hobbes had clearly questioned the efficacy of experimental proof, in this last passage he made an implicit appeal to Boyle's third criterion for theory acceptance and thus left his theory vulnerable to experimental refutation. If Hobbes had been content to argue as a mathematician, Boyle's subsequent response, wherein he used experi-mental evidence to defeat Hobbes's theory, would have been inappropriate, as

Shapin and Schaffer have claimed it was.[100] Hobbes's hubris had led him to make a much stronger claim, however, and he could not have it both ways. Either experiments are not evidence, in which case his alternative explanations would be beside the point, or experiments are evidence, in which case they could be used against his theory. Boyle noted this tension in Hobbes's criticisms and rightly saw that some might assume that Hobbes was following an experimental method. A response to Hobbes's contention that the air-pump experiments were inconclusive was therefore necessary. In particular, Boyle would use the concurrence notion to show, legitimately and easily, how Hobbes's explications were not consistent with the entire body of experimental evidence.

In his *New Experiments Physico-Mechanical* Boyle described experiments performed in an air-pump that had been constructed in such a way that the air could be evacuated from a glass vessel, called the receiver, in which various objects could be enclosed and manipulated. He did not, and never would, claim that a true vacuum was created in the receiver. Rather, the space in the receiver was "not a space, wherein there is no body at all, but such as is either altogether, or almost totally devoid of air."[101] His primary purpose in reporting the hundreds of experiments performed with this machine was to demonstrate that the "spring" of the air, its ability to be condensed and rarefied, was responsible for many phenomena that had previously been explicated by various plenist doctrines concerning nature's abhorrence of a vacuum.[102] Along the way, Boyle described two sets of experiments concerning the effects that his experimental vacuum had on the flames of candles and the lives of animals.[103] After repeating these experiments many times and varying the types of candles and animals enclosed, he noted that in all cases the flame or the life of the animal would be extinguished at least ten times faster when the receiver had been exhausted than when the objects were merely sealed up in the unexhausted receiver. Boyle therefore speculated that a "certain vital quintessence" in the air was necessary to support both respiration and combustion.[104] He also concluded that these results validated the integrity of his machine by supporting his claim that the atmospheric air had been pumped out of the receiver.

Hobbes disagreed. He maintained that when suction was applied to Boyle's machine, the air could not leave the receiver, since, the world being full, it would have nowhere to go.[105] Rather, what happened was that the air inside the receiver was put into a violent agitation by the pumping action, which increased the "consistency" of the enclosed air so that it approached the density of water. This heavier air circulating violently in the middle of the receiver caused the flame to be extinguished, much as the "damps" in mines could put out a flame.[106] In a similar manner, he maintained that the animals shut up in the receiver did not die for lack of air. If they did, he asked his interlocutor, "How then do they, who make a trade of diving, live under water, of whom there be some, who being accustomed from their childhood, have wanted air a whole hour?"[107] It was not a lack of air but a "compulsion of the air from all parts towards the center of the spherical glass, in which the animal is inclosed, and so may be seen to die stifled by the tenacity of the compressed air, as it were, with water."[108]

Boyle rejected Hobbes's alternative explanation for a number of reasons. First, the example of divers was not appropriate, since, as he noted, "his [Hobbes's] argument against my conjecture is, in the passage that proposes it, answered by himself." That is, the divers did not literally "lack air" but had been accustomed to holding their breath for long periods of time, something that it could not be assumed that the animals had been trained to do. Also, Hobbes's violent motion was supposed to be most operative in the center of the receiver, but Boyle was, he noted somewhat sarcastically, "wont to keep" the enclosed animals "not near the center of the receiver . . . but near the bottom of it, that the included animal might have something firm under his feet." In addition, Boyle observed that while it was true that a candle could be extinguished by the damps in mines, there were many other ways in which it could be extinguished, and there was simply "no sign of any damp or unusual thickness of the air" in the receiver.[109] He admitted that his machine was subject to leakage, but it was a quite different thing to maintain that not only was the receiver not empty but that the air inside had its density increased to that approaching water. Hobbes could have escaped these criticisms if he had consistently denied the efficacy of experimental proof, but as we have seen, he insisted that his theory could explain all of the phenomena produced by the air-pump. The most decisive part of Boyle's strategy was to show that while the individual explanations given by Hobbes might appear intelligible, taken together they were not.[110]

In the first edition of *New Experiments Physico-Mechanical* Boyle described an experiment that he had done in order to discover the density of the matter left in the receiver. He started a pendulum swinging inside the exhausted receiver at the same time as one was set outside in the open air. He had hoped that the enclosed pendulum would gradually swing at a quicker rate and thus give him some clue about the relationship between the enclosed medium and the atmospheric air. But there was no change—both pendulums kept time.[111] Hobbes took this experiment as confirmation of his plenist theory: "For the receiver was not, as you thought, emptier after the suction than before."[112] But surely this was no explanation. If the air was as dense as Hobbes claimed, so dense that the animals enclosed in an exhausted receiver could be described as having died by "drowning," then the pendulum's motion ought to have been retarded by such a medium. As Boyle noted in his response to Hobbes, "If the receiver be in our present case filled with a substance, whose consistence is so much nearer that of water than is our common air, as Mr. *Hobbes* would have it; how chance a pendulum should not move very sensibly slower in it, when in water the diadromes are so exceedingly much more slow?"[113]

In the second edition of *New Experiments* Boyle included reports of many other experiments, some of which were apparently designed specifically to refute Hobbes's theory, and in his *Continuation* of 1669 a particularly striking refutation was achieved when Boyle released a feather in the exhausted receiver and observed that "it descended like a dead weight."[114] While Hobbes might have been able to get around the argument of the pendulums, he could not account for this new phenomenon with a theory that maintained that a violent circular motion of thick air was present in the center of the receiver. Hobbes was a "failed experimentalist," and it was not illegitimate for Boyle to point this out. Hobbes had made an implicit appeal to the experimental standard of proof and had not been able to live up to it. He was *not* "able to show that the air-pump did not work in the manner claimed for it."[115]

Boyle's use of legal analogies shows that his notion of experience was much broader than an empiricist appeal to sense perception and that his standard of proof consisted of a complex reasoning process by which the rational acceptance of hypotheses would be established upon the basis of a concurrence of probabilities.[116] The spring of the air, for example, was not something that could be seen, but asserting its existence was the best hypothesis that could be constructed to explain all of the phenomena seen in Boyle's air-pump experiments. In an extremely eclectic fashion, Boyle constructed a moderate philosophy of science designed to possess the best elements from both empiricist and rationalist approaches to the study of nature. But this eclecticism also led him to produce a unique philosophy of science that ought to be evaluated in its own terms. I have argued that he modeled his philosophy in part upon the precepts set down by Bacon, and his use of legal analogies makes the influence of Bacon all the more apparent. But it must be stressed that while Boyle followed the earlier philosopher's larger vision of science, he expanded upon and modified that vision in its methodological details. He also gave a more sustained and sophisticated justification for the use of such methods based upon the experience he gained by an almost total immersion in experimental practices and the precedent he found in the work done by other experimentalists in the years following Bacon's death.

Three

THE EXPERIMENTAL TRADITION

WHEN BOYLE RETURNED TO ENGLAND in 1645, he spent a few months in London with his sister and the Hartlib circle and then retired to Stalbridge, his newly inherited estate at Dorset. While putting his affairs in order, he managed to find the time to continue his study of the new philosophy by reading, in addition to Bacon and Descartes, most of the ancient atomists as well as the recent account of their doctrine by Gassendi. Boyle wrote that he considered Gassendi "to be a very profound mathematician, as well as an excellent astronomer, and one that has collected a very ample treasury of numerous and accurate observations."[1] He also began to read the works of Mersenne. In one of them he found an account of a wind gun that led him to speculate about the construction of a pneumatic engine "to discover the weight of the air; which, for all the prattling of our book-philosophers, we must believe to be both heavy and ponderable, if we will not refute belief to our senses."[2] At the same time he continued to pursue his recently acquired interest in chemistry. After much difficulty, he had a laboratory set up at Stalbridge in the summer of 1649.[3] He began to record extensive notes on his readings, on the results of his experiments, and on observations from foreign correspondents. He also began to compile lists of subjects upon which he wished to write in the future. The first list contained a large number of ethical and theological themes but also included items such as "Of Natural Philosophy and Filosofers," "Of Chymistry and Chymists," "Of Universality of Opinions and Of Paradoxes," "Of Authority in Opinions," "Of Cold," "Of Atoms," and "Of Reasoning and Discourse."[4]

In 1652 Boyle once again moved his residence, this time to his newly acquired estates in Ireland.[5] He was unable to establish a chemical laboratory, but through renewed acquaintance with William Petty, who was serving as the general's physician in Dublin, he became involved in performing a number of anatomical dissections.[6] He had worked diligently at his studies and was beginning to receive recognition from the intellectual community for his mastery of the new philosophy. His circle of acquaintances had grown to include Kenelm Digby, the chemist George Stirke (Starkey), and Dr. Highmore, who dedicated his *History of Generation* to Boyle.[7] The esteem that others held for him can be seen, admittedly in exaggerated form, in a letter from John Wilkins, warden of Wadham College, Oxford, whom Boyle had met during a brief visit to England in 1653:

> I should exceedingly rejoice in your being stayed in *England* this winter, and the advantage of your conversation at *Oxford*, where you will be a means to quicken and direct us in our enquiries. And though a person so well accomplished as yourself, cannot expect to learn any thing amongst pedants, yet you will here meet with divers persons, who will truly love and honour you. . . . If I knew with what art to heighten those inclinations, which you intimate of coming to *Oxford*, into full resolutions, I would improve my utmost skill to that purpose; and shall be most ready to provide the best accommodations for you, that this place will afford.[8]

Boyle did not see Wilkins again until 1655, when he visited him at Oxford. He was quite impressed with this meeting, as he wrote to Hartlib:

> That which most endear'd to me my entertainment was the delight I had to find there a knot of such ingenious and free philosophers, who I can assure you do not only admit and entertain real learning but cherish and improve it, and have done and are like to do more toward the advancement of it than many of those pretenders that do most busy the press and presume to undervalue them, and during the little time I spent there I had the satisfaction to hear both the chief professors and heads of colleges maintain discourses and arguments in a way so far from servile that if all the new paradoxes have not found patrons there, 'tis not their dissent from the ancient or vulgar opinions but some juster cause that hinders their admission.[9]

Within a year Boyle had settled at Oxford and had become a member of the group that met weekly in Wilkins's chambers to advance experimental philosophy. Among this group were Christopher Wren, professor of astronomy; Richard Lower, physician and physiologist; Thomas Willis, physician; John Locke, medical student; and Robert Hooke, student and chemical assistant to Willis. Hooke, upon the recommendation of Willis, became Boyle's assistant around 1658. Shortly thereafter the two men constructed an air-pump and began the first of a long series of pneumatic experiments.

By the Restoration of 1660 Boyle was a well-respected and accomplished man of thirty-three. He had read most of the new works in natural philosophy, conducted experiments suggested by these works, and designed new experimental apparatuses and procedures to further the advancement of natural philosophy. He was also actively involved in constructing and justifying a new methodology. Aside from the general epistemological arguments in favor of experience derived from the legal and philosophical traditions, his defense of an experiential grounding for natural philosophy was primarily a product of his own intense and prolonged exposure to experimental practices. Therefore, in order to appreciate his philosophy of science fully, the experimental tradition within which he worked must be understood. Disciplinary boundaries were not as sharp as they are today, but some distinctions were drawn, and it will be helpful to look at the experimental tradition as comprising three areas of investigation: physical-mechanical, alchemical-chemical, and medical-anatomical. The following is not meant as a definitive study of the situation of these disciplines at that time, however, but as a discussion of Boyle's perceptions of their methodological details, particularly his criticisms of the way in which knowledge was being pursued within them and his recommendations for change.

The Physico-Mechanical Tradition

Boyle's early study of mechanics provided him with examples of the fruitfulness of experimental practice to the discovery of truths about nature. He was impressed by the successful work performed by Bacon on heat and by Mersenne on sound, but it was Galileo who remained for Boyle "the great master of mechanics" and whose work provided him with most of the examples that he would use for the promise of an experimental method in the physico-mechanical realm.[10]

Galileo's telescopic observations had significantly contributed to the successful overthrow of Aristotelian cosmology, and his experiments in mechanics were to change the science of motion.[11] Galileo did not attempt to set up a grand philosophical system but focused upon particular aspects of mechanics. As Boyle did later, he wrote that he would "be satisfied to belong to that class of less worthy workmen who procure from the quarry that marble" that later gifted sculptors may use to create "masterpieces."[12] Facts must be established before useful theories can be constructed. In mechanics, Galileo maintained that "anyone may invent an arbitrary type of motion and discuss its properties," but he wished to "consider the phenomena of bodies falling with an acceleration such as actually occurs in nature and to make this definition of accelerated motion exhibit the essential features of observed accelerated motions."[13] The properties of motion have to be "discovered by experiments." But as Galileo noted, this "pathway hitherto closed to minds of speculative turn" must be developed.[14]

Ordinary observation could mislead one about the actual acceleration of falling bodies, but so also could superficial experiments. There was "some difficulty" in the simple experiment of letting two bodies of unequal weight fall from a specific height, for example, because in small heights little or no difference could be discerned while in large heights the resistance of the air would have a greater effect upon the lighter body. To overcome this difficulty, Galileo presented a series of experiments employing pendulums and inclined planes, which reduced the effect of the resistance of the air, since the trials were performed over short distances, yet they allowed for measurement, since the intervals of time were "not only observable, but easily observable." Galileo maintained that the "absolute truth" of a hypothesis could only be "established when we find that inferences from it correspond to and agree perfectly with experiment."[15] His definition of uniformly accelerated motion, for example, was "confirmed mainly by the consideration that experimental results are seen to agree with and exactly correspond with those properties which have been . . . demonstrated by us."[16]

Boyle often used Galileo's work as an example of how experience could correct the judgments of reason. In *The Christian Virtuoso*, for example, he discussed a number of Galileo's discoveries, including the moons of Jupiter, the mountains on the Earth's moon, and the acceleration of falling bodies as cases "wherein we assent to experience, even when its information seems contrary to reason."[17] He noted that previous philosophers had been misled in their formulation of laws governing falling bodies because they had relied upon their reason without taking in the information of experience:

> Since gravity is the principle, that determines falling bodies to move towards the center of the earth; it seems very rational to believe, with the generality of philosophers, that therein follow *Aristotle*, that, in proportion as one body is more heavy than another, so it shall fall to the ground faster than the other. Whence it has been, especially by some of the peripatetic school, inferred, that of two homogeneous bodies, whereof one does, for example, weigh ten pounds, and the other but one pound, the former being let fall from the same height, and at the same time, with the latter, will reach the ground ten times sooner.[18]

But "notwithstanding this plausible ratiocination," Boyle argued that "experience shews us" that "bodies of very unequal weight, let fall together, will reach the ground at the same time, or so near it, that it is not easy to perceive any difference in the velocity of their descent."[19] He recognized that there was an element of idealization in this type of experiment, and he thought that the law of acceleration would probably not hold for very large or very small distances. But the results were significant all the same. One may not get a universal, metaphysical truth from such experiments, but one would get closer to the truth about the behavior of natural objects in the world than one could through rational analysis alone.[20]

In opposition to the Aristotelian doctrine that different bodies fall at different rates because of an innate difference in their natures, Galileo showed that the resistance of the medium was responsible for the observed phenomena. His work ushered in a new way of thinking about motion, and those who followed him employed the same techniques in their attempt to determine the nature of such mediums as air and water. In his two treatises on air and fluids, for example, Blaise Pascal appeared to follow Galileo's notion of "experimental proof" when he maintained that "in physical science experience is far more convincing than argument." In the search for cause-effect relationships, experiment is the "true master." "The experiment made on mountains has overthrown the universal belief in nature's abhorrence of a vacuum, and given the world the knowledge, never more to be lost, that nature has no abhorrence of a vacuum, nor does anything to avoid it; and that the weight of the mass of the air is the true cause of all the effects hitherto ascribed to that imaginary cause."[21] Pascal maintained that only through a long series of experiments would discoveries be made. Although "the secrets of nature are hidden," the "experiments which make her known to us are constantly being multiplied, and since they are the sole principles of physics, the results multiply in proportion." By building upon the work of previous philosophers, "we can discover things that were hidden from their view."[22]

For Boyle, work done in pneumatics and hydrostatics provided powerful examples of the way in which one could determine the unobservable properties of natural bodies through experimentation. Indeed, "we can scarce with a greater inducement to expect, that many new attributes may be discovered in the works of nature, if men's curiosity were duly set on work to make trials." He noted that mechanics, as a mathematical discipline, was a necessary component in this development: "Till geometry, mechanicks, opticks, and the like disciplines be more generally and skillfully applied to physical things, I cannot think otherwise, than that many of the attributes and applications of them will remain unknown; there being doubtless many properties and uses of natural things that are not like to be observed by those men, though otherwise never so learned, that are strangers to the mathematicks."[23]

It was necessary to learn about the mathematical regularities that hold between bodies, but once again, Boyle opposed the idea that mathematics alone would be sufficient for the development of natural philosophy. The "theorems and problems" of hydrostatics, for example, most of which are "pure and handsome productions of reason, duly exercised on attentively considered subjects," had to be expanded to include the physical properties of bodies. "Those mathematicians, that (like *Marinus Ghetaldus, Stevinius,* and *Galileo*) have added anything considerable to the Hydrostaticks, . . . have been wont to handle them rather as geometricians, than as philosophers, and without referring them to the explication of the phaenomena of nature."[24] Mathematics, while useful for the description of the action of bodies, could not provide natural philosophers with the reasons why bodies act as they do. When Boyle himself performed hydrostatic experiments, he was guided by the work of the mathematicians, but he wished to go beyond their accounts. His experiments concerning the "ingenious proposition (about floating bodies)," for example, that was "taught and proved, after the manner of mathematicians, by the most subtle *Archimedes*" were designed in order "to manifest the physical reason, why it must be true."[25] For Boyle, the importance of hydrostatic investigations extended beyond proving that certain regularities obtain in nature to an explanation of why they "ought to be so."[26] The first task was largely mathematical. The second was the true province of natural philosophy.

In his *Hydrostatical Paradoxes*, a critical review of Pascal's two treatises, Boyle described how experimental considerations could add to earlier productions of reason, while acknowledging that his mathematical readers "will not like, that I should offer for proofs such physical experiments, as do not always demonstrate the things, they would evince, with a mathematical certainty and accurateness; and much less will they approve, that I should annex such experiments to confirm the explications, as if suppositions and schemes, well reasoned on, were not sufficient to convince any rational man about matters hydrostatical."[27] He was willing to bear such criticism. Hydrostatics was not merely a mathematical discipline. It contained physical hypotheses that had to be proved by experience. When reviewing Pascal's first treatise, titled "On the Aequilibrium of Liquors," Boyle found that, despite the experimental rhetoric, Pascal had apparently been misled by an overconfidence in mathematics to set down experiments that he had not actually performed.[28] Indeed, the experiments described by Pascal could not have been performed, since "they require brass cylinders, or plugs, made with an exactness that, though easily supposed by a mathematician, will scarce be found obtainable from a tradesman." Boyle's criticism was not that Pascal had been dishonest in his reports but that he had relied upon reasoning that was not restrained by experience: "I remember not, that he expressly says, that he actually tried them, and therefore he might possibly have set them down, as things, that *must* happen, upon a just confidence, that he was not mistaken in his ratiocinations."[29]

Because "the experiments proposed by Monsieur *Pascal*" were "more ingenious than practicable," Boyle was "induced, on this occasion, to bethink myself of a far more expeditious way to make out, not only most of the conclusions wherein we agree, but others that he mentions not."[30] Pascal's conclusions were "consonant to the principles and laws of the Hydrostaticks," but because they were products of reason, they lacked the experimental basis necessary for adequate proof.[31] Thought experiments produced by mathematicians may be "speculatively true," but they "may oftentimes fail in practice."[32] As an example, Boyle discussed a "physico-mathematical" experiment of some "ingenious modern mathematicians" who, from the supposed hydrostatic principle that "warm water is lighter *in specie*, than cold . . . deduced, that wax, and other bodies, very near equiponderant with common water, will swim in that which is cold, and sink in that which is hot or luke-warm." Boyle doubted that any mathematicians had actually produced this experiment, because when he tried it he obtained a "quite contrary effect." He fashioned a ball of wax weighted with just enough bird shot to enable it to be submerged in cold water. He then placed the ball in warm water. Although it sank at first, after a short period of time it began to float. He repeated the experiment many times and concluded that an "unheeded physical circumstance" was responsible for the result contrary to the expectations of the mathematicians. That is, the texture of the wax and its being "somewhat (though not visibly) expanded" by the warm water gave it a "greater advantage towards floating, than the increased lightness of the water would give it disposition to sink." He then performed a like experiment with a glass bubble, in which he assumed that temperature would make "no considerable change of dimensions," and he got the desired result of the mathematicians.[33]

As both Bacon and Galileo had advised, Boyle insisted that actual experiments must be performed in order to learn about the physical circumstances that give rise to the mathematical regularities observed in nature. He gave numerous examples of the way in which mathematics failed to provide a sufficient ground for theories in natural philosophy, but he still met resistance from vocal critics of the new experimental program. Like Thomas Hobbes, Baruch Spinoza argued that experimentation could not yield the type of certainty that was required as a sign of knowledge. Spinoza wrote to Boyle's colleague Henry Oldenburg that he found the method employed by Boyle in his *Certain Physiological Essays* unsatisfactory and stated that "since Mr. Boyle does not put forward his proofs as mathematical, there will be no need to enquire whether they are altogether convincing."[34] Spinoza was not himself opposed to the performance of experiments, but he rejected the idea that experimental results could be used as proofs in natural philosophy: "I should think that notions derived from human senses, should by no means be numbered among the kinds of things . . . which are sure and which explain nature as it is in itself." Only mathematical reasoning could furnish philosophers with a knowledge of the first causes or the metaphysical natures of the things of this world, and such knowledge was, for Spinoza, necessarily prior to any speculation about natural processes: "One will never be able to prove this [the nature of corpuscles] by chemical or other experiments, but only by reason and calculation. For by reason and calculation we divide bodies infinitely, and consequently the forces which are required to move them; but we shall never be able to prove this by experiment."[35]

In his response, Oldenburg reflected the type of objections that Boyle had already made to mathematical reasoning: "Our Boyle is one of those who are distrustful enough of their reasoning to wish that the phenomena should agree with it." In addition, he explained that Spinoza's criticism of experiment was misplaced and would only apply to "ordinary experiments, where we do not know what nature contributes and what other factors intervene," not to "those experiments where it is known for certain what these factors are."[36] Spinoza replied that Boyle could not say that he knew what nature contributed, when he had not mathematically determined the nature of the corpuscles before-hand. Spinoza simply could not understand why Boyle had not begun in the Cartesian manner, by settling on definitions of the metaphysical first principles of nature, instead of attempting to prove by experiment certain truths, for ex-ample, that heat is a form of motion, which he believed had already been ade-quately proved by the reasoning of Descartes and Bacon.[37] Upon hearing Spi-noza's latest criticisms, Boyle apparently lost patience with the argument. Oldenburg, in his last response to Spinoza on the matter, wrote that Boyle "asks you to consult the preface to his experiments . . . , and you will then understand the real aim which he set himself in this work."[38]

Boyle recognized that he and Spinoza were talking past each other be-cause they had very different ideas about the goal of natural philosophy. Un-like Spinoza, Boyle had no intention of providing a description of the meta-physical nature of bodies, or of the first cause of their production, by mathematical *or* experimental means. His goal was "not so much to know, what is the general agent, that produces a phaenomenon, as, by what means, and after what manner, the phaenomenon is produced."[39] He agreed with Spinoza that the laws of matter and motion would provide the most satisfactory explica-tion of experimental trials, but these laws had to be determined by the trials themselves. As he had said in *Certain Physiological Essays*, Boyle accepted the intelligibility and fruitfulness of corpuscular explanations in general, but he was unwilling to accept the hypotheses of either the atomists or the Cartesians until he was "provided of experiments to help me to judge of them." In an essay in which he attempted to show how chemical experiments could be "useful to illustrate the notions of the Corpuscular Philosophy," he remarked that both atomism and Cartesianism explained natural processes as the result of "little bodies variously figured and moved," but they also had embedded within them "metaphysical notions" that were "more requisite to the explication of the first origin of the universe, than of the phaenomena of it, in the state wherein we now find it."[40] Boyle wished to avoid such metaphysical speculation until ade-quate information about the physical processes actually operative in nature had been discovered. It was his work in chemistry, in part, that led him to formulate a different version of the corpuscular philosophy and thus a different version of how explanations referring to corpuscular principles would be discovered and justified.[41]

The Alchemical Tradition

Boyle has been called the father of modern chemistry.[42] Because of his advocacy of the usefulness of chemical experiments to natural philosophy in the face of many critics who thought that the young man was wasting his great talents on such a mystical, if not fraudulent, pastime, he rightly deserves the title. Corpuscular hypotheses were to be based upon the discovery of the actions of Bacon's "real particles, such as really exist," not upon the abstract analyses of matter proposed by Descartes, Hobbes, and Spinoza that consisted almost entirely of "empty and extravagant speculations."[43] In order to go beyond speculation that was based upon common and superficial observations and "discover deep and unobvious truths," one had to perform "intricate and laborious experiments."[44]

The work by Mersenne and Galileo provided successful instances of how experiments could lead to discovery, yet their work was only a preliminary stage of inquiry that had to be extended to an investigation of the causes responsible for such regularities. This latter investigation would require that nature be "dissected" into its minute parts. For Boyle there were "scarce any experiments, that may better accommodate the Phaenician [corpuscular] principle, than those, that may be borrowed from the laboratories of chymists."[45] Instead of speculating about the makeup of a body from its manifest appearance, one could perform a chemical analysis to reveal differences in the heterogenous parts of bodies. Such trials make the bodies "more simple or uncompounded, than nature alone is wont to present them us."[46] These experiments were of the type to which Oldenburg had referred in his correspondence with Spinoza. From them one could learn not only about the material ingredients of bodies but also about the efficacy of the particular ingredients, because the circumstances of the trial were carefully controlled. Most were performed in closed transparent vessels, for example, so that one could know "what nature contributes" to the trials and therefore "better know what concurs to the effects produced, because adventitious bodies . . . are kept from intruding upon those, whose operations we have a mind to consider."[47] In a similar manner, the production of artificial compounds could provide important information about nature's processes. In the case of the production of "factitious vitriol," for example, "our knowing what ingredients we make use of, and how we put them together, enables us to judge very well how vitriol is produced."[48]

For Boyle, chemistry held the key to developments in the new philosophy, but its practice was difficult to master. Unlike the mathematical sciences, chemistry was a "science of signs."[49] In chemical experiments one gained the ability to manipulate a substance and thus learned how it was disposed to act and react in combination with other substances. This behavior would in turn provide hints about the powers or qualities possessed by the substance that made it able to effect changes in other bodies. When we are consistently able to produce such effects artificially, then it is a good sign that we have understood the particular powers inherent in the substance and thus have gained an insight into its nature or essence. Such knowledge was the promise of chemistry. Before this promise could be fulfilled, however, chemical practice had to be put on a more rational footing.

When Boyle began his chemical experiments at Stalbridge, he was relatively isolated and had to depend upon chemical treatises to learn the art. He found that because most of these works contained such an "obscure, ambiguous, and almost aenigmatical way of expressing what they pretend to teach," the experiments reported in them could not be repeated "without difficulty and hazardous trials."[50] One of his primary goals when he began publishing his chemical works in the 1660s was to bring the chemists' experiments "out of their dark and smoky laboratories," and he hoped that if he criticized their doctrines, the chemists themselves would "be obliged to speak plainer than hitherto" in response.[51] To make chemistry more rational, it was necessary to show "the way experiments are made, so that they lose their mystery."[52] It was also necessary to show the proper way in which to draw inferences from such experiments: "It is one thing to be able to help nature produce things, and another thing to understand well the nature of the things produced."[53] In *The Sceptical Chymist*, Boyle discussed these issues in detail.

He noted that many chemists, such as apothecaries, were mere "sooty empirics" who had "been much more happy in finding experiments than the causes of them."[54] At the other extreme, however, were the chemists who moved too quickly to form general theories upon their "substantial and noble experiments." Paracelsus and his followers, for example, had "fathered upon such excellent experiments" theories that were "phantastic and unintelligible." They were "either like peacocks feathers [that] make a great shew, but are neither solid nor useful; or else like apes, if they have some appearance of being rational, are blemished with some absurdity or other, that, when they are attentively considered, make them appear ridiculous."[55] The Paracelsians were better than the Aristotelians in their analysis of the production of qualities in bodies, and their experiments had provided sufficient evidence for a definitive refutation of the Aristotelian theory of four elements, but it was an "ill-grounded supposition of the chemists" to think that their theory of the *tria prima* (that all bodies are composed of three primary elements—salt, sulphur, and mercury) was confirmed because it could explain things not explained by Aristotle's followers. As Boyle noted, to make this "argumentation valid, it must be proved, (which I fear it never will be) that there are no other ways, by which those qualities may be explicated."[56]

Indeed, there had to be an alternative explanation, because, as Boyle's experimental work had shown, all bodies could not be analyzed into the *tria prima*.

The Spagyrists were only able to support their theory because of the "lax, indefinite, and almost arbitrary senses they employ the terms of salt, sulphur, and mercury."[57] If they were forced to make these terms definite, then they would see that the expressions lost explanatory force. The fault was not in their experiments but in the reasoning upon which their interpretations were based. Boyle admitted that sometimes one of the elements of the *tria prima* could be found in several bodies that all possessed the same quality, "yet . . . this may be no certain sign, that the proposed quality must flow from that ingredient." To illustrate the illegitimacy of this type of inference, Boyle produced a counterexample: tin added to other metals such as iron or gold would result in a brittle compound. One could not conclude from this case that tin was the sufficient cause of brittleness, however, because other bodies, such as glass, were brittle without the addition of tin.[58]

Of all the chemists, Boyle liked the Helmontians best, but they did not escape his criticism. Again, there was a problem with their advocacy of general theories that had been too quickly extrapolated from an insufficient number of experiments. Based upon his work on digestion, for example, Helmont had maintained that he could explain all natural processes by reference to the sympathy and antipathy of two material principles, acids and alkalies. Boyle thought that this theory was better than the Paracelsian *tria prima* and that it could be of some use to physicians, yet he could "not acquiesce in this hypothesis of alcali and acidum, in the lattitude, wherein I find it urged."[59] These chemical "duellists," like the Paracelsians and Aristotelians before them, were guilty of having "arbitrarily" assigned "offices to each of their two principles," and they only found these principles in all bodies because their theory prejudiced their interpretations of the experimental results.[60] Again, Boyle saw that his task was not to question the experiments themselves but rather to question "the truth of those very suppositions, which Chymists as well as Peripateticks, without proving, take for granted; and upon which depends the validity of the inferences they draw from their experiments."[61]

Boyle thought it "precarious to affirm" that in all bodies "acid and alcalizate parts are found; there not having been, that I know, any experimental induction made of particulars anything near numerous enough to make out so great an assertion."[62] In addition, the Helmontians had not as yet produced "any clear and determinate notion or sure marks" to distinguish between acids and alkalies, and thus it was not surprising that their "definitions given us of acidum and alcali should be but unaccurate and superficial."[63] Because the type of operational tests that they employed were not reliable indicators of the nature of the substances with which they worked, their conclusions were also not reliable. As Boyle explained, "To infer, as is usual, that, because a body dissolves another, which is dissoluble by this or that known acid," the new "solvent must also be acid" was "unsecure," since he had found "that filings of spelter [zinc] will be dissolved as well by some alcalies . . . as by acids."[64] Without a reliable indicator test, the Helmontians' claims to have found these two substances in every body could not be affirmed. It was not only the case that a large number of observations and experiments had to be collected, but these reports also had to be carefully verified before one could build any type of comprehensive theory. Although the Helmontians had a number of experiments, they could not be used for confirmation because of their faulty interpretive framework. Thus Boyle could say that he would withhold his "belief from their assertions, till their experiments exact it."[65]

For the most part, the result of Boyle's early chemical work in *The Sceptical Chymist* and essays was of a destructive nature. He provided experimental refutations of the Aristotelian, Paracelsian, and Helmontian doctrines, but he did not substitute any positive doctrine of his own. This was as it should have been. Chemistry was still in its preliminary stage, and there was not as yet a sufficient stock of reliable experiments upon which to ground a general theory. All of Boyle's refutations employed the same strategy. The theories were presented as general explications of all natural processes, but they could not in fact explicate all the phenomena and thus fell short of Boyle's standard for rational belief.[66] Boyle noted, in part because of his strong standard, that it was "far easier to frame objections against any proposed hypothesis, than to propose an hypothesis not liable to objections."[67] While he did not have a positive theory to replace those he criticized, he did have a general idea about what such a theory would look like.

Aside from their hasty theorizing and obscure manner of writing, the chemists were doomed to failure because they relied solely upon material principles that were "too few and narrow."[68] Their "way of judging, by material principles, hinders the foreknowledge of events from being certain," and "it much more hinders the assignation of causes from being satisfactory." Boyle's positive theory, when found, would include an account of the "coordination and contrivance" of the parts of bodies; that is, it would add mechanical considerations about the motion of matter to the material ingredients identified in the chemical experiments.[69] The material ingredients of bodies had to be discovered, but they would not be sufficient, since a "chemical ingredient itself . . . must owe its nature and other qualities to the union of insensible particles."[70] From his trials he had learned, for example, that the *tria prima* were "not primary" because salt, sulphur, and mercury are "each of them endowed with several qualities," which must be due to another, more primary cause. Boyle had no doubt that future trials would "oblige" the chemists "to have recourse to more catholic and comprehensive principles."[71] This last objection was directed not to any particular theory of the chemists but to their view of explanation in general.

The chemists' theories did "not . . . perform what may be justly expected from philosophical explications" because they only identified the agent and "not the manner of the operation" whereby the agent "produces the effect proposed, and it is this modus, that inquisitive naturalists chiefly desire to learn."[72] Because the efficient cause—that is, how the disposition that a body has to produce change can be communicated to another body—had been left out of the chemists' accounts, they had failed to be truly "philosophical."[73] The chemists proposed their elements as ways to explain all of the qualities possessed by bodies, but some qualities, such as gravity, "springiness," light, sound, electricity, and magnetism, could not be explained by a simple appeal to material ingredients.[74] How, for example, could a combination of salt, sulphur, and mercury, all of which are lighter "in specie" than gold, account for gold's specific gravity? Boyle noted, "I think it would much puzzle the chemists, to give us any examples of a compounded body that is specifically heavier than the heaviest of the ingredients it is made up of."[75] It was clear that the *tria prima* were "not the first and most simple principles of bodies." Indeed, they were not elements at all but really "primary concretions of corpuscles."[76] In the end, no matter how much they tried, the chemists would never be able to use their ingredients to provide the explanatory resources necessary for a comprehensive theory about the qualities of bodies. "The chemist and other materialists, if I may so call them, must (as indeed they are wont to do) leave the greatest part of the phaenomena of the universe unexplicated by the help of the ingredients (be they fewer or more than three) of bodies, without taking in the mechanical, and more comprehensive affections of matter, especially local motion."[77]

Boyle's work on the reconciliation of the corpuscularians with the chemists provides an example of a practical application of his standard of rational belief. Matter and motion are both relevant to the explanation of physical processes, and so information from both areas must be collected and compared before a general theory can be accepted. While he was highly critical of the chemists, Boyle nonetheless sought to make chemistry a part of natural philosophy. Once stripped of their mystical and extravagant speculations, chemical experiments could become "excellent tools in the hands of a natural philosopher."[78] Boyle noted, "And though I think, that many notions of *Paracelsus* and *Helmont*, and some other eminent Spagyrists, are unsolid, and not worthy the veneration, that their admirers cherish for them; yet divers of the experiments . . . deserve the curiosity, if not the esteem, of the industrious inquirers into nature's mysteries." Despite the fact that such experiments "may be misapplied by the erroneous reasonings of the artists" who perform them, they could still be "things of great use" for the "discovery or confirmation of solid theories," as well "as the production of new phaenomena, and beneficial effects," once they were put in the hands of those who would make rational reflections upon them.[79] Hypotheses could be discovered through reasoning on the results of chemical trials. Such experiments could also act as restraints upon the hypotheses already suggested by speculative philosophers. If theories about the nature of substances were correct, then they should lead to successful manipulations in the laboratory. If they did not, then this would be a sign of their doubtful validity.[80]

Boyle's interest in chemistry extended beyond the promise of value that he perceived it to have for natural philosophy in general, however. He was also interested in the production of "beneficial effects," particularly in the area of medicinal remedies, and here chemistry had produced some concrete results. He noted, for example, that while he distrusted chemical writers such as Paracelsus, who was not a "great logician or reasoner," he had to admit that Paracelsus had "attained to some such remedies" for the cure of specific maladies.[81] As Boyle had urged the corpuscularians to pay serious attention to the chemists, so too did he attempt to show how their work could benefit the practice of medicine. Reconciliation would not be easy, however, because most of the members of the medical community shared the natural philosophers' low opinion of chemical practices.

The Medical Tradition

Boyle's interest in medicine was in part a product of his own ill health. Throughout his life he suffered from frequent agues, weak vision, and repeated attacks of kidney stones, and he had at least one stroke. He was also motivated by self-defense. On two separate occasions in his youth, the "curative" potions given to him by physicians had had the opposite effect, and he had determined from that time to investigate the preparation of medicines himself so that he would have no need to rely upon physicians.[82] The medical profession had begun to derive benefit from the new learning, such as the discoveries by William Harvey and his predecessors at Padua, that had been made possible by the rejection of authorities and the return to actual anatomical dissections and physiological experiments. Boyle wanted chemistry to be included in the new learning, but few of the medical men welcomed the intrusion of the chemists into their domain.

Physicians were especially opposed to the extravagant claims made by the alchemists, and Boyle found that he had to play the role of an apologist for chemical medicine.[83] As he saw it, both sides were guilty of making dogmatic claims. In therapeutic procedures, for example, he found that the Paracelsians were too quick in their pronouncements that all diseases were curable, while the Galenic physicians were equally quick in pronouncing many diseases incurable.[84] He suggested a suspension of judgment as a way to arbitrate between the two groups. He defended the chemists by arguing that while not all diseases may be curable in every person, there was no benefit to be gained by announcing them incurable in principle: "The fault is rather in us, than either in nature or chymistry, that men do not, by the help of chymical experiments, discover more of the nature of divers medicaments than hitherto they seem to have so much as aimed at."[85] Yet the chemists were guilty of feeding the fires of the dispute, and Boyle could sympathize with the physicians who "do very much disapprove the indiscreet practice of our common Chymists and Helmontians, that bitterly and indiscriminately rail at the methodists [medical men], instead of candidly acquiescing in those manifest truths, their observations have enriched us with, and civilly and modestly shewing them their errors."[86]

Neither discipline was complete. Both Galen and Paracelsus had admitted that the complexities involved in medical research made the acquisition of certain knowledge difficult. Because of this, Boyle maintained that it is "no great presumption, if a man should attempt to innovate in any part of it [medicine], and consequently even in the *methodus medendi*."[87] The *methodus medendi*—the term is perhaps best translated as the "method of cure"—included both the therapeutic and diagnostic parts of medicine. Concerning the therapeutic part, Boyle was particularly interested in how a knowledge of chemistry could improve the pharmaceutical preparation of remedies that required the "skill of using the helps, that nature or art hath provided against diseases."[88] He argued that although most of the chemists' medicinal preparations had been cloaked in secrecy and their components, which often included precious stones and metals, were so expensive that the usefulness of the remedies up to that time had been quite limited, physicians could still benefit from the efforts of the chemists.[89] He did not denigrate the experience of the physicians. He admired the practical knowledge that they had built up, particularly about the standard courses of specific diseases, but he did maintain that they should broaden their experiential basis by taking account of the chemists' work. In a like manner, the chemists should take account of the experience of the physicians. Many drugs could be made safer and more economically, for example, if it could be discovered which of their many ingredients were efficacious. In order to gain such knowledge, however, a more complete understanding of the human body would have to be gained through anatomical and physiological studies.[90]

The reconciliation of chemistry and medicine could also aid diagnostics. One should know not only how the movement of bodily fluids is affected in pathological conditions but also how the chemical makeup of the fluids has been altered. In his *Usefulness of Natural Philosophy* Boyle devoted an essay to the "semeiotical part of physick." He discussed how a natural philosopher acquainted with chemical trials could "assist the physician to make more certain conjectures from the signs he discovers of the constitution and distempers of his patient." As Boyle explained, "He, that better knows the nature of the parts and juices of the body, will be better able to conjecture at the events of diseases." One such diagnostic tool discussed by Boyle was the practice of uroscopy, which in the Galenic tradition was the simple observation of the urine. In the reformed uroscopy of the alchemists, a "chemical dissection" was added to the procedure by which the components of the urine were separated by distillation. On the assumption that a chemical imbalance could be a cause of illness, then, the physician could learn through this process, which would yield the chemical composition of the urine, what sort of imbalance the patient was suffering from and thus could suggest the appropriate remedy to restore harmony and health. Boyle believed that fantastic elements were still involved in the process, but he did know of one "ancient chemist" who had an experiment by which he could make "rational predictions in some abstruse diseases, by a peculiar way of examining the patient's urine."[91]

As Boyle well knew from his association with a number of medical men, diagnostics is crucial to therapeutics. When he moved his residence permanently to London in 1668, he became close friends with his neighbor in Pall Mall, Dr. Thomas Sydenham.[92] In his *De arte medica* Sydenham stressed, as Boyle had done, that knowledge must be useful: "He that in Physick shall lay down fundamental maxims, and from thence drawing consequence and raising dispute shall reduce it into the regular form of a science has indeed done something to enlarge the art of talking and perhaps laid a foundation for endless disputes."[93] But "such a system" would not yield a "knowledge of the infirmities of men's bodies, the constitution, nature, signs, changes, and the history of diseases with the safe and direct way of their cure." Indeed, not only was such an "empty idle philosophy" useless, but because the "speculations in this subject, however curious or refined or seeming profound and solid," did not yield the discovery of any "new and useful invention," they "deserve not the name of knowledge."[94]

In order to be able to cure diseases, one first has to identify them properly. To advance this program, Sydenham compiled a vast history of diseases so that physicians could learn their various symptoms and by analogy with similar cases learn the best way in which they could be cured. Medicine, like chemistry, is a science of signs. The symptoms that a patient exhibits are the signs from which one has to infer the nature of the disease that is the cause of such effects. The physician's therapeutic decisions will only be as good as the diagnostic inferences upon which they are based. Because it is a science of signs, medicine cannot be expected to exhibit the character of mathematical demonstrations. Instead of arguing from causes to their effects, medical professionals make use of the inverted order of demonstration, whereby inferences are made from effects to their causes and then predictions, based upon the inferred causes, are made concerning the course of the disease and any possible cures.

Medicine is also like the law. Great care must be exercised when making inferences from the facts because, in both cases, human life depends upon them. It was said that in the law one had to take account of all of those circumstances that are "adjuncts of a fact, which make it more or less criminal, or make an accusation more or less probable."[95] So too in medicine: all of the attendant circumstances surrounding the patient must be taken into account and seen in their complexity. Such a set of circumstances, in both the legal and medical professions, composed the body of evidence that was to be used as the basis for reaching rational decisions. Both groups of professionals employed a practical form of reasoning that was concerned not with the establishment of metaphysical first principles but with the discovery of the actual events relevant to the cases under consideration. Chief Justice Hale made the analogy between law and medicine explicit when he wrote that the complexities involved in the law were similar to those encountered by physicians. Thus, he reasoned, the method required for the reasonable resolution of cases could not be founded upon purely abstract considerations: "The texture of human affairs is not unlike the texture of a diseased body. . . . it may be of so various natures that such physic as may be proper for the cure of one of the maladies may be destructive in relation to the other and the cure of one disease may be the death of the patient."[96]

Boyle also recognized the necessity of being aware of all of the complexities of a case before administering a remedy. He noted that a drug could be effective for the cure of a specific disease but could result in the death of the patient nonetheless because of some unforeseen or unheeded condition in the particular patient that made the drug either too strong or too weak to be effective. He also noted that a patient's expectations could often influence the effectiveness of the cure. All of the circumstances surrounding the case, including the patient's age, weight, and mental attitude, must be taken into consideration in order to determine whether the use of a drug would do more harm than good.[97] The more information that one could amass about the nature of the human body, in both its natural and pathological conditions, then the better would one's chances be of correctly diagnosing and treating the illnesses to which it is subject.

The information gathered from anatomical dissections and physiological experiments must be combined with that which chemical trials revealed about the makeup of the substances within the human body. In addition, a "knowledge of the nature of those things found outside man's body may well be supposed capable of illustrating many things in man's body" and may "occasion the discovery of the true genuine causes" of disease. Boyle speculated, for example, that the discovery of the cause of kidney stones could be aided by looking to other areas of nature where stones are produced. He hoped that if the cause of their production could be determined, then this would in turn suggest the best way to eliminate them and to prevent their future occurrence.[98] He believed that there were similarities between the human body and other natural bodies that justified the search for chemical remedies. He also maintained that practical experience showed that chemical remedies "really do sometimes succeed," so that "though sometimes they chance to fail, yet that possibility of their succeeding may sufficiently evince, that there are really in nature medicines, that work."[99]

Because medicine was a "low" science that consisted of conjectures based upon the interpretation of signs, some philosophers, such as Mersenne, viewed it as a defective science.[100] Boyle, on the other hand, saw the medical tradition as encapsulating the very mode of reasoning upon which the experimental philosophy was based. It was in this tradition that he found an example of a theory, both descriptive and causal, that had been indisputably established upon the basis of experimental evidence. Boyle would often cite Harvey's theory of the circulation of the blood as one of the most important discoveries of his century.[101] The theory was important in its own right for the contribution that it made to knowledge about the human body. The manner by which it had come to be confirmed was equally important, however, because it provided a paradigm of how future discoveries were to be made.

Although Galen had emphasized the need for "ocular demonstration" and practical experience as opposed to mere book learning, the education of physicians in the medical schools had paradoxically followed Galen's books and taught a system of anatomy and physiology with little reference to the experience that Galen himself had advocated.[102] In the generation preceeding Harvey, however, the principles of the Galenic system were being slowly eroded by a renewed stress upon Galen's original methodological principles.[103] New discoveries, such as the pulmonary transfer and the existence of valves in the heart and veins, called many of the earlier Galenic anatomical principles into question. The critical Galenists, especially at Padua, where Harvey was trained, were similar to Galileo and Bacon in their reliance upon observation and their negative attitude toward system building. Harvey followed this tradition, as can be seen in the methodological dictates set out in his works both on the circulation of the blood and on the generation of animals.

In his *De generatione* Harvey maintained that the particulars of sense observation were clear and distinct but that once the mind made a judgment of them, they became universal abstractions and thus lost their clearness. He cautioned that "without the due admonition of the senses, without frequent observation and reiterated experiment, our mind goes astray after phantoms and appearances. Diligent observation is therefore requisite in every science, and the senses are frequently to be appealed to." Before anything could be deduced from the phenomena, before it was permissible to "enter upon our second vintage," one first had to have a complete history of the subject.[104] Harvey was opposed to a purely rationalistic methodology, where "the eyesight is dazzled with the brilliancy of mere reasoning." But as with Bacon, his emphasis upon caution and sense observation did not preclude the use of reasoning or the discovery of unobservable processes. Both "the clearest reasonings, and the guarantee of experiments" are necessary for the discovery of facts about the world. Only by such "labour" could one "attain to the hidden things of truth."[105] The circulation of the blood was just such a hidden truth. It was not "seen" in any direct sense but had to be detected by the use of experiments expressly designed to reveal the process. The experiments were crucial to the validation of the theory, but the reasoning by which the experiments had been designed and the theory inferred was of equal importance. In the story that Boyle told of Harvey's discovery of the fact of the circulation of the blood and the subsequent confirmation of the causal theory that explained it, one can see that experimental proof is an extremely complex affair. It is not achieved by a single crucial experiment but must be composed of a number of experiments all concurring to evince the truth of a theory.

Harvey began with the experiential knowledge that had been gained by the critical Galenists at Padua. The discovery of the pulmonary transfer and the valves of the heart provided him with "hints" upon which he reasoned to the conclusion of the major circulation. As Boyle described it, there was a measure of teleological reasoning in the process. When an anatomist

> has learned the structure, use, and harmony of the parts of the body, he is able to discern that matchless engine to be admirably contrived, in order to the exercise of all the motions and functions, whereto it was designed. . . . Thus the circular motion of the blood, and structure of the valves of the heart and veins (the consideration whereof, as himself told me, first hinted the circulation to our famous *Harvey*) though now modern experiments have for the main (the modus seeming not yet so fully explicated) convinced us of them, we acknowledge them to be very expedient, and can admire God's wisdom in contriving them.[106]

An inference from the structure of a part to its function in the whole was the sign or hint that provided Harvey with an insight into what the part was designed to do. In this case the valves were designed to produce a unidirectional flow of the blood. It is not always clear, however, how Boyle, if not Harvey, could have advocated the use of teleological reasoning, since it seems to be inconsistent with a mechanistic philosophy and both Bacon and Descartes had been quite vocal in their rejection of any appeal to final causes.[107] But Boyle was not being inconsistent. He would agree with Bacon and Descartes that those vague and general explanations of natural phenomena that evoke the purposes or designs of a deity, or of nature itself, ought to be rejected.[108] The type of teleological reasoning he advocated differed in kind from such speculation.

In line with the principles of the new mechanical philosophy, Boyle noted that the body is not a "rude heap of limbs and liquors, but . . . an engine consisting of several parts so set together, that there is a strange and conspiring communication betwixt them."[109] The assumption that the parts of the body have a function by which they contribute to the good of the whole is a valuable heuristic from which to begin anatomical and physiological investigations. But the assumption is merely a general guide. Any teleological explanation obtained by the use of such an assumption would be acceptable only if it were able to fulfill a strong set of criteria whereby the use of a part of the body could be shown to be not only "manifest" but also "unique" and "necessary to an organism's welfare."[110] Not only living organisms have such relations between their parts.

> There are many things in nature, which to a superficial observer seem to have no relation to one another; whereas to a knowing naturalist, that is able to discern their secret correspondencies and alliances, these things, which seem to be altogether irrelative each to other, appear so proportionate and so harmonious both betwixt themselves, and in reference to the universe they are parts of, that they represent to him a very differing and incomparably better prospect than to another man.[111]

As long as one's teleological conjectures are subsequently tested and proved by experimental trials, then they have a positive heuristic role and can indeed be seen as necessary adjuncts to the mechanical philosophy.[112]

In Boyle's story of Harvey's discovery, reasoning upon the function of the valves had led Harvey to conjecture about their purpose, which in turn led him to speculate about the circulation and to design various ligature experiments by which it could be proved. Harvey had only gone part of the way in establishing his theory, however. As late as 1663 Boyle wrote that while the circulation of the blood was almost universally accepted, the manner by which the blood circulated was in dispute.[113] There was still a "controversy about the cause and manner of the heart's motion, betwixt those learned modern anatomists, that contend, some of them, for Dr. *Harvey's* opinion; and others, for that of the Cartesians."[114] While Harvey had maintained that the contraction of the heart was the principal cause of the circulation, Descartes had insisted that it was the expansion of the heart, caused by the blood's becoming heated and rarefied upon its entrance, that caused the motion of the blood from the heart to the body.[115]

Descartes's explanation rested in part upon his principle that at creation, God kindled in man's heart "one of those fires without light which I have already explained, and whose nature I understood to be no different from that of the fire which heats hay when it has been stored before it is dry, or which causes new wine to seethe when it is left to ferment from the crushed grapes." He believed that the fact that the arterial blood was brighter in color than the venous blood provided additional confirmatory evidence for the truth of his fermentation theory. It would be a consequence of the arterial blood's function of supplying heat to the body that it would be more rarefied, "livelier and warmer just after leaving it [the heart] (that is, when in the arteries) than a little before entering it (that is, when in the veins)."[116] Descartes's theory could account for the color of arterial blood, and expansion was certainly an available alternative for the causal process by which to explain the circulation, but his claim that the heart expanded because of its internal heat was, somewhat paradoxically, a teleological conjecture that, for Boyle, would have to be subjected to experimental trial.[117] When the appropriate trials were made, the conjecture was shown to be an unnecessary metaphysical assumption.

Several experimenters had to work over a number of years before Harvey's theory could be confirmed with certainty. The experiments that Boyle reported in his *Usefulness of Natural Philosophy* are representative of the type of tests that were required.[118] A consequence of Descartes's assumption that the heating of the blood in the heart caused it to expand was that if there was no blood in the heart, then its motion should cease. When Boyle removed the heart from a flounder and drained it of its blood, however, he "observed that for a considerable space of time, the severed and bloodless parts held on their former contraction and relaxation." In a similar experiment, he left the heart within a frog but emptied the blood from it. He noticed that here also the heart continued to contract and expand for a short while. These experiments clearly conflicted with Descartes's theory: the blood did not appear to be necessary for the heart's movement.[119] But the trials were not decisive.[120] Before Harvey's theory could be declared the winner, it would have to explain all of the phenomena associated with circulation, including the fact of the color change that was produced in the blood in its circulation through the heart.

The observation that was finally decisive was made in 1664 by Richard Lower, a colleague of Boyle's at Oxford. Apparently quite by accident, Lower noticed that the top layer of venous blood that he had left in an open dish became the brighter color of arterial blood after its brief exposure to the air. As he wrote in his *Tractatus de corde* of 1669, from this he was able to conclude that the "deep red coloration" of the arterial blood "must be attributed clearly to the lungs, as I have found that the blood, which enters the lungs completely venous and dark in colour, returns from them quite arterial and bright."[121] This discovery agreed well with the established fact of the pulmonary transfer— that the blood transversed the lungs in going from one part of the heart to the other—and eliminated the need to suppose any hidden flame, or fermentation process, within the heart itself.

The history of Harvey's theory showed how a set of "collateral experiments," none of which would be sufficient in themselves, could provide a "concurrence of probabilities" that would be decisive in the confirmation of a theory. According to Boyle, it had been shown beyond doubt that the contraction of the heart was the mechanism by which the circulation of the blood through the body was accomplished.

> When the circulation of the Blood, (who ever had any confused notion or gave any imperfect intimations of it before) was first clearly and almost fully, delivered by our justly famous Harvey; there appeared so many Opposers to this important Truth, and so many Objections were framed against it, by those that either envyed him the glory of so useful a discovery, or foresaw how much it would endanger, if not overthrow, divers of those received opinions of Physicians, which their reputations or prejudices made them solicitous to maintain, that the Circulation of the Blood continued for many years doubted of, or was confidently rejected, in several Parts of Europe: and perhaps would never have prevailed, if this so strongly opposed truth had not been vigorously seconded by several collateral and subsequent Experiments, that were made and inforced by ingenious and dextrous men who by the differing Tryals they made, supplied impartial men with so many *mediums* conspiring to prove the same conclusion, that the truth, though much opposed, could not be supprest, or hindered from being at length triumphant.[122]

Harvey's case typified the experimental program and became the paradigm that Boyle would follow in all of his investigations. In his discussion immediately following the above passage, for example, he noted that the Torricellian and air-pump experiments were at the moment producing the same type of concurrence of probabilities in support of his theory about the weight and spring of the air:

> 'Tis also well known that the Doctrine of the weight and Springiness of the Air, met for a long time with so much opposition and so many differing objections, that if a considerable variety as well as number of Experiments had not been brought to the assistance of that unpopular Doctrine, it would never have got so much advantage of the ancient and received Perpatetick notions, as now it has obtained in the judgment of the most skilfull and impartial sort of philosophizers.[123]

The doctrine of the "spring of the air" had found much more support from experimental trial than could be had for the Aristotelian notion that nature abhorred a vacuum. But the case was not as simple here as it had been for the circulation of the blood. The decision between Harvey and Descartes was a straightforward choice: either the primary motion of the heart was expansion or it was contraction. Most natural phenomena, however, are open to more than two possible explanations, as Boyle had noted when he criticized the Paracelsians for their faulty reasoning in assuming that because their theory was superior to Aristotle's it was therefore also true. Yet the concurrence achieved in the proof of Harvey's theory could still be a paradigm for determining the truth in harder cases. Put simply, each possible alternative was to be tested and eliminated until only one was left. As theories increased in complexity and comprehensiveness, so too would the tests required to decide between them, but eventually one theory should emerge that would be able to account for all of the phenomena.

Boyle's advocacy of the Baconian inverted order of demonstration was clearly in part a result of his extensive work in such "low" sciences as chemistry and medicine. Because the law was also a low science, his belief in the appropriateness of legal analogies to illustrate his method is understandable. But a question remains about why it would be appropriate to use the low sciences as a model for natural philosophy. Boyle was certainly aware of the numerous critics who believed this tactic to be inappropriate. In addition, the "experimental philosophy," as Boyle admitted, "is a study, if duly prosecuted, so difficult, so chargeable, and so toilsome" that it would certainly not be unreasonable for someone to prefer the "natural philosophy, wont to be taught in schools," that was not "very difficult to be learned."[124] If the experimental method is "to be performed as it ought to be," it will "in many cases, besides some dexterity scarce to be gained but by practice, require sometimes more diligence, and oftentimes too more cost than most are willing, or than many are able to bestow."[125] And Boyle did nothing to make the practice easier. Indeed, by maintaining that all areas of study had to be included within an experimental investigation of nature, he had made the practice much more complex.[126] In the end, in order to understand why he would advocate such a difficult and time-consuming method, it is necessary to understand how his ontological views concerning the nature of the object being investigated influenced, and ultimately justified, his methodological choices.

Part Two

BEING A CHRISTIAN VIRTUOSO

Four

Natural Theology

Boyle had become a virtuoso of the new philosophy, but as he stressed on a number of occasions, he desired to live as a "Christian virtuoso." His theological beliefs, particularly about God's act of creation, played a crucial role in his ultimate formulation of a definition of nature, which in turn influenced his general epistemological position. Because one's view of what the world is like will, at least in part, restrict the methodological options suitable for its study, any discussion of Boyle's construction and justification of an experimental philosophy must make reference to his ontological and theological commitments.

Much has been written about the relationship between theology and the new philosophy in the seventeenth century. Some studies have focused upon how the theological views of voluntarists and necessarians led to divergent metaphysical positions concerning the contingent or deterministic nature of the world's processes.[1] Others have written about how various organized religions had different beliefs about the ability of the human rational faculty and different attitudes toward the ethic of work and grace that influenced the methodological positions of their members.[2] Finally, issues involving the wider political realm have been introduced by those who have studied the social implications of various theological doctrines, such as predestination and free will, and the polemics, particularly in England, against enthusiasts and atheists that reflected in some measure concurrent discussions concerning the legitimacy of existing social structures.[3]

A cursory glance at this literature will show that the identification of Boyle's position within these debates is not an easy matter. Indeed, because there were so many permutations within the various doctrines and so many philosophers who chose a middle ground between the extremes expressed in the dichotomies, it is difficult to categorize most of the historical figures. Even when such a categorization is achieved, no straightforward inference to a particular epistemological position can be made. The metaphysical distinction between necessarian and voluntarist doctrines, for example, has often been associated with the epistemological distinction between classical rationalism and empiricism. Necessarians are characterized as advocates of a rationalist, and primarily mathematical, approach to the study of nature because they viewed the world, created by the wisdom of God, as a self-contained mechanism subject to deterministic and universal laws of motion. Voluntarists chose instead to emphasize the free power of God who was actively involved in all aspects of his created object. Because God is the immediate and sole cause of all activity in nature, a voluntarist would claim that the best that one could hope to achieve at the epistemological level would be an empirical science of the regular occurrences of phenomenal appearances.[4]

Such strict dichotomies give rise to a number of interpretive problems. Descartes, for example, clearly advocated a rationalist methodology, yet he also supported a type of voluntarist conception of creation, as did Leibniz, who, although he spoke of the preestablished harmony of the universe, was severely critical of Spinoza's necessarian approach.[5] Mersenne, on the other hand, was an empiricist who had a strictly voluntarist conception of creation yet advocated the superiority of mathematical methods for discovering the universal harmony of the world.[6] Given what we have seen already, it should not come as a surprise to find that Boyle also had an eclectic attitude toward these metaphysical and theological categories and that he produced his own particular vision of the world by a synthesis of the elements embedded within them. Indeed, his eclecticism in part explains why there has been, and still is, considerable controversy about which categories best capture his beliefs. He had a rather robust sense of physical causality that would distance him from a strictly voluntarist conception of nature, for example, yet he did at times stress the free power of God in his creation.[7]

A similar ambiguity arises in attempts to determine Boyle's affiliation with organized religion, because his eclecticism and diffidence led him to refuse extreme positions here also. The group to which he was most closely aligned, the "latitudinarians," primarily consisted of Anglicans who, at the Restoration, called for religious toleration and unity through understanding, rather than through force.[8] But it would be misleading to attempt to assimilate Boyle with this group completely and infer his theological views from such an association.[9] In the following discussion of his "philosophical worship" of God, for example, a marked Calvinist-Puritan influence is evident. My concern in this chapter, therefore, will not be to categorize Boyle but rather to treat him as an individual and see what he explicitly had to say about his theological and ontological commitments.[10] The epistemological and methodological inferences that he actually drew from his beliefs will be discussed in chapter 5.

The Book of Nature and Philosophical Worship

At an early age Boyle determined that his life should be devoted to the service of God.[11] While Boyle was still in his teens, Samuel Hartlib advised him that he could best serve God by joining with a small group of persons who had recently instituted a program for the advancement of learning upon Baconian precepts. Boyle believed that he could contribute to man's spiritual well-being by "discovering to others the perfections of God displayed in his creatures," because such knowledge "excites devotion" and "increases admiration" for God and his works.[12] This was no mere side benefit of natural philosophy. Rather, the study of nature is "the first act of religion." Through the study of the "sensible representations" of God in his works, one can gain an understanding of the divine attributes, which is "the true use of all the discoveries of nature."[13]

There were criticisms of this new emphasis upon the study of nature, however, from "divines" who "out of a holy jealousy (as they think) for religion, labour to deter men from addicting themselves to serious and thorough inquiries into nature, as from a study unsafe for a Christian, and likely to end in atheism."[14] Boyle believed that the divines were well intentioned, but "the prejudice that might redound from their doctrine (if generally received) both to the glory of God from the creatures, and to the empire of man over them, forbids me to leave their opinion unanswered."[15] In response, he constructed a complex concurrence argument designed to show his critics that their "severity" was not "befriended, either by Scripture, reason, or experience."[16] He could use these three areas of information for his argument because all were relevant to his purpose.

> Provided the information be such, as a man has just cause to believe, and perceives, that he clearly understands, it will not alter the case, whether we have it by reason, as that is taken for the faculty furnished but with its inbred notions, and the more common observations, or by some philosophical theory, or by experiments purposely devised, or by testimony human or divine, which last we call revelation. For all these are but differing ways of informing the understanding, and of signifying to it the same thing.[17]

In the first volume of *The Usefulness of Natural Philosophy*, published in 1663, Boyle argued that reason alone tells us that the study of God's work will not lead one to atheism. Rather, a Christian is obliged to study the book of nature because it was "written for man's instruction" and God "deserves, to be honoured in all our faculties, and consequently to be glorified and acknowledged by the acts of reason, as well as by those of faith." Indeed, there is "a great disparity betwixt the general, confused, and lazy idea we commonly have of his power and wisdom, and the distinct, rational, affecting notions of those attributes, which are formed by an attentive inspection of those creatures, in which they are most legible, and which are made chiefly for that very end."[18] According to Boyle, God gave us reason for a purpose, so that we would be able to understand his power and glory in a way that common beasts cannot. The neglect of the study of nature would, therefore, be an act of impiety. We are housed in a complex body, for example, the study of which should heighten, not lessen, our admiration for its creator:

> It seems to me, not only highly dishonourable for a reasonable soul to live in so divinely built a mansion, as the body she resides in, altogether unacquainted with the exquisite structure of it; but I am confident, it is a great obstacle to our rendering God the praises due to him, for his having so excellently lodged us, that we are so ignorant of the curious workmanship of the mansions our souls live in.[19]

Second, there was historical evidence that those who had studied nature had not become atheists: "The universal experience of all ages manifests, that the contemplation of the world has been much more prevalent to make those, that have addicted themselves to it, believers, than deniers of a Deity."[20] That the book of nature was intended to provide material for reflection upon divinity had been "not only embraced by Christians, but assented to even by Jews and Heathens." Plato taught that "the world is God's epistle written to mankind," and Menasseh ben Israel had labored "to prove it by scripture and tradition."[21] Many of the "heathen philosophers" had been led by their "contemplation of nature" to "acts of religion."[22]

The study of nature, as a religious practice, was called by Boyle a "philosophical worship" of God. There was a long tradition wherein "philosophers of almost all religions have been, by the contemplation of the world, moved to consider it under the notion of a temple."[23] Among "the greatest celebrators of God" had been the "Indian gymnosophists, the Persian magi, the Egyptian sacrificers, and the old Gauls druids," who "were to their people both philosophers and priests."[24] Plutarch, Seneca, the "Jewish Philosopher" Philo, Macrobius, Mercurius Trismegistus, and almost all Christian philosophers had also accepted the notion that the world was a temple wherein the philosopher was required to act as a priest and offer worship to its divine architect. For Boyle, the meaning of this metaphor was clear: "If the world be a temple, man sure must be the priest, ordained (by being qualified) to celebrate divine service not only in it, but for it."[25] Man as the "priest of nature" is obliged and "bound to return thanks and praises to his Maker, not only for himself, but for the whole creation." As the "representer," he has the duty to "present with his own adorations the homages of all the creatures to their Creator, though they be ignorant of what is done, as infants under the law were of the sacrifices offered on their account."[26]

The philosophical worship of God through the study of his works was a primary act of religious devotion for Boyle. As he said, "I dare not confine the acts of devotion to those, which most men suppose to comprize the whole exercise of it." Rather, "I esteem, that God may be also acceptably (and perhaps more nobly) served and glorified by our entertaining of high, rational, and as much as our nature is capable of, worthy notions, attended with a profound and proportionable admiration of those divine attributes and prerogatives, for whose manifesting he was pleased to construct this vast fabrick."[27] The historical precedent for this attitude was not limited to philosophers per se but also encompassed the practitioners of the low sciences. Paracelsus had maintained that to study nature *"is to walk in the ways of God,"* and Galen had written that piety does not "consist in sacrificing" but in first knowing and then declaring "what his wisdom, power, providence, and goodness is: the ignorance of which, not the abstaining from sacrifice, is the greatest impiety."[28]

Finally, in addition to the evidence provided by reason and experience, Boyle also found support for his view in the Bible. Indeed, the "neglect of this philosophical worship of God" was in direct opposition to "that invitation of the Psalmist, *to sing praises to God with understanding."* Knowledge is a "gift of God, intrusted to us to glorify the giver with it," and those who neglect "imploying it gratefully" are culpable.[29] Solomon, "who was pronounced the wisest of men by their omniscient Author," did, according to Boyle, "not only justify the study of natural philosophy, by addicting himself to it, but ennobled it by teaching it, and purposely composing of it, those matchless records of nature."[30] And Augustine, as a leading church father and authority on biblical interpretation, had noted the same when he advised his reader "not to use your eyes as a brute, only to take notice of provisions for your belly, and not for your mind; use them as a man; pry up into heaven; see the things made, and enquire the Maker; look upon those things you can see, and seek after him, whom you cannot see."[31]

Boyle followed "Saint Austin" in the belief that one had to use one's senses "as a man." In fact, he noted that he would be "sometimes angry with them" who "never find the leisure to discharge that primitive and natural obligation," who "both worship God so barely as Catholick or Protestants, Anabaptists or Socinians," and who "live so wholey as lords or counsellors, Londoners or Parisians," that they "never find the leisure, or consider not, that it concerns them to worship and live as men."[32] The Christian has an obligation to investigate and learn from God's work:

The works of God are not like the tricks of jugglers, or the pageants, that entertain princes, where concealment is requisite to wonder; but the knowledge of the works of God proportions our admiration of them, they participating and disclosing so much of the inexhausted perfections of their author, that the further we contemplate them, the more footsteps and impressions we discover of the perfections of their Creator; and our utmost science can but give us a juster veneration of his omniscience.[33]

God is the "author of nature," and he may "declare truths to men, and instruct them, by his creatures and his actions, as well as by his words."[34] In this context, Boyle's support of teleological reasoning can be seen not merely as a valuable heuristic device but also as an integral part of a Christian's philosophical understanding and worship of God.

Boyle admitted that it would be a "presumption" if he were "peremptorily to define all the ends and aims of the omniscient God." One could suppose, however, that "two of God's principal ends were, the manifestation of his own glory, and the good of men."[35] That is, one could not determine a priori the specific purposes of God, but one could suppose that whatever the purposes, the things of this world were "designed to instruct us" about the attributes of God, especially "his power, his wisdom, and his goodness." So that his reader would not think that this proposition was merely "affirmed gratis," Boyle subjoined a number of instances from scientific practice where these attributes of God were manifest. From astronomy, for example, one could learn of God's power. While the Earth seems great in its variety and intricacy from the perspective of those who inhabit it, astronomers taught that the Earth is "but a point in comparison of the immensity of heaven."[36] In a similar manner, chemical practice, by which seemingly magical effects could be wrought by mere man, taught how great the wisdom of God was in providing the materials for such awe-inspiring results. As the Paracelsian Thomas Tymme wrote in 1612, "The wisdome of Natures book men commonly call Naturall Philosophie, which serveth to allure to the contemplation of that great and incomprehensible God, that we might glorifie him in the greatness of his works."[37]

For Boyle, the most perspicuous examples of God's wisdom, power, and goodness could be had from his creation of living organisms. Experience of nature showed him that God provided different species with instincts beneficial for their survival. Silkworms, for example, did not learn to spin their cocoons by imitation but were born with the inbred knowledge of how to do so, given them by their "wise preserver."[38] Such wisdom and providence could also be seen in the process of generation. All who, like Aristotle, Dr. Highmore, and Dr. Harvey, had made the effort to "watch and diligently observe, from time to time, the admirable progress of nature in the formation of a chick" had, unlike the "ordinary eaters" of eggs, been led to make "philosophical reflections" upon the process and offer praise to the designer of it.[39] By studying all aspects of nature, from astronomy and chemistry to the anatomy of the least created being, one would be led to a knowledge of the "Author of things, to whom alone such excellent productions . . . may be ascribed."[40] Because we owe to God "that glory, praise, and admiration, he both expects and merits, from such a contemplation of the creatures," a "true knowledge of their nature and properties" will be requisite.[41] It would therefore be incumbent upon a good Christian to design and perfect a method that would achieve such knowledge.

Natural philosophy and divinity are not "at such variance, as the divines we deal with would persuade us."[42] Boyle, quoting Bacon, maintained that even though a "little or superficial taste of philosophy, may, perchance, induce the mind of man to atheism," a "full draught thereof" would bring "the mind back again to religion."[43] In nature, one is able to behold "the dependency, continuation, and confederacy of causes, and the works of providence," and Boyle's ultimate goal was to discover and present these causes to mankind.[44] Throughout his life he published works with this end in mind. Works such as *The Usefulness of Natural Philosophy* (1663), *Occasional Reflections* (1665), *The Reconcileableness of Reason and Religion* (1675), *Of the High Veneration Man's Intellect Owes to God* (1685), *A Free Inquiry into the Vulgarly Received Notion of Nature* (1686), and *A Disquisition about Final Causes* (1688) all exhibit Boyle's constant concern to show his contemporaries how it was possible to be a "Christian virtuoso."[45] Theology and natural philosophy were both studies of the books of God, and the two would be compatible once "rightly understood."[46] But there was a real danger to theology from those natural philosophers who did not rightly understand the text of nature. In large part, *A Free Inquiry into the Vulgarly Received Notion of Nature*, although written twenty years earlier, was finally published by Boyle in the 1680s because, despite the work done in the new philosophy, many still had a confused and prejudicial conception of nature.

A Free Inquiry

On the first page of *A Free Inquiry*, Boyle stated that his purpose in the work was to discuss whether nature "be that almost divine thing whose works, among others, we are; or a notional thing, that in some sense is rather to be reckoned among our works, as owing its being to human intellects."[47] He began his inquiry with an attempt to clear up the ambiguities that resulted from contemporary usages of the term *nature*. Sometimes, he noted, *nature* was meant to refer to God directly, as when philosophers employed the phrase *natura naturans*. At other times, it had a physical sense, when it was meant to refer either to the general course of nature that resulted from the activity of God as "the universe, or the system of the corporeal works of God," or to the particular nature of individual bodies as that "on whose account a thing is what it is," such as when one spoke of "what belongs to a living creature at its nativity."[48] Finally, from the custom of speaking of the activity of nature, some philosophers, both ancient and modern, had come to use the term as though it referred to "a goddess, or a kind of semi-deity."[49]

To combat the latter and most insidious use of the term, Boyle went to great lengths to argue that "nature" is not a "true, physical, and distinct or separate efficient." Simply because it is acceptable linguistic usage to say "that nature does this or that, we ought not to suppose that the effect is produced by a distinct or separate being."[50] He used an analogous case from the social sphere to argue this point. He noted that although it was common practice to speak of the law of the land as doing things, such as when one would say that "the law punishes murder with death," it was plain to all that the law "cannot, in a physical sense, be said to perform these things." Rather, "they are really performed by judges, officers, executioners, and other men, acting according to that rule."[51] The linguistic confusion that gave rise to the idea of nature as a type of purposive agent was a theological threat, because it could result in a failure to appreciate that, by his unlimited power, God had created the world and all of its processes without the assistance of a "vice-gerent." It was also "no small impediment to the progress of sound philosophy."[52]

Because of the "great ambiguity" surrounding the term *nature* and "the little or no care which those, that use it, are wont to take to distinguish its different acceptions," a "great deal of darkness and confusedness" and a "multitude of controversies" had been generated, "wherein men do but wrangle about words, whilst they think they dispute of things."[53] Boyle hoped to make advances in natural philosophy by clearing up the ambiguities of the term *nature* and thus displaying the illegitimate inferences that had been based upon careless linguistic practice. In place of the "custom of assigning, as true causes of physical effects, imaginary things or perhaps arbitrary names," he wished to discover the true locus of causality, which would have to make reference to physical agents.[54] If a cause proposed by a hypothesis "be not intelligible and physical, it can never physically explain the phaenomena."[55] One could still speak of nature in philosophical discourse, but the word's use would have to be carefully delineated and restricted to the two purely physical senses, and its meaning in these contexts would have to be carefully defined.

The most general of the physical senses, where *nature* was used to denote the universe as a whole, should be understood to mean "the aggregate of the bodies, that make up the world, framed as it is, considered as a principle, by virtue whereof they act and suffer, according to the laws of motion prescribed by the Author of things."[56] Boyle did not mean to imply in this passage that nature is merely an aggregate of unrelated parts whose actions are guided by laws externally imposed upon them, as J. E. McGuire and others have argued.

In an immediate paraphrase of his new definition, Boyle explained that the discrete bodies of which nature is composed have their properties determined by the internal relations that result from their placement in the whole: "Nature, in general, is the result of the universal matter, or corporeal substance of the universe, *considered as it is contrived in the present structure and constitution of the world,* whereby all the bodies, that compose it, are enabled to act upon, and fitted to suffer from one another, according to the settled laws of motion."[57] The world is a set of relations, and it is the structure of the world as a whole that gives it definition. As with a mechanism such as a clock, where the parts must be related in such a way as to produce the final product, so too could nature, in its general sense, be referred to by the "more compendious" expression as a "cosmical mechanism."[58]

In a similar manner, Boyle argued that it was appropriate to retain the term *nature* to refer to the particular essences of bodies, those that make them distinct, identifiable entities, but this sense also must be redefined and "be conceived to signify a complex or convention of all the essential properties, or necessary qualities that belong to a body of that species."[59] The nature of an individual body, that which makes it a particular body different from other bodies, "consists in a convention of the mechanical affections . . . of its parts." These affections are "sufficient to constitute" bodies as members of "particular species or denominations," and so the "more compendious" expression to denote such a nature would be the "individual mechanisms of that body."[60]

It was in the context of his discussion about the particular natures of individual bodies that Boyle inserted a short digression on the notion of a scientific law. As we have seen, he wanted to eliminate nature as a causal agent in order to clear the way for finding the real mechanisms operative in the world that give rise to "the aggregate of powers belonging to a body." He noted that Helmont had suggested that the Aristotelian notion of the nature of a body could be replaced with that of a "law, that it receives from the creator, and according to which it acts on all occasions."[61] Boyle clearly found this construction to be better than that of the "naturists," yet he cautioned that it could also create problems in philosophical discussion if it came to be used carelessly.

> But to speak strictly, (as becomes philosophers, in so weighty a matter) to say, that the nature of this or that body is but *the law of God prescribed to it,* is but an improper and figurative expression: for, besides that this gives us but a very defective idea of nature, since it omits the general fabric of the world, and the contrivances of particular bodies, which yet are as well necessary, as local motion itself, to the production of particular effects and phaenomena; besides this, I say, and other imperfections of this notion of nature, that I shall not here insist on, I must freely observe that,

to speak properly, a law being but a *notional rule of acting according to the declared will of a superior,* it is plain, that nothing but an intellectual Being can be properly capable of receiving and acting by a law.[62]

Boyle was making two separate points about natural laws in this passage. The one in the middle of the paragraph, about the way in which laws provide a defective idea of nature, is similar to his criticisms of mathematicians and chemists. The other point, which begins and ends the paragraph, on the improper and figurative sense in which the term *law* is used to describe nature, has given rise to some confusion in the interpretation of Boyle's conception of the lawlike behavior of bodies.

Boyle had described nature as a "notional entity" when he meant to imply that *nature* was nothing more than a "summary appellation" constructed by the human understanding to refer to the universe. It may seem that in the above passage he is referring to the laws of nature as "notional rules" and therefore that he found them also to be mere constructions imposed upon nature by the human understanding.[63] Actually, Boyle was arguing the exact opposite. Laws of nature are not notional rules. Laws, "to speak strictly," are the rules instituted by a government to moderate the behavior of its citizens. In the social realm, these laws are "notional" in the sense that for them to be effective, the citizens must have some notion, or idea, of the law in order to be able to act in conformity with it.[64] When laws are spoken of in the natural realm, they cannot be considered as notional rules, because a "body devoid of understanding and sense" cannot "moderate and determinate its own motions, especially so, as to make them conformable to laws."[65] Laws of nature must be understood figuratively. It is "by virtue of the original frame of things, and established laws of motion" that "bodies are necessarily determined to act on such occasions after the manner they would do if they had really an aim."[66]

Boyle entered upon this digression in order to warn his contemporaries against confusing the two senses of *law* and imagining that matter in some sense is able to "obey" the laws of nature. The laws of nature are not notional rules. Neither are they imposed upon nature by human understanding. Humans may learn about such laws from the observation of regularly occurring phenomena, but the laws themselves, as properties of bodies, exist independent of our ability to recognize or conceptualize them. Immediately following the passage about laws quoted above, Boyle stated that while it was inappropriate to speak of inanimate bodies following notional rules of conduct, it was "intelligible" that God should "at the beginning impress determinate motions upon the parts of matter, and guide them, as he thought requisite, for the primordial constitution of things; and that ever since he should, by his ordinary and general concourse, maintain these powers, which he gave the parts of matter, to transmit their motion thus and thus to one another."[67] According to Boyle, there are laws of nature, originally imposed by God, that give rise to causal processes in the world. They are the impresses on natural things that define their essences and give them the power to act and be acted upon by other bodies. Laws of nature are absolutely necessary for the determinate fabric of the world in its present condition, and they constitute the world, wherein "all bodies are governed by the laws of the universe."[68]

Natural laws refer to the "course of nature," where things are "brought to pass by their proper and immediate causes, according to the wonted manner and series or order of their acting." In addition, Boyle noted that there were two types of natural laws. The first, universal laws, hold absolutely for all bodies; the other, less general laws, could be called "customs of nature" because they belong not to all bodies but "to this or that particular sorts of bodies."[69] He had made this distinction earlier in his "Cosmical Suspicions," printed as an appendix to his *History of Particular Qualities*, in which he had criticized his contemporaries for neglecting the second type of law in favor of the more general. According to Boyle, the customs of nature "may have a greater influence on many phaenomena of nature, than we are wont to imagine." Instead of focusing upon the more general laws, he urged his readers to seek out these lesser laws, as intermediate stages in their inquiries. In "Cosmical Suspicions" he also cautioned naturalists not to impose their own ideas of regularity upon nature. In astronomy, for example, he believed that there might be "more accurateness fancied than there really is" about the "limits within which some great masses of matter are supposed to perform their motions."[70] On the other hand, naturalists should not assume that because a regularity is not easily discernible by us, there is no law to be discovered:

> There are cases wherein I am not quite out of doubt, but that we may sometimes take such things for deviations and exorbitancies from the settled course of nature, as, if long and attentively enough observed, may be found to be but periodical phaenomena, that have very long intervals between them; but because men have not skill and curiosity enough to observe them, nor longaevity enough to be able to take notice of a competent number of them, they readily conclude them to be but accidental extravagancies, that spring not from any settled and durable causes.[71]

Boyle had a nominalist critique of those philosophers who tended to reify concepts, but he did not deny the reality of natural laws and physical causality.[72] His conception of causality can be seen in the other objection that he expressed concerning the dangers inherent in the use of the term *law*. Boyle believed that the discovery of laws and mathematical regularities was not sufficient for the establishment of a solid natural philosophy, and here he noted that laws are defective expressions because they leave out reference to "the general fabric of the world, and the contrivances of particular bodies."[73] In

"Cosmical Suspicions" he made the same point when he wrote that while he admired the "industry of astronomers and geographers," yet

> they have hitherto presented us, rather a mathematical hypothesis of the universe, than a physical, having been careful to shew us the magnitudes, situations, and motions of the great globes, such as the fixed stars and the planets (under which one may comprize the earth) without being solicitous to declare what simpler bodies, and what compounded ones, the terrestrial globe we inhabit does or may consist of.[74]

The aim of his corpuscular philosophy extended beyond the determination of regularities to the discovery of the hidden mechanisms that would define the essences of the bodies that compose the great cosmic mechanism of the world.

The Corpuscular Philosophy and Physical Causality

As Boyle stressed repeatedly, his version of the corpuscular philosophy differed from those formulations proposed by his contemporaries. In "The Excellency and Grounds of the Corpuscular or Mechanical Hypothesis" he stated explicitly, and at some length, that he differed from the others because of his beliefs about how the world had been created:

> When I speak of the corpuscular or mechanical philosophy, I am far from meaning with the Epicureans, that atoms, meeting together by chance in an infinite vacuum, are able of themselves to produce the world, and all its phaenomena; nor with some modern philosophers, that, supposing God to have put into the whole mass of matter such an invariable quantity of motion, he needed do no more to make the world, the material parts being able by their own unguided motions, to cast themselves into such a system (as we call by that name:) but I plead only for such a philosophy, as reaches but to things purely corporeal, and distinguishing between the first original of things, and the subsequent course of nature, teaches, concerning the former, not only that God gave motion to matter, but that in the beginning he so guided the various motions of the parts of it, as to contrive them into the world he designed they should compose, (furnished with the seminal principles and structures, or models of living creatures,) and established those rules of motion, and that order amongst things corporeal, which we are wont to call the laws of nature. And having told this as to the former, it may be allowed as to the latter to teach, that the universe being once framed by God, and the laws of motion being settled and all upheld by his incessant concourse and general providence, the phaenomena of the world thus constituted are physically produced by the mechanical affections of the parts of matter, and what they operate upon one another according to mechanical laws.[75]

Such a strict demarcation between the original creation of the world and its subsequent established order was crucial for achieving a right understanding in natural philosophy. It would be illegitimate to assume either that because there is physical necessity in the world as now constituted, it had therefore been created from necessity, or that because the world had been created by an omnipotent free agent, it is therefore radically contingent in its present constitution. As Boyle said, "There are divers errours that men fall into for not being aware, or for not considering, the difference between the world such as it *now* is, (after the course of nature has been instituted) and *before* it was created and framed, and the universal matter was contrived into the world."[76] In the world "constituted as it now is," manifest effects are not produced by the immediate action of God but are "physically produced" by the powers God invested in his creation.[77]

Boyle was not a strict voluntarist like Mersenne and Gassendi, who settled for a "positivistic-pragmatic conception of knowledge" where the best one can do is learn about the regular conjunctions between the observable phenomena of nature.[78] His voluntarist conception extended only to the creation of the world, not to its present state. There is a physical necessity in the world because, in a sense not always easy to understand, God's creative act is already complete. The Scripture "affirmed, *That all his [God's] works are known to him from the beginning.*" God had "resolved, before the creation, to make such a world as this of ours," and "the phaenomena, which he intended should appear in the universe, must as orderly follow, and be exhibited by the bodies necessarily acting according to those impressions or laws, though they understand them not at all."[79] From this initial contingent act of creation, a world of physical necessity resulted. In *A Free Inquiry* Boyle described the type of necessity that would flow from God's omniscience: "I ascribe to the wisdom of God in the first fabric of the universe, which he so admirably contrived, that, if he but continue his ordinary and general concourse, there will be no necessity of extraordinary interpositions, which may reduce him to seem, as it were, to play after-games; all those exigencies . . . being foreseen and provided for in the first fabric of the world."[80]

The presence of physical determinism in the world did not in any way decrease the power that God had over his creation. Rather, determinism was an actual manifestation of power. Those who assumed that God must immediately act in the world viewed his creation "after the nature of a puppet," wherein "almost every particular motion the artificer is fain . . . to guide and oftentimes over-rule the actions of the engine."[81] In opposition to this view, Boyle maintained that the world "is like a rare clock," where

all things are so skilfully contrived, that the engine being once set a moving, all things proceed, according to the artificer's first design, and the motions . . . do not require, like those of puppets, the peculiar interposing of the artificer, or any intelligent agent, employed by him, but perform their functions upon particular occasions, by virtue of the general and primitive contrivance of the whole engine.[82]

Boyle would "suppose no other efficient of the universe, but God himself," who by "his infinite wisdom" had framed the corporeal world "according to the divine ideas, which he had, as well most freely, as most wisely determined to conform them to."[83] But while God was the only efficient *of* the universe, he was not the only efficient *in* it. The world is a vast interconnected system, and all natural processes are related in such a way that the divine plan will be realized. God constructed the world according to his own design and foresaw all that would happen by virtue of the laws that he instituted. For Boyle, the world is "not a moveless or undigested mass of matter" to be explained by some type of "billiard-ball" model of causality. Rather, the world is a "self-moving engine," and all bodies in it, "however they first came by it, are moved by an internal principle."[84] Bodies have been empowered to act and react in particular situations, and it is the "motive force" that they possess that gives rise to the ultimate structure of the universe.

Boyle acknowledged that his discussions of his version of the mechanical philosophy had not always been clear when he noted that some "visitants" at his lodgings had supposed that he believed in the absolute rest of bodies from a mistaken interpretation "of the true meaning of a passage or two" in his *History of Fluidity and Firmness* (1661). In order to clear up such misunderstandings, he had "An Essay of the Intestine Motions of the Particles of Quiescent Solids" published as an appendix to a new edition of the *History* in 1669.[85] He noted that it was most likely that even those bodies that appeared to be completely at rest to our senses of vision or touch retained an internal motion of their minute parts that was responsible for the qualities that the macroscopic bodies possessed. These "intestine motions of the corpuscles of hard bodies need not be solely, nor perhaps principally ascribed to those obvious external agents, to which we are wont to refer them, since these may but excite or assist the more principal or internal causes of the motions we speak of." While it is "not easy for us, who are wont (perhaps too much) to follow our eyes for guides in judging of things corporeal," to discover such motions, experiments fitly made could reveal them.[86]

Leibniz, who is often considered to have developed a natural philosophy much different from Boyle's, used his essay on intestine motions "Where the Absolute Rest of Bodies Is Called in Question" in a planned critique of Locke's contention that "bodies can be without motion."[87] Leibniz also acknowledged his intellectual debt to Boyle in his 1698 essay "On Nature Itself," a work in which he argued against the "occasionalist tendencies" of some Cartesians and presented his positive metaphysical view that "activity, is in the body itself, and not merely in God." Indeed, Leibniz borrowed the title of this essay from the Latin translation of Boyle's *Free Inquiry, Tractatus de ipsa natura* (1688), and explicitly approved of the definition that Boyle had given there of nature as "the very mechanism of bodies itself."[88] A closer reading of *A Free Inquiry* reveals that in this work Boyle had set out the details of an ontology that was similar to Leibniz's notion of a preestablished harmony, and thus a brief comparison of the two philosophers will be useful for achieving a better understanding of the complexity of Boyle's ontological position.

Boyle frequently spoke of the world as God's created text. The "book of nature" was a common and somewhat paradoxical metaphor in the seventeenth century, and its meaning for any individual user is not always easy to ascertain. Given Boyle's discussions in *A Free Inquiry* and elsewhere, however, it is clear that for him the metaphor was not only a device for justifying his pursuit of science to theologians. It also played an important role in the development of his ontology. Considered as a text, the world is a coherent, albeit extremely complex whole, written all at once, where all of the parts are perfectly suited to the purposes for which the completed object was designed.[89] Boyle's conception of the universal harmony inherent in nature is revealed most perspicuously in his discussions of the perfection of the world.

Boyle thought it appropriate to ask "*Whether the world, and the creatures, that compose it, are as perfect as they could be made?*" Even though the world is "an admirable piece of workmanship" and God has "immense power and unexhausted wisdom," it does not follow that "the divine architect could not have bettered it."[90] Actually, it had to be admitted that individual creatures could have been made more perfect. An oyster, for example, "could have been made better since it can neither hear nor see, nor walk, nor swim, nor fly, etc." But "if the question be better proposed," and one asks

> not whether God could have made more perfect creatures, than many of those he has made, for that it is plain he could do so, because he has done it; but whether the creatures were not so seriously and skilfully made, that it was scarce possible they could have been better made, with due regard to all the wise ends he may be supposed to have had in making them, it will be hard to prove a negative answer.[91]

When considered as parts of a coherent whole, each creature is perfect in its kind. The world itself is also not perfect in any absolute sense, but it is a perfect vehicle for the ends for which it was created.

According to Boyle, because God foresaw all that would happen when he instituted the particular processes of nature, we can be assured that they provide the best means for producing his all-wise aims, even if we cannot know what these aims are. Boyle used this point, in much the same way that Leibniz would later, to respond to the apparent presence of evil in the world. Because God has an "infinite understanding, to which all things are at once in a manner present," he clearly discerned "what would happen, in consequence of the laws by Him established, in all the possible combinations of them, and in all the junctures of circumstances, wherein the creatures concerned in them may be found." When "all these things were in his prospect," God "settled among his corporeal works general and standing laws of motion suited to his most wise ends."[92] It would be

> very congruous to his wisdom, to prefer . . . catholic laws and higher ends, before subordinate ones, and uniformity in his conduct before making changes in it according to every sort of particular emergencies; and consequently, not to recede from the general laws He at first most wisely established, to comply with the appetites or the needs of particular creatures, or to prevent some seeming irregularities (such as earthquakes, floods, famines, etc.) incommodious to them, which are no other, than such, as He foresaw would happen . . . and thought fit to ordain, or to permit, as not unsuitable to some or other of those wise ends, which He may have in his all-pervading view.[93]

Even those events that appear to be disasters from our standpoint "are unfit to be censured by us dim-sighted mortals." God's providence is, for the most part, a general one.[94] He "subordinated his care" of individual creatures "to his care of maintaining the universal system and primitive scheme or contrivance of his works." Whenever "such a concourse of circumstances" occurs that either "particular bodies . . . must suffer, or else the settled frame, or the usual course of things must be altered," the "welfare and interest of man himself (as an animal) and much more that of inferior animals . . . must give way to the care, that providence takes of things of a more general and important nature or condition."[95] Given God's omniscience and omnipotence, all events that occur in the world must be for the best even if we cannot discern the good in them. It "became the divine Author of the universe" to act consistently in his creation and to provide for the "private portions" of it only to the extent that their welfare would be "consistent with the general laws settled by God in the universe, and with such of those ends, that he proposed to Himself in framing it."[96]

Boyle's theological beliefs, particularly about God's act of creation, strongly influenced his ontology, but the traditional historical categories of "voluntarism," "nominalism," and "occasionalism" are not of much help in understanding the complexity of his beliefs. For Boyle, a strict determinism prevailed in the world, where physical bodies were empowered to act in such a manner that certain preplanned results would be realized. Because bodies are actual "corporeal agents," causality exists at the physical level. It is the task of natural philosophy to discover these true causes. Boyle's conception of determinism differed from that of most of the atomists of his day, however, because of his emphasis upon the complex interrelations in the world. This explains why he was critical of the atomists' almost exclusive focus upon the geometrical properties of bodies and why he allowed for the appropriateness of teleological reasoning.[97] The world is a coherent whole. All of its parts are fitted for the ends designed by its creator. In order to understand the whole, then, a knowledge of the relations that hold between its parts will be required.

Causal Relations and the Essences of Bodies

In *The Origin of Forms and Qualities* (1666), described by Boyle as an introduction to his version of the corpuscular philosophy, he discussed many of the same themes that we have already seen.[98] He noted, for example, that his view of creation in part separated his philosophy from that of his contemporaries: "The Author of nature did not only put matter into motion, but, when he resolved to make the world, did so regulate and guide the motions of the small parts of the universal matter, as to reduce the greater systems of them into the order they were to continue in; and did more particularly contrive some portions of that matter into seminal rudiments or principles, lodged in convenient receptacles."[99] He also urged his fellow philosophers not only to know "the general laws and course of nature, but to inquire into the particular structure of the bodies they are conversant with as that wherein, for the most part, their power of acting and disposition to be acted on does depend."[100] The true locus of causality resides in the contrivance of matter.

Boyle wanted to find the physical agents responsible for the manifest effects of nature in opposition to those supposed entities of the Scholastics that were nothing more than reified concepts. But he stressed that while these agents must be physical, they did not have to be sensible. He criticized the "vulgar of philosophers," who because they were "accustomed to converse with visible objects, and to conceive grossly of things, cannot easily imagine any other agents in nature than those that they can see" and were thus forced to "confess themselves utterly at a loss" when they were unable to find such "sensible" causes. The "school-philosophers," however, "run too far to the other side, and have their recourse to agents that are not only invisible but inconceivable," as when "they ascribe all abstruse effects to certain substantial forms." Boyle maintained that "betwixt visible bodies and spiritual beings there is a middle sort of agents, invisible corpuscles; by which a great part of the difficulter phaenomena of nature are produced, and by which may intelligibly be explicated those phaenomena, which it were absurd to refer to the former, and precarious to attribute to the latter."[101]

In his *Origin of Forms and Qualities* Boyle argued that these invisible particles of matter are the true causal agents responsible for the effects that sensible bodies are able to produce both in us and in other bodies. Their particular contrivances give rise to the qualities that bodies have, by virtue of which we "distinguish any one body from others, and refer it to this or that species of bodies." Boyle rejected the Aristotelian notion of a "substantial form," but he would "for brevity's sake retain the word Form" as long as he was "understood to mean by it, not a real substance distinct from matter, but only the matter itself of a natural body, considered with its peculiar manner of existence." A form consists in the "concurrence of all those qualities which men commonly agree to be necessary and sufficient to denominate the body which hath them, either a metal or a stone, or the like."[102] The "first and universal, though *not immediate* cause of forms, is none other but God," but "among second causes, the grand efficient of forms [is] local motion, which, by variously dividing, sequestering, transposing, and so connecting the parts of matter, produces in them those accidents and qualities, upon whose account the portion of matter they diversify, comes to belong to this or that determinate species of natural bodies."[103] Boyle's corpuscular philosophy was unique, as both he and his publisher noted, because of this emphasis upon discovering the forms and qualities of bodies.[104] Unlike Mersenne, Gassendi, Hobbes, or Descartes, Boyle had a goal similar to that of the Aristotelians. He desired to know the true essences of bodies and merely rejected the Scholastic manner of discovering them.

Although qualities are accidents of matter and a form is "nothing but an aggregate or convention of such accidents," it does not follow that bodies themselves are devoid of essences. Boyle referred to the modifications of matter that gave rise to the qualities and forms of bodies as accidental in order to avoid the mistakes of the Aristotelians, who imagined a form "to be a very substance." As he explained, however, the term *accident* could be used in two ways: "Nor need we think that qualities being but accidents, they cannot be essential to a natural body; for accident, as I formerly noted, is sometimes opposed to substance, and sometimes to essence. And though an accident cannot but be accidental to matter, as it is a substantial thing, yet it may be essential to this or that particular body."[105] Qualities are accidental to matter. They are not substances in their own right, but they are still essential to the bodies that possess them.

Forms are also essential to bodies. He went on to explain: "This convention of essential accidents being taken (not any of them apart, but all) together for the specific difference that constitutes the body and discriminates it from all other sorts of bodies, is by one name, because considered as one collective thing, called its form . . . or, if I may so name it, an essential modification." A form is a modification "because it is indeed but a determinate manner of existence of the matter," yet it is essential "because that though the concurrent qualities be but accidental to matter . . . yet they are essentially necessary to the particular body, which without those accidents would not be a body of that denomination."[106] In Boyle's most explicit statement about the nature of these forms and qualities, he wrote: "Though we do not admit substantial forms, yet we need not admit natural bodies to be *entia per accidens*; because in them the several things that concur to constitute the body, as matter, shape, situation, and motion, *ordinantur per se & intrinsicè* to constitute one natural body."[107]

According to Boyle, the investigation of the forms and qualities of bodies is "one of the most important and useful that the naturalist can pitch upon for his contemplation." As the principle of individuation, a form is that which can explain the qualitative changes—of generation, corruption, and alteration— that take place in natural bodies. Because qualities are the "concurring physical causes" of forms, we must learn about the qualities of bodies, as the accidents that accrue to them by virtue of the structure of their minute parts.[108] We saw that in *A Free Inquiry* Boyle insisted that "the contrivances of particular bodies" are "as well necessary, as local motion itself, to the production of particular effects."[109] In *The Origin of Forms and Qualities* he explained this notion more completely: "As it is by their qualities that bodies act immediately upon our senses, so it is by virtue of those attributes likewise that they act upon other bodies, and by that action produce in them, and oftentimes in themselves, those changes that sometimes we call alterations, and sometimes generation or corruption."[110]

Qualities are the causal powers that bodies possess to effect changes in other bodies. Qualities are not "like to the ideas they occasion in us." The ideas that we have of the attributes of bodies could by extension be called "sensible qualities," but Boyle used the names "primary qualities" and "secondary qualities" to refer to those qualities actually possessed by matter and bodies respectively.[111] Because the qualities of bodies "have an absolute being irrelative to us," causation for Boyle did not refer merely to a category of the human understanding. A "glowing coal," for example, "would be hot, though there were no man or any other animal in the world." Heat is a quality that does not only "work upon our senses, but upon other, and those inanimate bodies; as the coal will not only heat or burn a man's hand if he touch it, but would likewise heat wax . . . and thaw ice into water, although all the men and sensitive beings in the world were annihilated."[112] Indeed, even if these bodies were "placed *in vacuo*" or in one of those "imaginary spaces which divers of the schoolmen fancy to be beyond the bounds of our universe, they would retain many of the qualities they are now endowed with, yet they would not have them all; but by being restored to their former places in this world, would regain a new set of faculties (or powers) and dispositions."[113]

Boyle was often vague and sometimes changed his mind about which type of contrivance was responsible for a particular quality. The importance of his corpuscular philosophy did not consist in its specific theories, however, but in the general program for the interpretation of nature. His goal was the discovery of true essences, which are the forms of bodies that result from the set of qualities that they possess. These qualities are in turn the powers that bodies have to effect changes, and powers are, by definition, relational. In the world as it is presently constituted there are "great multitudes of corpuscles mingled among themselves." How these corpuscles come to be related to each other will determine which qualities are produced. In generation, for example, parts of matter "that did indeed before pre-exist" are "now brought together and disposed of after the manner requisite to entitle the body that results from them to a new denomination, and make it appertain to such a determinate species of natural bodies, so that no new substance is in general produced, but only that which was pre-existent obtains a new modification or manner of existence."[114] In a change of form, or essence, new qualities are taken on by the matter because of its differing manner of existence. In a similar way, there are also "*de facto* in the world certain sensible and rational beings" whose ideas are products of "the relation that happens to be betwixt those primary accidents of the sensible object and the peculiar texture of the organ it affects."[115] And one must discover "what changes happen in the objects themselves" in order to understand how they are able to "cause in us a perception sometimes of one quality and sometimes of another."[116]

Boyle criticized his contemporaries for ignoring the qualitative aspects of matter because, as he said, "qualities do as well seem to belong to natural bodies generally considered, as place, time, motion, and those other things, which upon that account are wont to be treated of in the general part of natural philosophy."[117] The new philosophers were understandably irritated by the endless Scholastic disputations about forms and qualities, but they had gone too far in their total rejection of these notions. Boyle agreed with them that it was a mistake to think of qualities or forms as actual physical entities, but he believed that it was legitimate to treat them as actual physical relations. In a section of *The Origin of Forms and Qualities* entitled "An Excursion about the Relative Nature of Physical Qualities," he maintained that an appreciation of the relative nature of qualities would be "of no small importance towards the avoiding of the grand mistake that hath hitherto obtained about the nature of qualities."[118] This idea also appears in his "Cosmical Qualities of Things" and "Cosmical Suspicions." In the "Excursion" he opened his discussion with an analogy, frequently used elsewhere, of a lock and key. When the "smith that invented" them made the first locks and keys, they were nothing but pieces of iron contrived into particular shapes,

> but in regard that these two pieces of iron might now be applied to one
> another after a certain manner, and that there was a congruity betwixt

the wards of the lock and those of the key, the lock and the key did each of them now obtain a new capacity; and it became a main part of the notion and description of a lock, that it was capable of being made to lock or unlock by that piece of iron we call a key, and it was looked upon as a peculiar faculty and power in the key, that it was fitted to open and shut the lock; and yet by these new attributes there was not added any real or physical entity either to the lock or to the key, each of them remaining indeed nothing but the same piece of iron, just so shaped, as it was before.[119]

The qualities of natural objects should be understood in the same way in which the qualities of man-made objects are. Those qualities, for example, "which we call sensible" because of "a certain congruity or incongruity in point of figure . . . to our sensories" are by "the portions of matter they modify . . . enabled to produce various effects, upon whose account we make bodies to be endowed with qualities." But just as with the lock and key, these qualities "are not in the bodies that are endowed with them, any real or distinct entities, or differing from the matter itself, furnished with such a determinate bigness, shape, or other mechanical modification." The yellow color that is perceived when a piece of gold is presented to a human sensory, for example, is a product of the internal configuration of the metal to act in a specific way when externally related to an organ of sense, just as its disposition to dissolve in aqua regia is a product of its configuration in relation to the configuration of the dissolving agent.[120] Boyle wished to investigate the processes of this world, where only a few of the qualities of bodies are products of internal configurations alone. Most are products of the relations that obtain between their bodies and other bodies. The latter type of qualities may be called "cosmical" or "systematical," because "they depend upon some unheeded relations and impressions which these bodies owe to the determinate fabric of the grand system or world they are parts of."[121] To discover the processes that give rise to these qualities, one must become acquainted with the universe as a whole, because "a portion of matter that is indeed endowed but with a very few mechanical affections, as such a determinate texture and motion, but is placed among a multitude of other bodies that differ in those attributes from it and one another, should be capable of having a great number and variety of relations to those other bodies."[122]

Boyle's project was to discover those qualities that bodies have because of their position in the universe that in turn define their essences. These "cosmical qualities" are not "chimaeras." They "are not merely fictitious qualities, but such whose existence I can manifest, not only by considerations not absurd, but also by real experiments and physical phaenomena."[123] As such, the goal of Boyle's experimental philosophy went beyond that of an empiricist science of appearances. He sought to discover the essences of bodies by investigating the complex causal relations that obtain in a coherent and harmonious universe.[124] His was an extremely difficult task, made more so by the fact that there was "not any one quality of which any author has yet given us an anything competent history."[125] To fulfill his explanatory goal, then, he would find it necessary to perfect the experimental philosophy. He would need to make the method sophisticated enough to capture the complexity exhibited by the book of nature.

Five

BIBLICAL HERMENEUTICS

BOYLE'S NATURAL THEOLOGY provided him with an ontological conception that led him to formulate a natural philosophy, the goal of which was to discover the ways that interrelated causal processes produce the manifest effects of nature. His theological beliefs would also provide another crucial argument for his justification of the experimental method as the best means for discovering such causes. Boyle drew upon the Baconian, legal, mechanical, chemical, and medical traditions in part to compile a set of "collateral arguments" in support of his program. While none would be sufficient in themselves, taken together they could be convincing, but more so for those who already had an inclination to experiment. Boyle was well aware that his arguments did not convince those philosophers, such as Hobbes and Spinoza, who had no such inclination.

It should be clear, for example, that while Bacon provided a significant precedent, the experimental method could not be justified simply by an appeal to such an authority. Indeed, the motto of the Royal Society, "Nullius in verba," understood in the Latin context from which it was taken, stated that all such appeals to authority were illegitimate.[1] Nor could the experimental practices of his predecessors serve as conclusive proof of the superiority of his method, because in fact the successes of the experimental program up to his time were not overwhelming. It was not apparent, for example, whether Galileo's work rested upon actual experimental trials or whether it resulted from an abstract mathematical analysis.[2] Also, while he had disapproved of Pascal's method of "thought experiments," it is significant to note that Boyle's actual experiments led to the same conclusions that had been reached earlier by his French contemporary.[3] Harvey's theory of the circulation of the blood provided a better case for Boyle's purposes, but it is still not clear that actual experimental work had been absolutely necessary for the theory's acceptance. Hobbes, for example, accepted the theory even though, as we saw in chapter 2, he was highly critical of experimental modes of proof.[4] Boyle's practical arguments from the success of experiment, therefore, were not altogether convincing, and it is reasonable to see how some could argue that the results achieved through experiment were not worth the great amount of labor and expense involved, since such results could be had in an easier manner.

Boyle's appeal to the legal standard of rational assent is also problematic. First, the standard itself was rather vague. To say that all of the information relevant to a conclusion must be included in the decision-making process was arbitrary at best, because one could always question the relevance of the evidence. We have seen that Hobbes held that only evidence from mathematical analysis, whereby one settled on definitions and then derived consequences from them, was relevant to the pursuit of natural philosophy. On the other hand, one could argue that Boyle did not include all of the relevant information since he refused to include metaphysical first principles as evidence. Spinoza was as critical of Boyle as he had been of Descartes and Bacon for having thus "strayed so far from the knowledge of the first cause and the origin of all things."[5]

In addition, a further objection, although not explicitly made in Boyle's time, could be formulated against the legitimacy of his use of legal analogies. While such analogies clarify, in a descriptive way, the historical role of experience in his experimental method, it is not completely clear why a procedure followed in the legal realm, which concerns the contingent actions of persons possessed with free will, would be appropriate for the natural realm, which concerns the lawlike behavior of inanimate matter. Boyle did draw such a distinction between the natural and human worlds. He noted, for example, that because humans are free agents, "the skill of ruling nations" is more difficult than other arts that work only with "inanimate materials."[6] And he insisted that such "inanimate agents act not by choice, but by a necessary impulse."[7] Even more emphatic was his criticism of the Helmontian theory of acids and alkalies, which assumed that a special relation held between these substances by which there was a "supposed hostility" between them, as between two warring tribes, while there was a "sympathy with bodies belonging to the same tribe." These notions of "amity and enmity" were "affections of intelligent beings," not of material substances.[8] Against such chemical "duellists," Boyle stated, in no uncertain terms, that there was "one thing, which I would gladly recommend and inculcate to you, namely, that 'those hypotheses do not a little hinder the progress of human knowledge, that introduce morals and politicks into the explications of corporeal nature, where all things are indeed transacted according to laws mechanical.'"[9] If it is not legitimate to introduce morals and politics into theorizing, could it not also be said that it is inappropriate to introduce a method that is concerned with such human areas into the investigation of nature?

To understand why Boyle could use the limited success of experiment and the legal standard of rational assent, it is necessary to see these arguments in relation to his ontological view. Nature is a "book" written by an omniscient and omnipotent author. As such, it is a fait accompli. The study of nature involves, then, the attempt to understand the completed act of a free agent and thus requires the same stringent appeals to experience and evidence as those employed in the legal realm. One cannot reason on purely a priori grounds about such a divinely created product, because God's reason and power extend far beyond human faculties. Rather, one must look at nature—read the text—in order to determine what was actually done. The world *is* like a text. It is a coherent, albeit extremely complex, whole. To understand any part of the great cosmic mechanism, the relations that hold between that part and the rest of the whole have to be known.

As Boyle explained in his *Usefulness of Natural Philosophy*, "Each page in the great volume of nature is full of real hieroglyphicks, where (by an inverted way of expression) things stand for words, and their qualities for letters." [10] As a particular sequence of letters make up a word, so a particular set of qualities make up a body of a particular type. One must find these "letters" or qualities first in order to understand the things, or words, of nature. Because the world is such a complex text, an alphabet would be required, large enough to ensure the decipherment of all of the words written. Neither the mathematical points of the geometers nor the material principles of the chemists would suffice. The evidence relevant to the truth of a corpuscular explanation would have to include the particular configurations of matter, plus the motion that these least parts have had impressed upon them. The alchemists lacked such "comprehensive principles," and thus their attempts to explain the qualities of bodies were doomed to failure, as Boyle explained, using the text metaphor to illustrate this point: "Methinks, a chemist, who by the help of his *tria prima*, takes upon him to interpret that book of nature, of which the qualities of bodies make a great part, acts at but a little better rate than he, that seeing a great book written in a cypher, whereof he were acquainted but with three letters, should undertake to decypher the whole piece." Such a chemist would be kept "from decyphering a good part of those very words," because "it is more than probable, that a great part of the book would consist of words wherein none of his three letters were to be found." [11]

Because the book of nature "was written for man's instruction," its complexity does not preclude our understanding it. [12] But its complexity does require that our successful understanding of nature will depend upon knowledge of a vast number of particulars and upon our ability to reason correctly about the relations that hold between them. The world is a coherent whole, but we are not able to comprehend it all at once. Only by employing a method of proof that has the flexibility exhibited by moral demonstration will progress in our knowledge be assured. As we learn about the particular processes of nature, we are able to postulate theoretical constructs that join the particulars together into a coherent account. These constructs are probable (worthy of belief) given the knowledge thus far gained about nature, but they should never be accepted dogmatically. It is always possible that a further increase in our knowledge may lead to a revision of what we formerly believed. Boyle stated that his view of nature led him to see the intelligibility of corpuscular principles and that his acceptance of corpuscularianism led him to see the necessity of an experimental approach.[13] For Boyle, the experimental method was a means by which one could "interpret" the book of nature.

In order to understand fully the details of his method and how it was justified, it is necessary to see that the experimental philosophy was designed as a method of interpretation. But then, it is also necessary to understand what Boyle meant by *interpretation*. In addition to offering well-known arguments in natural theology, Boyle was a student of Scripture who used biblical exegesis in support of his views. What has been often overlooked, however, is that he was also interested in what we would today call biblical hermeneutics. At the same time that he was working out his experimental approach for the interpretation of nature, he was also working out the details of how best to interpret Scripture. From his early days on the Continent, where he met such thinkers as Menasseh ben Israel and Jean Diodati, to his days at Oxford, where his circle of acquaintance included not only the experimentalists but also theologians, classicists, and linguists, such as Thomas Barlow, Thomas Hyde, Edward Pococke, and Samuel Clarke, Boyle was concerned with finding the right way of understanding both books.[14] As early as 1652 he had written "Essay of the Holy Scripture," parts of which were later incorporated into his published works, particularly *Considerations Touching the Style of the Holy Scriptures* (1661) and *The Excellency of Theology* (1674).[15] In these works he clearly expressed what he meant by the interpretation of a text and how it could best be accomplished.

The Two Books

We saw in the last chapter that Boyle certainly believed that science and religion were compatible. Indeed, since both are studies of divine works, they would have to be so. He allowed that some theological assumptions could act as guiding principles for a general ontology. He also believed that the book of Scripture and the book of nature could be studied in much the same way. It must be stressed, however, that, except at the most general level, the content of the two books had to remain separate.

When he quoted Bacon on the compatibility of science and religion in his *Usefulness of Natural Philosophy*, Boyle also noted that Bacon had cautioned his readers to "*beware*" that "*they do not unwisely mingle and confound these distinct learnings of theology and philosophy and their several waters together.*"[16] He was opposed to any "unwholesome mixture" of the two disciplines. The two books could be used to shed light on each other, but care was required so as not to confound them.

The study of the book of nature can guide a knowledge of theology, because a knowledge of the "created book" is "both conducive to the belief, and necessary to the understanding of his written one." Of the "qualities of the creatures," for example, that are used in Scripture to illustrate virtues and vices, "there are divers not to be fully understood without the assistance of more penetrating indagations of the abstrusities of nature, and the more unobvious properties of things."[17]

On the other side, theology could shed light on the more general aspects of nature. The fundamental tenet of creation, for example, could guide a Christian to see how atheistic formulations of atomism, such as those proposed by "those great denyers of creation and providence, *Epicurus,* and his paraphrast *Lucretius,*" were based upon an incoherent foundation. Even if these "very subtile philosophers" could explain all by matter and motion, "it would not thence necessarily follow, that, at the first production of the world, there was no need of a most powerful and intelligent being" who had established "the universal and conspiring harmony of things; and especially to connect those atoms into such various seminal contextures, upon which most of the more abstruse operations, and elaborate productions of nature appear to depend."[18] The belief in divine creation could act as a guiding principle in deciding on the best formulation of a general cosmology. But it was only a guide, and Boyle could point to non-Christians who had found ancient atomism unacceptable. Thales, Anaxagoras, and Aristotle, for example, had all recognized that matter must be moved by a "creating power." Indeed, Boyle based his argument for the necessity of God's initial creation of the particular motive forces in the world upon Aristotle's definition of essence, "as on account of which a thing is what it is," to conclude that motion "is no way necessary to the essence of matter, . . . for matter is no less matter when it rests, than when it is in motion."[19] Of the two choices for a general ontology, Boyle noted that it was easier to conceive of an all-powerful God who had existed from eternity and was the original cause of the motion of matter than it was to conceive of matter's having existed from eternity and somehow being the cause of its own motion.[20]

Although major tenets could be used as a guide, Boyle noted that many passages in the Bible could not be used for the establishment of physical truths because they were "spoken of rather in a popular than accurate manner."[21] The two books are distinct. Scripture is "designed to teach us rather divinity than philosophy."[22] And the book of nature, although it reveals something of the glory, power, and wisdom of God, was not specifically designed to teach the particulars of theology. Because Boyle believed that "nature is only one way to learn about God," it is a mistake to view his theology as founded purely upon a natural basis.[23] Divine revelation, as delivered in the Bible, was a necessary component of his rational belief.[24] In fact, in his *Excellency of Theology* Boyle produced an extended argument in support of the intellectual superiority of scriptural studies over natural theological studies for learning about God and his works.[25] He criticized Descartes, for example, for believing that he had proved the immortality of the human soul by natural means. Descartes may have proved that the soul was immaterial, but only from revelation could one learn that it was also immortal.[26]

Nor was it legitimate to attempt to explain natural phenomena by reference to spiritual agents. In order to explain physical processes, one must appeal to physical causes. An "immaterial principle or agent" will "not enable us to explain the phaenomena," because "we cannot conceive, how it should produce changes in a body, without the help of mechanical principles, especially local motion." Natural philosophy is concerned with "things purely corporeal."[27] "General answers," such as "that it pleased the author of the universe" to create things in the manner found, are not explanatory at the physical level.[28] Those who used such illegitimate teleological explanations merely masked their ignorance of the true causes operative in the world.[29] When working as a "naturalist," Boyle sought to "discourse of natural things" only, without "intermeddling with supernatural mysteries."[30]

There is a compatibility and an overlap between the two books, but Boyle noted that one must be extremely cautious in how they were put together. God may have given some "hints" in Scripture about the nature of the world, and he may also aid philosophers in their attempts to understand the world, "partly by protecting their attempts from those unlucky accidents, which often make ingenuous and industrious endeavours miscarry," and partly, "or rather principally, by directing them to those happy and pregnant hints, which an ordinary skill and industry may so improve, as to do such things, and make such discoveries by virtue of them, as both others, and the person himself, whose knowledge is thus increased, would scarce have imagined to be possible." It is only through our own rational and practical activity, however, that our knowledge of nature will increase.[31]

Boyle expended much effort in his defense of the pursuit of science against the criticisms of theologians. At the same time, he also defended Christianity against "libertines" who believed that the new science could be used to disprove the rationality of religion.[32] Both of his antagonists' arguments were premature at best, because neither text was as yet fully enough understood to be capable of being used for such polemical purposes. As he explained, it would "not be so easy a matter" to say that the principles of natural philosophy could explain the world without recourse to the original creator of it: "For we are yet, for aught I can find, far enough from being able to explicate all the phaenomena of nature by any principles whatsoever. And even of the atomical philosophers, whose sect seems to have the most ingeniously attempted it, some of the eminentist have themselves freely acknowledged to me, their being unable to do it convincingly to others, or so much as satisfactorily to themselves."[33]

The explication of all the "various phaenomena," even of a "single inanimate, and seemingly homogenous body" as mercury, "is like to prove a task capable of defeating the industry and attempts, I say not of more than one philosopher, but of more than one age."[34] Because of the complex interrelations in nature, knowledge of it will only be gained by a slow process that will demand the combined effort of a number of workers over many years. Neither debates between science and religion nor debates within the disciplines would be appropriate at such an early stage of inquiry. Boyle's "irenic personality" was not simply a product of his desire for political peace among quarreling partisans but also a product of his belief that true knowledge could only be achieved through a tolerant, cooperative effort of sincere searchers after it.[35] Just as with natural philosophy, so too should theology be practiced as a tolerant and undogmatic process of discovery: "You should as little think, that there are no more mysteries in the books of scripture, besides those, that the school-divines and vulgar commentators have taken notice of, and unfolded; as that there are no other mysteries in the book of nature, than those, which the same schoolmen (who have taken upon them to interpret *Aristotle* and nature too) have observed and explained."[36]

Belief in the divine authorship of both books led Boyle to have a reverent and extremely humble attitude toward both Scripture and nature. It also gave him hope that God would assist him in coming to learn something of these works if he actively and sincerely pursued their study. The contents of the two books were distinct for the most part, but Boyle's general attitude toward the learning process through which one would discover truth was the same for both. This is not to say that Boyle developed a hermeneutic based upon experimental principles or that his experimental philosophy influenced the way in which he studied Scripture. There are parallels between the two studies, however, and in both Boyle exhibited the same dynamic approach to knowledge acquisition. His epistemological conception of the progressive nature of knowledge entailed the belief that it could only be achieved through a complex process of interpretation and the reconciliation of truths from all areas of learning.

The Interpretation of Scripture

Boyle developed a rational approach to religious belief, and the moderating and reconciling attitude that went with it, while still in his teens. The religious controversies that he encountered during his stay in Geneva, a home for some of the more radical sects, caused him to have doubts about his beliefs. His anxiety proved fruitful, however, as he wrote in his third-person autobiography, since it led him to "the advantage of groundedness in his religion: for the perplexity his doubts created obliged him, to remove them, to be seriously inquisitive of the truth of the very fundamentals of Christianity, and to hear what both Turks, and Jewes, and the chief sects of Christians could alledge for their several opinions."[37] Upon his return home to England, he encountered yet more controversy, which greatly disturbed him, because he saw such religious disputes as being detrimental to Christianity in general.[38] The religious partisans were tearing Christianity apart by "quarrelling for a few trifling opinions" rather than attempting "to embrace one another for those many fundamental truths wherein they agree."[39] As he wrote to John Dury in 1647, he hoped that a "unity of peace" could be achieved through a "moderate and satisfactory reconcilement" of the major sects, while the numerous "upstart sectaries" would be "smitten at the root with the worm of their irrationality" and "be as sudden in their decay, as they were hasty in their growth."[40]

Aside from the perceived threat to Christianity, Boyle noted that the proliferation of sectarian controversy was also leading to a devaluation of the Bible itself. Many critics of the Bible cited the number of inconsistent doctrines apparently justified by the same text as support for the claim that the Bible was "immethodical" and "contradictory to itself" and thus not worthy of study.[41] In response, Boyle argued that it was not any fault in the text, but in its readers, that led to interpretive difficulties, because "we often impute to the scripture our own faults and deficiencies."[42] The charges of obscurity and lack of method, for example, were misguided because they were based upon the "presumption" of many critics who saw their "own abilities as the measure of all discourses" and were thus led "to call all that transcends their apprehensions, dark, and all that equals it not, trivial."[43] But as Boyle noted, God's manner of working is different from ours. The "book of grace doth but therein resemble the book of nature," and it is only from our limited standpoint that the text appears to lack a rational method: "It became not the majesty of God to suffer himself to be fettered to human laws of method, which, devised only for your own narrow and low conceptions, would sometimes be improper for, and injurious to his, who may well say (as he doth in the Prophet), that his thoughts are so far from being ours, that, *As the heavens are higher than the earth, so are his thoughts higher than our thoughts.*"[44]

The critics were guilty of attempting to impose a human standard of rationality upon the divine author and of looking for "apodictical" arguments where such were not appropriate.[45] The Bible is not a philosophical text composed of syllogisms that are clear upon a first reading. Rather, it is a complex text designed to reveal supernatural mysteries oftentimes by digressions "which do not readily seem pertinent to the series of the discourse, but are extremely so to some scope of the author, and afford much light and excellent hints to the reader" who is attentive.[46] We should not "suffer our not understanding the full meaning at first to deter us from endeavouring to find it out by further study."[47] That is, we may not achieve the absolute certainty so much sought for in the philosophical tradition, but we can achieve some insight into the truth of Scripture.

Boyle admitted that there are "seeming contradictions in Scripture, whose reconcilement is a taske very uneasy."[48] Most of these apparent contradictions arise from the misplaced belief that the true meaning of the text has been exhausted. We often "deceive ourselves by presuming we understand it [the Bible] when indeed we do not."[49] We have concluded too quickly that we know what the text says, and therefore when contradictions appear they are attributed to the text itself. But "we may be easily mistaken" in our interpretation, and thus the disagreement between expositors does not mean that the Bible itself is contradictory.[50] Rather, the contradictions are a direct result of the expositors who allow their particular preconceptions to influence their interpretations:

> It is not oftentimes so much the various aspects of the texts, as the divers prepossessions and interests of the expositors, that make books seem replenished with interfering passages and contradictions. For if once the theme treated of do highly concern men's interests, let the book be as clear as it can, subtile and engaged persons on both sides, perusing it with forestalled judgments of biassed passions, will be sure to wrest many passages to countenance their prejudices, and serve their ends, though they make the texts never so fiercely fall out with one another, to reconcile them to their partial glosses.[51]

Because of the presence of these types of prejudices, one could not take for granted the interpretation of Scripture of that "society of Christians" one chances "to be born and bred in."[52] Rather, one has to study the Scripture itself. Theology should not be accepted as a static set of dogmas delivered as the precepts of an established church but pursued as an open-ended enterprise of discovery about God and his mysteries:

> The greater reverence I owe to the scripture itself, than to its expositors, prevails upon me to tell you freely, that you will not do right, either to theology, or (the greatest repository of its truths) the bible, if you imagine, that there are no considerable additions to be made to the theological discoveries we have already, nor no clearer expositions of many texts of scripture, or better reflections on that matchless book, than are to be met with in the generality of commentators, or of preachers, without excepting the antient fathers themselves.[53]

Boyle acknowledged that particular interests, metaphysical notions, and sectarian doctrines would always guide the "heated spirits of men" and thus that scriptural interpretation would always be problematic.[54] But in works such as *Considerations Touching the Style of the Holy Scriptures, The Excellency of Theology,* and

The Christian Virtuoso, he suggested guidelines designed to lessen the influence of these forestalled judgments.

To begin, we should not attempt to use the Bible as an "arsenal, to be resorted to only for arms and weapons to defend this party, or defeat its enemies." Rather, we should "make such observations, as may solidly justify . . . a reverence for the scripture itself, and Christianity in general."[55] To make such observations, the first thing we must do is recognize that we do not yet fully understand the Bible. This recognition ought to lead us back to a more serious examination of the text itself, which would include a critical investigation of the way that it has been translated. According to Boyle, most of the partisans of the disputes "take it for granted that the texts of Scripture are rightly translated, but that is oftentimes an ill grounded supposition." He argued that "since according to the laws of the best logic, to make a contradiction strictly so called, the terms must be understood in the same sense according to all necessary circumstances," we cannot conclude that passages are contradictory "if we do not understand the meaning of those words and phrases" of which the passages are composed.[56] Before the interpretive task can begin, the right translation of passages must be determined, and a knowledge of the original languages of the text will be necessary. But as Boyle found when he taught himself Hebrew, oftentimes "the same word or phrase may have had diverse other significations, than interpreters have taken notice of."[57] It is, therefore, "probable, that many of those texts, whose expressions, as they are rendered in our translations, seem flat, or improper, or incoherent with the context, would appear much otherwise, if we were acquainted with all the significations of words and phrases, that were known in the times, when the Hebrew language flourished, and the sacred books were written."[58]

Learning the original languages of Scripture involves much more than simply learning their grammar. It is not only "our want of Knowledge of the Originall Tongues, and their several dialects," but also "our ignorance of the History, Geography and customes of foreign times, and nations," that causes us to "father upon the Scripture, Absurdities which are indeed the issue of our own Ignorance."[59] We need to learn about a "multitude of particulars relating to the topography, history, rites, opinions, fashions, customs, etc. of the antient Jews and neighbouring nations," so that we may understand idiomatic phrases, for example, whose literal translation would be meaningless.[60] We cannot always get this historical knowledge from the Bible itself, because the ancient penmen would have assumed that their readers were familiar with many things that the modern reader is not: "It is not to be expected, that out of those books we should be able to collect and comprehend, either complete ideas of Israelitish government, civil and ecclesiastical, or the true state of their several sects, opinions and affairs in matters of religion: and yet without the knowledge of those it cannot be, but that many texts will seem obscure to us, which were not at all so to them." Many "Mosaical texts," for example, seem obscure only because of a lack of knowledge about the "antient Zabians, in opposition of whose magical worship and superstitions, I am apt to think divers ceremonies of the ritual law of the Jews to have been instituted." Similarly, many passages in the New Testament may be obscure simply because of "our ignorance or want of taking notice of the persuasions and practices of the Gnosticks, Carpocratians, and the sects allied to theirs."[61]

Some historical sources, such as "the writings of those Jewish Rabbies, that lived about our Saviour's and his Apostles times," had already been used by biblical scholars to clear up the meaning of obscure passages. Boyle sought to promote this type of scholarship because he believed that "higher and valuable attainments in that kind of learning" would "disperse that obscurity, which yet dwells upon divers other texts" and show the "groundlessness" of "our too fierce contentions about them."[62] The more observations of this type that are made, the better our chances will be of correctly understanding the true meaning of Scripture: "The more knowing its [the Bible's] pious studiers have been, the greater store of excellent truths they have met with in it; the scripture being indeed like heaven, where the better our eyes and telescopes are, the more lights we discover."[63] Linguistic and historical scholarship is required as a way to provide the firm informational basis necessary for biblical study, but it is only the beginning of the larger interpretive task.

When attempting to determine the true meaning of anything, the type of proposition within which it is expressed must be taken into consideration. In *The Christian Virtuoso* Boyle distinguished between two types of propositions: absolute and conditional. Absolute propositions express truths such as "those theoretical principles and axioms, which are the foundations of our reasonings, such as are, two contradictories cannot both be true," and "the definitions of our more simple mental ideas, such as the clear conceptions we have of a triangle, a square, a circle, a cube, a cylinder, etc." These are "self-evident principles" and "may be called eternal truths," because "there neither has, nor will be any time, wherein these principles of knowledge and ratiocinations may not be safely assented to, without any relation to contingent circumstances."[64] Conditional propositions, on the other hand, express "truths upon supposition."[65] The two types of propositions must be kept distinct because the justification procedures for the two are radically different: "It is far from being all one, to consider a proposition as an absolute one, and to consider it as part of a system, where it is linked, and as it were enchaced by others that are manifestly true, or granted, or demonstrable."[66] Absolute propositions can be known immediately by the light of reason in isolation from any contingent circumstances, whereas conditional propositions are justified by an appeal to our knowledge of other propositions to which they are related. Also, unlike the truth expressed by absolute propositions, the truth of conditional ones can be known to varying degrees. When dealing with such propositions, we must assess the amount of information upon which they are based to determine whether they are gradual or complete. The latter category is reserved for those "judgments, that are duly formed upon a perfect information, or at least such a one, as is sufficient to ground a right judgment upon."[67]

The Bible, as the product of a sole omnipotent author, is made up of conditional propositions. Boyle believed that because of its divine authorship, the Bible must be true. From this it follows that the Bible cannot be contradictory to itself, for God, "as the author of our reason, cannot be supposed to oblige us to believe contradictions."[68] But then, because the Scripture is a coherent whole that "loses much by not being considered as a system," internal evidence from the text itself is required in order to constrain the prejudicial and superficial readings that result in the faulty translations and interpretations from which the apparent contradictions arise. We must attempt to "make the scripture coherent, or discursive; and then, for our opinions, rather to conform them to the sense of scripture, than wrest the words of scripture to them."[69] When a seeming contradiction arises, we should not take it as a sign of a fault in the text, but rather as a sign of a fault in our interpretation, and we must seek a reinterpretation that will eliminate the contradiction.[70]

At the second level, then, translations, guided by historical information, must be made internally coherent. Translators must "let the context and the speaker's scope regulate their choice, amongst all the various, though not equally obvious, significations of ambiguous words and phrases." As an example, Boyle noted that the "Hebrew conjunction copulative Vau, or Vaf, (as it is diversly pronounced by the Jews)" may signify not only "AND, but hath also . . . four or five and twenty other significations (as *that, but, or, so, when, therefore, yet, then, because, now, as, though,* etc.) and that the sense [of the text] only gives it this great diversity of acceptions."[71] In the end, however, the task of translation and interpretation is much more complex. The text as a whole, the books of both the Old and the New Testaments, must be made coherent.

> Those must derogate hugely from the scripture, who only consider the sense of the particular sections, or even books of it: for I conceive, that . . . he, that shall attentively survey that whole body of canonical writings we now call the Bible, and shall judiciously in their system compare and confer them to each other, may discern, upon the whole matter, so admirable a contexture and disposition, as may manifest that book to be the work of the same wisdom, that so accurately composed the book of nature, and so divinely contrived this vast fabrick of the world.[72]

All of the books give light to each other, not merely those that are close together in space or time or in sense. There are "mutual irradiations and secret references" throughout the entire text, and one must use these as helps to

achieve the proper interpretation. "The books of scripture illustrate and expound each other; *Genesis* and the *Apocalypse* are in some things reciprocal commentaries; (as in trigonometry the distantest side and angle use best to help us
to the knowledge one of the other:) . . . so do some texts of scripture guide us
to the intelligence of others, from which they are widely distant in the Bible,
and seem so in the sense."[73]

Biblical interpretation must proceed by way of an active process of reconciliation whereby apparent inconsistencies within the text or between the text
and other known truths are eliminated.[74] The process consists roughly of three
stages. First we must gather historical and linguistic information as a basis for
correct translation and interpretation. Then we must look at the sense of the
passage as a whole to determine whether we have given the right meaning to
the words found there. Finally, we must compare these newly interpreted passages with others that we previously believed ourselves to have rightly interpreted. If a contradiction arises at some point, it is a good sign that there is a
fault in at least one of our interpretations. Reason must then be employed to
reconcile the apparent conflict. The effort will often result in a higher level of
understanding, where we not only learn a new truth but also gain a new insight
that will aid us in the acquisition of more truth: "The texts of the Bible inter-
change light with one another, and every new degree of scripture-knowledge
is not only an acquist of so much, but an instrument to acquire more."[75]

By this process of interpretation, we gradually clear up the obscurities in
the Bible that were initially a "mistaken discouragement from reading it."[76] It
is our "carelessness, or ignorance," that keeps us from "discovering, as well in
the *written*, as in the *created* book of God . . . signatures of the divine Author's
wisdom and goodness." Interpretive difficulties can be cleared up as long as
one is not a superficial reader who, like a butcher, "kills and divides into pieces,
sheep and oxen, perhaps a thousand times, . . . without ever discovering anything of that wonderful contrivance, that a good anatomist will, by dissecting
the same animals, take notice of." To be successful, we must be like the "diligent
and devout" reader who, "furnished with the original languages, and other useful parts of learning, by attentively and assiduously reading those excellent
writings, and carefully comparing place with place, phrase with phrase, and in
short, adding one help of interpretation to another, may discover excellent and
mysterious truths, that are wholly missed by vulgar readers."[77]

The truths that we gain through this active process of reconciliation are
gradual in degree. We will never be able to rest content in the satisfaction of
having achieved the formal certainty of mathematics or logic, and it is quite
likely that we will also never achieve complete truth. But this should not cause
us to be like the critics who, "because they cannot understand all of it [the
Bible], they will not endeavour to learn any thing from it."[78] We do learn. The
eradication of prejudice that is achieved through the reconciliation of apparent

inconsistencies is an indication that we are getting closer to the whole truth. Every seeming contradiction has a solution, if we have enough wit to discern it. In *Things Said to Transcend Reason* Boyle used another example from natural philosophy to illustrate this point.

> Now if one, that had observed *Venus* only in the mornings, should have affirmed, that besides the six known planets, there was but a seventh (namely the phosphorus) which preceded the rising sun; and another, (that had taken notice of her only in the evenings) should assert, that besides the same six known ones, the only seventh was that called *Hesperus*, which sometimes appeared after his setting; a by-stander would presently have concluded, that their assertions were not reconcileable, either to one another, or to the truth, which (in his judgment) was, that there must be no less than eight visible planets: and yet *Pythagoras*, who had more skill, and more piercing wit, did, (as was lately noted) discern and teach, that these two phaenomena were produced by one and the same planet *Venus*, determined by its peculiar motion (about the sun) to shew itself near our horizon, sometimes before he ascends it, and sometimes after he had left it.[79]

Reconciliation is not always easy. Boyle used examples like the above to "perswade us from being too forward to reject every proposition, that we see not how to reconcile to what we take for a truth."[80] We must remain flexible. It is crucial to the progress of our understanding that we be willing to change our opinions as new information relevant to our interpretation is discovered.

Boyle believed that the interpretive task might be difficult, but the goal of understanding the scriptural message was of such importance that the great amount of labor involved in the effort was justified. The goal of understanding nature, as God's production, was also important. And as we will now see, because of its great complexity, understanding nature would involve the same type of arduous task of interpretation and thus would require the same type of hermeneutic principles that were employed for an actual text, as constraints upon the theories that we construct for the "explication" of nature's processes.[81]

The Interpretation of Nature

Boyle's choice of method was guided by his ontological view of nature as a divine text. In broad outline, his experimental philosophy was designed as a means of gathering as much information about natural processes as possible in order to be able to construct plausible hypotheses that would provide the most coherent interpretations of how the particulars of nature are connected into one grand cosmic mechanism. The experimental acquisition of knowledge was to be a gradual, and admittedly fallible, process of discovery. But this was as it should be, given the nature of the object being studied. "The book of nature is a fine and large piece of tapestry rolled up, which we are not able to see all at once, but must be content to wait for the discovery of its beauty, and symmetry, little by little, as it gradually comes to be more and more unfolded, or displayed."[82] Knowledge must be grounded "historically." As we gradually uncover more information about the world, we will accumulate background knowledge, all of which will, in the end, be assimilated into one consistent narrative of nature. Because the experimental philosophy is a method of interpretation, however, it should be clear that reason will provide the final warrant for the knowledge claims produced by it. Boyle only opposed the use of unaided reason when he rejected those who attempted to build science upon the metaphysical modalities of a priori ratiocinations.

In a manner reminiscent of his attitude toward scriptural study, Boyle maintained that "the right understanding of the Book of nature" is "a thing of that worth to a philosophical mind, and of that importance to mankind, that it ought to be preferred to the reputation of any of those that have taken upon them to Interpret it."[83] But he would not therefore reject all that had come before. While previous systems of natural philosophy had been erected prematurely, they might still contain some useful pieces of information, or subtle insights into nature, that could form part of the informational basis for a truer kind of philosophy. Indeed, "every Sect, and almost every Philosopher, has contributed something to the knowledge of nature," even though "none has furnished us with *all* that is necessary, to the compleating of Physics."[84] Boyle had an eclectic, reconciling attitude toward the systems of the past as well as those of the present, because the "true philosophy" is

> a thing of a more noble nature, and of greater extent, than the hypotheses of any one sect of philosophers, being indeed a comprehension of all the sciences, arts, disciplines, and other considerable parts of useful knowledge, that the rational mind can attain to . . . by reason, that is improved by meditation, literature, exercise, experience, and any other help to knowledge.[85]

Boyle's effort to reconcile the corpuscularians, chemists, and physicians formed a part of his eclectic approach to knowledge that was based upon his belief that all physical processes are interrelated.[86] This attitude was especially apparent in his discussions about complex living organisms. In the first volume of *The Usefulness of Natural Philosophy* he noted, for example, that among the areas of expertise required for a full understanding of human bodies he would include "a skill in divers parts of physiology, and especially chymistry," to gain a "competent knowledge of the nature of those juices that . . . pass through them"; skill "in the principles of mechanics and in the nature and properties of levers, pulleys, etc.," to understand the "origination, shape, bulk, length, progress and insertion of each particular muscle"; skill in optics, in order to "shew the wisdom of God in making the crystalline humour of the eyes"; and the skill of "statuaries and painters," to discover the "certain harmonious proportion betwixt the parts of a human body, in reference both to the whole, and to one another."[87] He continued to stress the need for an eclectic and reconciling approach to knowledge in volume 2, in which he compiled a number of instances from different disciplines in order to show how such a collection

> may serve to beget a confederacy, and an union between parts of learning, whose possessors have hitherto kept their respective skills strangers to one another; and by that means may bring great variety of observations and experiments of differing kinds into the notice of one man, or of the same persons; which how advantageous it may prove towards the increase of knowledge, our illustrious Verulam has somewhere taught us.[88]

The possessors of individual parts of learning, such as the mathematicians who focused exclusively upon the motion of geometrical bodies or the chemists who restricted their attention to the material ingredients of bodies, had made their inquiry too narrow and superficial and had thus left the way open for faulty interpretations of nature prejudiced by the biases of their own disciplines.[89] According to Boyle, "Nature never contradicts herself."[90] Because it is a divine text, any contradictions that seem to arise in the propositions about nature must be taken as a sign that we do not as yet fully understand it. The prejudice of "local reason" is responsible for seeming conflicts. As with any text, care must be taken not to impose one's own particular prejudices and idols of the mind upon the words of the author.[91] An experimental approach to nature can help to eradicate such prejudices because it is the way by which one's vision of the world can be expanded. By including all disciplines within its compass and actively searching for the fullest information possible, the mind becomes aware of those things that would never have occurred to it through superficial observation and reason alone: "He, that seeks for knowledge only within himself, shall be sure to be quite ignorant of far the greatest parts of things."[92]

One must collect information about natural processes into an accurate history before attempting to interpret the book of nature, but Boyle's goal was not a simple "ordering of phenomenal experience."[93] As we have seen, he did not deny that we can know the inner natures of things. He merely believed that before we can speculate about the hidden causes operative in the world, we first have to find the effects that are actually produced. He was "uneasy" about elaborate philosophical systems because it is difficult "to erect such theories" before the "tenth part of those phaenomena, that are to be explicated," have been taken notice of.[94] It is as prejudicial to natural philosophy to base theories upon faulty data as it is to theology to base scriptural interpretations upon faulty translations. The "collection of materials" that Boyle compiled and published throughout his life was meant to be the foundation upon which future theories, especially those that would "explicate qualities" of bodies, were to be built.[95] As he said in his essay on the "production of qualities," his experiments could "serve for confirmation" of corpuscular explanations of qualities as well as "contribute to the natural history of them."[96]

Boyle himself was more concerned with setting up a "program for interpretation" than he was with advancing his own ideas about particular interpretations.[97] The number of experiments that must be performed varies in direct "proportion to the comprehensiveness of the theory to be erected on them." The corpuscular philosophy was a very comprehensive theory. When he speculated at the theoretical level, Boyle did so in a tentative manner and noted that "I should, if it were not for mere shame, speak yet more diffidently than I have been wont to do" because he had "not unfrequently" found that "what pleased me for a while, as fairly comporting with the observations, on which such notions were grounded, was soon after disgraced by some further or new experiment."[98] His diffidence extended only to the substance of particular theories, however, not to the manner by which such theories were to be discovered and justified. Indeed, his diffidence at the substantive level was a direct result of his confidence in having found the correct method for interpreting nature. In order to discover the "particular structure" of bodies upon which their power to act and react depends, "inquirers must take notice of abundance of minute circumstances."[99]

The inductive establishment of facts is a necessary prerequisite to theory construction, but Boyle's experimental philosophy was ultimately designed as a way to construct corpuscular hypotheses, and the senses could only provide indirect evidence of the action of such minute particles of matter. Reason must be used to draw inferences from this indirect evidence. "The eye or the imagination can never reach to so small an object as an atom . . . [but] there is no necessity . . . that visibility to a human eye should be necessary to the existence of an atom, or of a corpuscle of air, or of the effluviums of a loadstone."[100] Corpuscles are unobservable in principle. We learn about them by using our "illative knowledge," not our "apprehensive."[101] Our senses provide us with information upon which to reason, and experiments can increase the quality of that information better than the methods used by those philosophers who "go no further than the outside of things, without penetrating into the recesses of them."[102] But such information is merely a restraint upon our reason and is not to be taken as decisive:

They that are wont, . . . to trust too much to the negative informations of their senses, without sufficiently consulting their reason, . . . imagine that such minute corpuscles (if they grant that there are such) as are not, for the most part of them, capable to work upon . . . the sight, cannot have any considerable operation upon other bodies. But I take this to be an error, which, as it very little becomes philosophers, so it has done no little prejudice to philosophy itself.[103]

In order to "search out and discover deep and unobvious truths," we must not be like Augustine's "brutes."[104] We must go beyond a mere passive viewing of nature and actively seek to interpret it with the rational faculty given to us. Experiments are required because we need to determine not only how bodies act upon our sensories but also how they act upon other bodies. These operations are the "signs" from which we can learn about the powers, the qualities, that bodies possess. Heat, for example, is a quality, produced by the motion of the matter in a body, that not only is sensible to us but also affects bodies such as wax and ice. Similar relations have to be sought for the other qualities of bodies before one can say that they are completely understood: "A body is not to be considered barely in itself, but as it is placed in, and is a portion of the universe."[105]

Reason must be used in the construction of corpuscular hypotheses because "it is part of reason to judge, what conclusion may, and what cannot be safely grounded on the information of the senses, and the testimony of experience."[106] In addition, there are "secret correspondences and alliances" in nature, and the mind must actively seek to reconcile nature's apparently diverse phenomena into a coherent account.[107] It was not merely Boyle's view of nature as a text but also his own experience that had revealed to him the "strange relations . . . betwixt natural things" that made it "extreamly difficult to understand any one of them perfectly without understanding also diverse others."[108] It was because of such relations that he would "dare not pretend to assign the true and adequate cause of more than very few even of the familiar phaenomena of Nature."[109] But it was also the existence of these relations that justified his appeal to the strong criterion of theory acceptance that he employed against Hobbes (see chapter 2). Just as the propositions of Scripture are of a conditional type, so too are those about nature. They must be judged by "their fitness to solve the phaenomena, for which they were devised, without crossing any known observation or law of nature."[110] Hypotheses of this type are worthy of approbation, but one does not stop at that point. Further testing must be done to see if the hypothesis can also explain "all other phaenomena of nature, as well as those it is framed to explicate."[111] If there are errors in our theoretical interpretations, they will be revealed at some point by the inability of our hypotheses to provide consistent explications of the phenomena.

Boyle's stress upon the importance of consistency shows that reason is the primary warrant for hypotheses. Indeed, "Philosophy, when it deserves that name, is but reason."[112] It should be noted, however, that reason is also the warrant for the factual propositions that form the informational basis upon which hypotheses are built. Because it is difficult "to make and relate an observation accurately and faithfully enough for a naturalist to rely on," one cannot simply take the information of the senses at face value.[113] "Whereas . . . the sensories may deceive us, if the requisites of sensation be wanting, as when a square tower appears round at a great distance, and a strait stick half in the water appears crooked, because of the double medium; it is part of reason, not sense, to judge, whether none of the requisites of sensation be wanting."[114] The informational basis must be built up in a rational and historical manner, by actively comparing together, and correcting, the appearances made to the senses by manifest phenomena. As Boyle said in his draft outline for "The Diffident Naturalist," "Our senses alone are not safely to be trusted in the informations they are believed to give of the natures of things."[115] While experience is required, reason is fundamental to the process of building up a base of information about nature. It is merely an "improper way of speaking" to say that "experience corrects reason," since it is "reason itself, that upon the information of experience, corrects the judgments she had made before."[116]

As one should expect with the interpretation of a text, a "reconciling disposition" will be required at every stage of inquiry, from the discovery of particular effects to the description of the regular occurrence of these effects and the construction of causal explanations for these regularities.[117] Boyle's experimental philosophy consists of an extremely complex method of discovery by which facts are verified and then collected and reconciled in such a way as to allow for the construction of a coherent narrative about the world. The book metaphor legitimates Boyle's use of legal analogies in support of his experimental program. The best method for the investigation of nature is one that includes a rational adjudication procedure whereby as much information as possible, from all areas, is collected and critically compared. The grounding of our knowledge claims must be historical: knowledge will be a product of the experience built up over many years by a number of workers actively investigating nature, and "the more things one knows by Experiment, the greater number he has of different Touch-stones (if I may so speak) or ways of *discovering* and *judging*; or (which is the result of both) of *knowing*, the nature of any thing proposed, that relates to it."[118]

While some knowledge is possible, Boyle remained diffident because, as he explained, "We know perfectly and fully no truth; because we know not all its connections and respects."[119] Knowledge should never be static or dogmatic. Given his dynamic conception of natural philosophy and his willingness to change his opinion about the acceptability of particular hypotheses, Boyle may appear inconsistent at times, but this type of inconsistency should not be viewed as a failed attempt to construct a philosophical system. Rather, it was a product of his epistemological dictates. "'Tis not a Fickleness but a laudable constitution of the Humane Understanding, that it changes Opinions according to the Prevalency of the Reasons, or the Arguments that determine it this way or the other."[120]

Boyle's justification of his experimental philosophy was based upon a complex set of epistemological and ontological considerations. He presented his program as an alternative mode of reasoning, but it is equally important to recognize that his experimental philosophy also contained a set of practical activities. Throughout his published writings, he gave detailed accounts of how an experimentalist should act, amply illustrated by the successes and failures that he encountered in his own daily work. The dynamic epistemological conception upon which Boyle's experimental philosophy was grounded was balanced by an equally dynamic methodological conception that contained three distinct types of activities: observing, experimenting, and writing. While these three were interconnected at the level of reasoning (because they would all contribute to the factual foundation upon which inferences were to be made), they will be discussed as three separate modes of acting in the following chapters, and the details of Boyle's advice on how to proceed in each area will be examined.

Part Three

ACTING EXPERIMENTALLY

Six

OBSERVING

BECAUSE OF THE COMPLEX RELATIONS in the created world, "we know very little *à priori*," and thus a methodological priority must be given to the construction of a solid factual foundation.[1] Despite the work begun by Bacon, Descartes, and others, Boyle maintained that natural philosophers were still "far from having such a stock of experiments and observations, as I judge requisite" for the construction of theories about the processes operative in the natural world.[2] In order to overcome the imperfections of natural philosophy, imperfections in the history of nature had to be corrected. Boyle's initial project, then, would involve the amassing of a great amount of information from all areas of inquiry for the construction of a new factual foundation broad and deep enough to support the construction of theory.[3]

Constructing the Factual Foundation

Before turning directly to a discussion of Boyle's collection of observations, we must examine what he meant by a *fact*. In the philosophical literature of the twentieth century, the factual has frequently been described as a linguistic category composed of singular synthetic statements. Because Boyle frequently contrasted the factual with the hypothetical, it may seem that he followed this linguistic usage. Indeed, among the historians who have turned their attention to the introduction of the factual category in the seventeenth century, most have taken the twentieth-century conception of the factual as their starting point. Peter Dear, for example, has represented Boyle's factual category as being characterized by the "historical report of a particular event."[4] The factual was a new category in the seventeenth century, however, and its usage was not clearly delineated at that time. While Boyle did not write explicitly about the category, one can see from the items that he identified as factual that his idea of it was not the same as today's.

In the most general sense, "matters of fact" for Boyle were statements that referred to natural effects, the existence of which were to be explained by theories about the causes responsible for them. Because the domain of the factual is composed of claims about existence, singular synthetic statements belong to this category, but Boyle's category was much more inclusive. Among the "facts" that he presented was that on average animals "may be killed in our engine by the withdrawing of the air in about one minute of an hour." This is a statement about a regularly occurring effect that was arrived at not by a singular experience but from repeated observations of experimental results.[5] In addition, facts necessarily involve reference to activity because they are effects produced either directly by nature or with the help of human agency. In a draft for a type of dictionary that recorded the usage of words, Boyle listed one entry under *fact*: "to confess the f[act]."[6] This legal sense of the term *fact*, which referred to a thing done or an action performed, was introduced into the philosophical realm by Bacon to refer to actions performed by nature and clearly influenced the way in which Boyle used the term. In his *Experimental History of Colours*, for example, he wrote that he hoped to "excite" his readers "by the delivery of matters of fact, such as you may for the most part try with much ease."[7]

Facts differ from hypotheses, doctrines, and notions not by their degree of generality or by the absence of human agency in their construction, but by the degree of evidence upon which they are based. It is not the case that some statements belong to the category of the factual by virtue of their linguistic structure. Rather, statements *become* factual when the evidence for them provides a sufficient degree of certainty upon which to base one's acceptance of them as containing terms that refer to actual processes operative in nature. In *Leviathan and the Air-Pump* Shapin and Schaffer recognize the epistemic dimension of the factual, but their subsequent analysis focuses upon what they term the "procedural boundary" constructed by Boyle between "speech of matters of fact" and "speech of the physical causes of these facts." Thus they reduce the factual to a linguistic category composed of descriptive statements.[8] The imposition of such a rigid linguistic distinction on Boyle's discussions of the factual and the hypothetical can only lead to confusion. If, for example, the factual is made coextensive with descriptive language and facts are the only items that possess moral certainty, then it would seem that Boyle did not believe in the possibility of discovering true causal statements about natural processes. But as we have seen, Boyle's programmatic statements clearly indicate that he did believe in the possibility of causal knowledge. His cautious attitude toward hypotheses was not based upon a prior belief about the probabilistic nature of causal knowledge as opposed to factual knowledge. Rather, hypotheses, by definition, are speculative claims that as yet are not sufficiently grounded by the evidence.

The factual and the hypothetical are mutually exclusive epistemic categories, but the factual and the causal are not mutually exclusive linguistic categories. Factual statements may refer to causes, while hypothetical statements, such as those of the mathematicians, may be purely descriptive. The way in which the distinction between facts and hypotheses actually worked in Boyle's philosophy can be seen from his discussions of the "spring of the air." When he began his *New Experiments Physico-Mechanical,* Boyle discussed how the spring or pressure of the air was a "notion" useful for the explication of the phenomena produced by his engine. But as the work proceeded, he began to refer to the spring as a fact, even though it clearly referred to a claim about a causal property of the air, while he remained tentative toward hypotheses that would describe the structure of the particles of air that would be causally responsible for its springlike property.[9] Boyle did not detail the way in which the spring, but not its cause, had been transferred from the hypothetical to the factual domain in his *New Experiments,* because, as he said, he did not believe that the work was the appropriate place to elaborate upon general methodological precepts.[10] In the following year, 1661, however, he did elaborate upon this topic in *Certain Physiological Essays,* and in this work it is clear that there were two ways in which the spring differed from hypotheses about its cause: its place within the chain of causes, and the degree of evidence upon which the notion of a spring was based. Taken together, these two criteria allowed causal statements referring to the spring of the air to change categories and become factual instead of hypothetical.

Following Bacon's advice about the inverted order of demonstration, Boyle maintained that one should first seek to discover the "subordinate causes" in nature that would have immediate practical applications and would also serve as helpful guides for further investigations into more general causes. Philosophers should not neglect these lower-level causes. As he said in his "Proëmial Essay" in *Certain Physiological Essays,* "In the search after natural causes, every new measure of discovery does both instruct and gratify the understanding."[11] Among the examples he used to illustrate his conception of subordinate causes, Boyle noted that one could use the notion of specific gravity to explain why gold would not "swim in quicksilver," or one could appeal to the laxative and astringent qualities of rhubarb to explain "why it purges choler." In the same place, he also used the spring of the air as an example of such an intermediate cause:

> So if it be demanded, why, if the sides of a blown bladder be somewhat squeezed betwixt one's hands, they will, upon the removal of that which compressed them, fly out again, and restore the bladder to its former figure and dimensions; it is not saying nothing to the purpose, to say, that this happens from the spring of those aerial particles, wherewith the bladder is filled, though he, that says this, be not perhaps able to declare, whence proceeds the motion of restitution, either in a particle of compressed air, or any other bent spring.[12]

Given Boyle's conception of the order of causes, his discussion of the factual status of the spring in *New Experiments* is understandable. He noted that he wished "only to manifest, that the air hath a spring, and to relate some of its effects." This clearly indicates that the spring is a causal notion. But at the same time he observed that it was also an effect when he said that he would not attempt "to assign the adequate cause of the spring of the air."[13] A hypothetical construct about an intermediate causal process can become a matter of fact if there is sufficient evidence that the process exists, because such causes are also effects, the existence of which will be explained by higher-level causes. As we saw in chapter 3, Boyle maintained that the spring's factual status had been established in much the same way as the status of the circulation of the blood had been. As early as his second edition of *New Experiments* in 1662, he predicted that the truths delivered there concerning the spring of the air would, "in time, in spite of opposition, establish themselves in the minds of men, as the circulation of the blood, and other, formerly, much contested truths, have already done."[14] According to Boyle, the evidence for the existence of a springlike quality in the air was overwhelming.

The *New Experiments* consisted of forty-three numbered "experiments," but each title comprised a series of experiments and each series compiled and compared the evidence from "several other of the experiments afforded by our engine" and thus produced a concurrence of probabilities to prove "that we have not mis-assigned the cause of this phaenomena."[15] The hundreds of experiments reported in the first edition, together with the additional experimental reports compiled in the second edition and two subsequent "continuations," all of which Boyle maintained, rightly or wrongly, could be explicated by the causal notion of a spring, allowed him to conclude that the air's spring could be accepted as a factual claim about an actual process operative in nature. Boyle did not present any single experiment as crucial. There were problems with all of them, as he freely acknowledged in his conclusion.[16] But taken together, his experiments provided a set of collateral arguments and thus produced a moral demonstration, that is,

> a complex of arguments which do not pretend each of them to the grand conclusion but to be either like the *Lemmata* of Mathematicians previous or preparatory truths or else to be collateral Arguments whereof one infers this conclusion and another that, and a third perhaps answers an objection, so that when all these partial and some of them perhaps indirect arguments or ratiocinations are taken together, they make and become one total Argument, if I may so speak, that directly proves the grand and ultimate conclusion.[17]

The epistemological status of the spring should be clear. While a causal claim, it is also a "matter of fact" because it is an effect in the chain of causes whose existence has been established by a concurrence of evidence. As Boyle described the process of the generation of factual knowledge, it is by our "illative" faculty that we are able to construct "positive proofs" that such intermediate causes exist.[18] When we discern that effects, such as those exhibited by his engine, "must proceed from such a cause," we may "conclude, that such a cause there is, though we do not particularly conceive, how, or by what operation it is able to produce the acknowledged effect."[19]

Factual claims are inferred from and justified by a concurrence of evidence, but especially when the facts referred to unobservable causal processes, Boyle would also use analogical reasoning to support his inferences. Because the spring of the air was inferred indirectly from the empirical evidence and was based upon analogous instances of the behavior of macroscopic bodies, Boyle's concept of the spring remained somewhat vague. He introduced the concept at the beginning of his *New Experiments*:

> That which I mean is this; that our air either consists of, or at least abounds with, parts of such a nature, that in case they be bent or compressed by the weight of the incumbent part of the atmosphere, or by any other body, they do endeavour . . . to free themselves from that pressure, by bearing against the contiguous bodies that keep them bent; and, as soon as those bodies are removed, . . . by presently unbending and stretching out themselves, . . . and thereby expanding the whole parcel of air, these elastical bodies compose.[20]

He was never able to produce a clearer formulation of the spring, in part because that would depend upon a knowledge of its cause, that is, of the configuration of the particles of air that gave it its springlike quality, and his experimental techniques were not advanced enough to yield such information. As he wrote in a letter to Henry Oldenburg that was later incorporated into his posthumously published *History of Air*, the springlike quality of the air was factual, but he could not decide what type of configuration was necessary for this quality since too many "comparisons" could serve to explicate it. The "springy particles" could be "like the very thin shavings of wood, that carpenters and joiners are wont to take off with their plainers" or like "curled hairs of wool" or "like extremely slender wires." Any of these shapes may illustrate "the action and reaction" of the parts of air, and he admitted that "one may fancy several other shapes (and perhaps fitter than these we have mentioned) for these springy corpuscles, about whose structure I shall not now particularly discourse, because of the variety of probable conjectures, that I think may be proposed concerning it."[21] The evidence was incomplete. The discovery of the true nature of the air's spring would have to wait until further research yielded more decisive information.

To ask that Boyle be more specific about the cause of the spring, or that he be more rigorous in his use of terms at this early stage of inquiry, would be to ask him to be inconsistent with the open-ended and dynamic way in which he characterized the nature of his experimental program. In addition, the spring of the air functioned in large part as a model for thought in Boyle's writing, and models need a certain amount of vagueness to be successful. We have seen that at the methodological level Boyle often used models and analogies, such as those from the legal and theological realms. He also found them useful at the descriptive and explanatory levels within science. In an unpublished notebook, Boyle wrote that there were two reasons why he used "similitudes or comparisons" in his works:

> [1] comparisons fitly chosen and well applied, may on many occasions usefully serve to illustrate the notions for whose sake they are brought and by placing them in a true light help men to conceive them far better than otherwise they would do.
> [2] apposite comparisons do not only give light but strength to the passages they belong to since they are not bare pictures or resemblances but a kind of Arguments, being if I may so call them analogous instances which do declare the nature or way of operating of the thing they relate to, and by that means do in a sort prove that as 'tis possible, so it is not improbable, that the thing may be such as 'tis represented.[22]

The fact that there were so many comparisons by which to explicate the spring created a problem for deciding which one might represent its actual cause, but at the same time the numerous comparisons made it all the more probable that there was such a springlike quality.

In this chapter we will see how Boyle used comparisons and analogical reasoning, not only to arrive at hypotheses but also to establish factual items for inclusion in his natural histories. In part because of the way that they are discovered and established, "matters of fact" do not compose a rigid category delineated by linguistic structure. Facts are highly confirmed items of knowledge that may refer to singular effects, regular occurrences, or causal processes. The category of the factual, while foundational, is also dynamic. The material included within the informational basis remains open to revision in light of further discoveries. The spring of the air provides an example of a fact that was generated primarily from experimental evidence, but there are two ways in which facts are discovered, and while Boyle called his program the experimental philosophy, it is important to remember that observing remained a crucial element in the construction of the factual foundation.

Some of the problems that Boyle identified as arising from the fallibility of our senses could be solved by the practical activity of experimentation, yet he maintained that observations were important in their own right and sometimes were even more important than intricate experiments. As he wrote to

Samuel Hartlib, because "things of the greatest consequence do oft-times depend upon the most common observations, . . . I would have no man, who
hath leisure, opportunity, and time, to think it a slight thing to busy himself in
collecting observations of this nature."[23] Before one can speculate about the
causes that give rise to the particular qualities of bodies, one must first accurately ascertain which qualities bodies actually possess. In his "Proëmial Essay"
Boyle noted that observations were as useful as experiments in constructing
such a history of the qualities of bodies. As he said, "If we did but diligently
turn our eyes to the observations, wherewith even neighbouring and familiar
objects would, if duly consulted, present us," we might be relieved of "that
ignorance" by which philosophy was "prejudiced." Indeed, he was quite "confident, that very much may be done towards the improvement of philosophy
by a due consideration of, and reflexion on, the obvious phaenomena of nature,
and those things, which are almost in every body's power to know, if he pleases
but seriously to heed them; and I make account, that attention alone might
quickly furnish us with one half of the history of nature."[24]

Boyle often grouped observation reports together with reports of experimental results in his published works because, as he said, both are "historical,"
that is, they concern "matters of fact," and both are often produced by reason
and design rather than by chance.[25] In addition, observations and experiments
are not differentiated by the function that they serve in the factual foundation,
since either one may subsequently be used for the discovery, confirmation, or
refutation of theoretical claims. While some observations are made simply for
the compilation of data, others are made to test a generally accepted theory,
and experiments need not be confined to a testing role but may be simply
"exploratory."[26] Both observational and experimental reports are designed to
increase the informational basis, and they are not distinguished by either the
information that they yield or the roles that they play. They do differ, however,
at the level of activity, and because the experimental philosophy is action-
oriented, this difference is significant and justifies their separate treatment.

As activities, observing and experimenting differ first in their level of complexity. As Boyle noted in his *Origin of Forms and Qualities*, observations are made
"of what nature does, without being over-ruled by the power and skill of man,"
while experiments are made when "nature is guided, and as it were, mastered
by art."[27] Boyle had stressed years earlier in his "Proëmial Essay," however, that
it is difficult "to make and relate an observation accurately and faithfully
enough for a naturalist to rely on."[28] Observing requires "either skill or curiosity, or both, in the observer." This circumstance explained why so many predecessors had failed to notice the significance of those "simple common observations" that formed the basis of Archimedes' work in hydrostatics and Galileo's
in kinematics.[29]

In addition, the activity of observing will often precede experimenting because accurate observations of nature are needed to provide us with the initial ideas for the design of more intricate experiments. Boyle noted, for example, that when faced with a new substance or a new sample of a seemingly familiar substance, one should first make observations by sight, touch, and taste about its apparent properties before proceeding to perform elaborate experiments. In *History of Blood* he described this preliminary general survey of an object of study:

> It may suffice for the first time, that the mind do, as it were, walk round the object it is to contemplate, and view it on every side, observing what differing prospects it will that way afford, (as when a painter, or an anatomist, looks upon a man's body, first when the face and belly are towards him, then when the back and other hinder parts are so) and that it takes notice of the limits and boundaries of it, and of the most essential and considerable parts or other things that belong to it.[30]

Because of their controlled conditions, experiments may be more reliable, but common observations are also required for the informational basis of the new philosophy. Boyle expended enormous effort in collecting and recording a vast number and variety of observations.

Collecting Observations

Boyle was indeed "le plus curieux".[31] His thoughts were full of questions about glaciation, echoes, electrical bodies, lime, water, air, blood, light, tin, and the customs and climates of foreign countries.[32] He would sometimes complain about the "slavery of sleep" and the "designless visits" of idle nobility and neighboring gentry that stole his time away from the "useful employments" that he desired to follow to find the answers to his questions.[33] Early on, observing things became a way of life for Boyle. He would often visit the shops of apothecaries and the "work-houses of mechanicians" and other tradesmen to see firsthand the practical knowledge that they had, and he urged others not to believe that such converse with tradesmen was beneath their dignity.[34] When his move to Stalbridge in 1645 caused an initial interruption in his chemical work, he made the most of the situation by turning it into an opportunity to learn about horticulture and husbandry.[35] Indeed, he could turn almost any occasion into one for making an observation. One day while visiting the royal court, for example, he had a lengthy discussion with "two very fair ladies" about the curious electrical properties that their hair exhibited in particular types of weather.[36]

At least ten years before he began to publish his observations, Boyle began the practice of recording them, first in "philosophical diaries" and later on loose sheets.[37] His earliest surviving collection, begun at Stalbridge and entitled "Memorials Philosophical beginning this Newyear's day 1649/50 & to End with the Year," primarily comprised chemical and medical observations and experiments.[38] In the same year, on his twenty-third birthday, he began another type of collection, entitled "Materials and Addenda Designed towards the Structure and Completing of Treatises already begun or written January the 25th 1649/50." Among these were listed a group of twenty-six "essays" on such topics as philosophy, chemistry, idleness, atoms, antiquity, and authority.[39] During Boyle's early years he evidently began a new diary annually, although the only ones still preserved among his papers are those that appear to correspond roughly with the years in which he changed his residence. The next diary by date, entitled "Memorials Philosophical Beginning this First Day of the Year 1651/52," was written about the time that Boyle moved to Ireland and contains his first recorded entries of chemical preparations that he had received from George Starkey, who had newly arrived in London from America.[40] The next diaries preserved date from the time that Boyle was back in England. "A Philosophical Diary Begun this First of January 1654/55" contains his first attempt to record entries in numbered "centuries"—groups of one hundred—a practice that he continued the following year in "Private Philosophical Diary Begun this first day of January 1655/6," which coincides approximately with the time that he moved his residence to Oxford.[41]

The last diary preserved, "Begun this First of Jan. 1656/7," probably dates from about the time that Boyle began to keep his observations and notes on loose sheets.[42] As he wrote to Henry Oldenburg, he had been "so troubled and discouraged by having had the number of the first books that I writ, lessened by more than one, that were surreptitiously got away from me; that I afterwards, to secure myself against the like losses of a whole treatise at a time, resolved to write in loose and unpaged sheets."[43] Dated and undated observations and experiments covering the next thirty years are scattered throughout Boyle's papers on loose sheets and pasted-up scraps of paper. Some of the confusion of these papers may be due to the way in which the volumes of papers came to be bound after his death, but Boyle himself would admit that the jumble was in part caused by his own lack of organization. Sometimes he had the information from the loose sheets transcribed into sets of observations about particular topics, but most often even the transcriptions were not done systematically. He noted that in one collection dating from the mid-1660s he had "Philosophical Entrys and Memorialls (of all sorts) . . . confusedly thrown together: to be Hence transferred to the Severall Treatises whereto they belong."[44] Periodically he made attempts to introduce more system into his notes by constructing indexes of his published and planned treatises and prefixing numbers to those entries that were to be used as marginal notations for his loose observations, indicating to which works the observations belonged. Although most of the loose observations recorded in his papers do have such numbers attached to them, the presence of more than one index, with different numbering schemes, makes it difficult to determine exactly the references of these numbers.[45] The first index, entitled "The order of my several Treatises" and compiled sometime between 1665 and 1670, shows that his curiosity and breadth of interest had continued unabated:

1. The Engine Book
2. Physiological Essays about Contingent Experiments, The History of Fluidity and Firmness
3. The Scepticale Chymist
4. The Defence against Hobbes & Linus
5. The Usefulness of Experimental Philosophie
6. Of Colours
7. Of Cold
8. The unpublished part of the Usefulness
9. Appendix to the Engine Book
10. Hydrostatical Paradoxes
11. The Origin of Forms and Qualities
12. The History of Qualities
13. Spontaneous Generations
14. Of The Compiling of a Natural Historie
15. Improbable Truths
16. The Scepticale Naturalist
17. Concerning Sensation in general
18. Considerations and Experiments touching Occult Qualities
19. of the Origin of Minerals, especially gems
20. Historia Naturalis or Promiscuous Experiments and Observations
21. Miscellania Physica
22. Communicated observations physiological and medicall
23. [Varia Lectiones] Physicae
24. Topica Particularia.[46]

The most extensive collection of observations and experiments composed by Boyle, entitled "Promiscuous Paralipomena," takes up over two hundred pages in his papers, but as can be seen from his preliminary remarks to this work, it was intended to be much longer. The "first tome" of it was to comprise 500 pages of 3,000 observations, and the second tome was to begin with page 501, with the first observation there numbered 3,001.[47] According to Boyle, this collection was intended as a contribution to the *"materials, that are gathering in this Industrious Age, towards the History of Nature."*[48] The index provided at the beginning of this work for "the Chapters of the Paralipomena" had numbers and letters that were to be used as marginal notations for the observations that followed to indicate to which of his works they belonged:

1. A. Particulars for the most part referable to the Discernment of suppositions
2. B. Pts. belonging chiefly to the Essay of Improbable truths
3. C. Pts. for the most part referable to the Origin of gems
4. D. Pts. referable chiefly to the changes of colours
5. E. Pts. belonging chiefly to the memoirs of the History of Tin
6. F. Pts. belonging to the Letter about the Tryal of Ores
7. G. Pts. referable chiefly to the Origin of Minerals
8. H. Pts. many of them referable to the Dialogue of the Melioration and Transmutation of Metals
9. I. Pts. that may be referable chiefly to the Production of Inflammability
10. K. Pts. most of them referable to the Examen and Improvement of *Materia Medica*
11. L. Pts. referable to the danger of too soon forsaking Experiments upon probable reasons
12. M. Pts. Referable to the Treatise of various observations about the air
13. N. Pts. containing some loose memoirs belonging to the History of flame.
14. O. Pts. belonging to the Essay of strange narratives
15. P. Pts. belonging to the surface of contiguous fluids
16. Q. Pts. belonging to the Essay about the Porosity of Bodies
17. R. Pts. belonging to the Essay of the Pores and figures of Corpuscles
18. S. Pts. belonging to the Tracts of Effluviums
19. T. Pts. belonging to the various effects of Fire as differingly applied
20. U. Pts. belonging to the Mutual Usefulness of Speculative and Practical Physiology
21. V. Pts. belonging to the Essays about the Physiological, Pathological and dietetical parts of Physick
22. X. Pts. belonging to the therapeutical part of Physick
23. Y. Experimenta et Observationes Physicae.[49]

Following this index, Boyle discussed how the collection was to be organized and the significance of the numbers and letters prefixed to his titles:

> The ordinal numbers placed in the margent of this Chaos, are set down for the conveniency of references. And for the same reason, 'tis thought fit, that those numbers be retained and kept unvaried; though all this is contained in the Paragraph that number relates to, should be transferred to this or that among my other writings whereto it belongs. . . . The

alphabetical letter that accompanies each number, shows to what title the Particular where to the number is prefixed, does belong. And if there be more letters than one that accompany the same number, they denote that, although the Particular belongs primarily to the Title signified by the first letter, yet 'tis also referable to the Titles pointed to by the other letter or letters.[50]

Some of the works from the index were published in Boyle's lifetime, such as *The Origin of Gems* and *Experimenta et observationes physicae*, but most of them appeared again in the last list of his unpublished works, "set down July the 3rd, 1691," a few months before his death, with slightly altered titles. Entry 8.H., for example, became "Conferences about Melioration and transmutation of Metals"; 9.I. became "Notes about the Mechanical Origin and History of Inflammability and Flame"; and 12.M. became "Some Incoherent Memoirs about the Natural History of the Air." All of these works have been lost to us with the exception of *The History of Air*, which was prepared for publication after Boyle's death by John Locke.[51]

While Boyle was actively engaged in making observations himself, he also collected a large number of observations from other sources. As Bacon had advised, only by the use of such sources would one be able to amass the amount of information required for a complete and accurate history of nature.[52] Boyle's early diaries from Stalbridge and Ireland, for example, contained a number of entries from his alchemical and medical acquaintances, including Kenelm Digby, Samuel Hartlib, Frederick Clod, George Starkey, Dr. Butler, Dr. Mayerne, Dr. Herbert, Dr. Boat, Mr. Bardon, and Mr. Smart.[53] While at Oxford, Boyle continued to collect observations, and Henry Oldenburg corresponded with him, keeping him informed about new developments in London and on the Continent and procuring the numerous books that Boyle required in order to learn more about the operations of nature in foreign lands.[54] Scattered throughout his papers are notes that he made from works such as a "History of Barbados," *Greenlandia*, "Purchase's Pilgrimage," Gulielmus Piso's *De medicina brasiliensis*, "Linschoten's Voyage," Alvaro Semmedo's *History of China*, "DuTertre Hist. Nat. des Antilles," and "Monsieur Tavernier's Reports of Travels about a Tribe in Africa."[55]

As time went on Boyle's circle of acquaintances grew as a result of his newly acquired business contacts. In 1661 he was appointed to the Council for Foreign Plantations, in 1662 he was chosen by Charles II to be the first governor of the Corporation for Propagating the Gospel in New England and the Parts Adjacent in America, and in 1664 he was elected into the company of the Royal Mines.[56] He valued these and other associations, such as those he had with the East India and Hudson's Bay Companies, for the access they gave him to merchants, navigators, and others who related to him the observations they made in their travels. In his *Experimenta et observationes physicae,* for example, Boyle noted that his membership on the committee of the East India Company had been in part motivated by his desire to learn more about the properties of diamonds and other gems:

> The opportunity I had of being one of the committee or directors of the English *East-India* company (whereto the desire of knowledge, not profit, drew me) allowed me in some measure to gratify my curiosity about them [diamonds], by adding, to some observations of my own, the answers I had to the questions I propounded to some *East-India* merchants and jewellers, that had opportunity to deal much with those gems.[57]

Boyle was particularly interested in travelers' observations about "odd" phenomena, and in the 1660s he began keeping a separate record of these strange reports in his "Outlandish Book."[58]

The first pages of the Outlandish Book contain a list of Boyle's treatises, similar to the first index described above, except that letters were used instead of numbers for the first twenty-four entries and the entire index was set in verse. The entry for the letter Z, corresponding with the earlier entry at number 24, the "Topica Particularia," was now listed as "Our topics doth particularize." Under this last heading thirty-eight numbered titles were included, beginning with "the first tract the Free Inquiry comes, the next that brutes may be but engines dooms, the third does Suppositions discern." Some titles, such as *A Free Inquiry,* would be published in Boyle's lifetime, while others, such as the "Discernment of Suppositions," would remain on the 1691 list of his unpublished works.[59] Among the strange reports recorded are those concerning "perfume from a dung hill," the "odd phenomenon consequent upon the loss of an arm," "Paracelsus on plaisters," a "cure of an odd distemper," and an "odd observation about seed."[60] As the title of the notebook suggests, a large number of the observations concerned the oddities reported to Boyle from travelers to foreign places, such as Jamaica, Mount Sinai, the Cape of Good Hope, Egypt, China, Japan, Russia, Ceylon, the Congo, Goa, Brazil, Virginia, Massachusetts, Maine, and the "Springs of Mara in the East."[61]

Boyle sought to encourage others to observe and report the peculiarities of the countries they visited in a short work that first appeared in *Philosophical Transactions* as "General Heads for the Natural History of a Country, Great or Small; Drawn Out for the Use of Travellers and Navigators."[62] He divided the things to be observed according to whether they were "supraterraneous," "terrestrial," or "subterraneous." A complete account should be taken of the positions of the "fixed stars," of the temperature, weight, clearness, and "refractive variations" of the air, of the composition of the waters and the types of fish to be found in them, and of the dimensions and figure of the land, including any particulars relating to the land's human inhabitants:

both natives and strangers, that have been long settled there: and in particular, their stature, shape, colour, features, strength, agility, beauty (or
the want of it) complexions, hair, diet, inclinations, and customs that
seem not due to education. As to their women (besides the other things)
may be observed their fruitfulness or barrenness, their hard or easy labour, etc. and both in women and men must be taken notice of what
diseases they are subject to, and in these whether there be any symptom,
or any other circumstance, that is unusual and remarkable.[63]

Throughout this work he stressed the need to search out the unusual. For example, an observer should note "what peculiarities are observable" in the plants
grown in a particular country and "whether it have any animals that are not
common, or any thing that is peculiar in those that are so."[64] In addition to
simple observations about the physical characteristics of a country, Boyle
noted that one should also make "1. Enquiries about traditions concerning all
particular things relating to that country, as either peculiar to it, or at least
uncommon elsewhere. 2. Enquiries, that require learning or skill in the an-
swerer: to which should be subjoined proposals of ways, to enable men to give
answers to those more difficult enquiries."[65]

Oddities of nature constituted a special class of observations collected not
merely as a way to expand the informational basis but also as a way to test
claims that had already been accepted as part of that basis.[66] As we saw in
chapter 4, Boyle believed that the study of nature would reveal the "divine
wisdom" that in the beginning ordered things in such a way that God's purposes would be fulfilled by the gradual unfolding of the intricate processes that
make up the grand cosmic mechanism. So-called oddities are unusual occurrences only in the sense that they appear to be opposed to the regular course
of nature.[67] But as Boyle noted repeatedly, philosophers should remember that
the theories and laws they have constructed about regularities are based upon an
imperfect history of nature. There can be no contradictions in the physical world
as divinely ordered. Only our limited perspective makes oddities appear unintelligible. In his *Free Inquiry* Boyle discussed how the world could be likened to "an
excellent letter," some parts of which "were written in plain characters, others in
cyphers, besides a third sort of clauses, wherein both kinds of writing were variously mixed."[68] If a "very intelligent person" found upon reading the letter

that those passages, that he can understand, are excellently suited to the
scopes, that appear to be intended in them, it is rational as well as equit-
able in him to conclude, that the passages or clauses of the third sort, if
any of them seem to be insignificant, or even to make an incongruous
sense, do it but because of the illegible words; and that both these passages, and those written altogether in cyphers, would be found no less
worthy of the excellent writer, than the plainest parts of the epistle, if

the particular purposes, they were designed for, were as clearly discernable by the reader.[69]

Oddities are like cyphers in the sense that they are meaningful parts of the natural order even though we are not as yet able to discern their significance. To recognize phenomena as odd, one must have an initial idea about the regular course of nature.[70] But once collected, oddities provide severe tests of the acceptability of those initial ideas. They provide a type of check on "local reason," that is, on the partiality and prejudice that arises from our experience's being limited to one geographic area and from the "superficial account given us of things by their obvious appearances and qualities."[71] Rather than searching only for those observations that will confirm a favored theory, one must actively search out observations that appear to refute it. A "thousand experiments" or observations made to confirm a theory do not have the force of one made "to prove the contrary."[72] An oddity need not completely overturn an idea about a regularity in nature, but it may show how that idea needs to be revised. We learn from foreign countries, for example, that even familiar bodies may have different properties in different climates.[73] Because oddities are natural productions, when we attempt to reconcile them with our earlier ideas of nature, we can achieve a deeper and more comprehensive understanding of natural processes.

Boyle's diffidence was not merely a result of his views about the probabilistic status of hypotheses but also the result of his belief in the flexibility of the factual foundation upon which hypotheses were to be based. Occurrences considered unusual in England may represent the usual course of nature in other parts of the world. Extreme degrees of heat and cold not experienced in England but common elsewhere, for example, could radically alter ideas about the effects that these qualities are able to produce. The customs and practices of the inhabitants of foreign lands could also lead to "additions and alterations" in the factual foundation.[74] Boyle noted, for example, that while the practice of medicine in China was quite different from that in the West, it was apparently successful, and a further investigation into the practice could lead to a revision of common beliefs about what was healthful.[75] A significant number of the facts that were to compose the informational basis of the new philosophy—both about regularities and about oddities—did not require the artificial manipulation of nature but could be gathered simply by observing that which occurs in one's surroundings. Before they could be admitted into the factual category, however, observation reports had to be assessed for their credibility.

Assessing Credibility

Facts are highly confirmed items of knowledge about the effects produced in nature. Because our senses can deceive us, we should not depend upon perceptual experience alone for the establishment of these facts. Boyle was careful to note, however, that while the use of reason would therefore be primary in the construction of the factual foundation, reason could also be a source of error. It should not be assumed that the mind is "but a spectator." While "almost all" causes and effects are external to the mind, the mind remains active in how it discerns such events.[76] Indeed, "in many of the reports of meer senses as they are presumed to be, there is . . . couched or mingled though unsuspectedly, somewhat of Rationalism; and in such cases (which are very frequent) if the *Ratiocination* be erroneous, that which we call *Sense* may mislead us."[77]

In his planned work on the "discernment of suppositions," he noted that while it may "be thought strange," not only is it the case that "the greater part of men do usually believe more than they should, but even learned men (and perhaps philosophers too) do sometimes believe more than they *think they do*."[78] Often it is the case that "through custom or inadvertancy" many propositions include a supposition that when taken "for granted, makes men too prodigal of their Assent; which in such cases is *always* illgrounded and *oftentimes* given to things that are not true."[79] Among the topics of inquiry listed for this planned work, Boyle noted the number of circumstances surrounding an observation that must be taken into account in order to guard against the presumptions of reason that occur when

> Men presume that themselves and their own organs of sense, continue in the same state, when indeed they are not.
>
> Men presume that a medium or instrument used in sensation or otherwise, is in the same state when 'tis not.
>
> Men are not sufficiently aware how far in some cases a seemingly light circumstance or a slight variation may go, towards the producing of considerable effects.
>
> Men presume that time within which a thing is performed observed is not material a circumstance as in divers cases it is.[80]

While the presumptions of reason may create problems, accurate observations cannot be made without the use of reason. Indeed, for Boyle, there was no such thing as a pure, unbiased sense perception. From the notes compiled throughout his life on sensation and from the preface he prepared for his never-published work "Concerning Sensation in General," it is clear that Boyle arrived at his conclusions about the "imbecility of the visive faculty" and the role of reason in sensation from his study of nature. He wrote that even in seemingly simple sensations there is a complexity involved that includes

1. the change made by the impression of the object on the exterior organ
2. the conveyance of that impression to some internal parts of the brain (or seat of common sense)

3. the reception of the impressions there, on which follows *in man,*
4. the perception of the mind and the judgment etc. it makes on occasion of that perception.[81]

All sensations involve a judgment of reason. It is not reason per se, however, but the "want of the due exercise of it" that leads to errors in observation.[82] In the first instance, it is necessary to recognize that, because of the role of reason in sensation, one cannot treat perceptual experience as incorrigible. In addition, it will be necessary to expand, as fully as possible, one's experiences of the world in order to have a significant number and variety of observations with which one can compare new observations and by which one can correct faulty ideas constructed from a too-limited experience about the ordinary course of nature.

Recognition of the presumptions of reason leads to the search for an expanded experiential basis that in turn leads to improved theoretical knowledge about natural processes that can be used to correct defects in perceptual experience. Skill, expertise, and theoretical knowledge are needed to judge "whether the reports of the senses have all the requisite conditions" and thus to ensure that the "misinformations of our senses" do not "delude us."[83] An observer ought to make an accurate survey of "all the due conditions of sensation, relating to the organ, the object and the *medium*" before relating the observation as factual.[84] To guard against optical effects, for example, one ought to vary the amount and type of light in which the phenomenon is observed, and one also ought to seek other instances of the same type of phenomenon, since a single observation is not as reliable as an observation obtained from repeated experience.[85] While some of the controls that reason may exercise over the faculty of sensation can be performed without recourse to artificial manipulations of nature, Boyle often resorted to experiment as the best way to test the accuracy of his own observations. But sometimes experimental tests are not possible. Boyle collected a number of observations, especially about odd phenomena to which he had no direct access. As Peter Shaw wrote in his abridgment of Boyle's works in the early eighteenth century, he was always most "inquisitive into the foundations and evidence of the narratives he receives from others."[86] He would not take the observation reports of others on trust but would judge them based upon the credibility of the witnesses who related them and the credibility of the observations so related.[87]

Before assessing the accuracy of reported observations, one first has to determine if they have been related by "credible persons" upon "their own knowledge."[88] Witnesses are assessed by characteristics such as truthfulness and credulity, of course, but their credibility also depends upon whether they are "competent judges" in the particular areas about which they report.[89] When Boyle published an observation made about the odd properties of a "piece of glass, . . . which was red and pretty transparent like glass of antimony made *per se*," for example, he prefaced the account by noting that it was related to him by "a very intelligent person, well versed in chemistry, not credulous, and in a word very well worthy of credit, who assured me, that he had himself seen" the substance.[90] Competence is a trait acquired either through extensive reading and university training or through prolonged practical experience in a certain area. Witnesses could therefore be drawn from one of two broad groups in society: the learned and the illiterate. Among the first group were professional scholars who, as a result of their learning, possessed knowledge about the laws of nature that would enable them to judge the accuracy of their own observations. When Boyle discussed some strange reports concerning "vitiated sight," for example, he included one from a "mathematician, eminent for his skill in optics, and therefore a very competent relator of phaenomena belonging to that science."[91]

The group of literate witnesses was, of course, socially constituted. It was composed of those who had access to university training or who, like Boyle, had the leisure to learn on their own, including the younger sons of nobility and members of the growing merchant class. While social class would be an identifying characteristic of this group, it was not their class but their competence that gave credence to their reports. When Boyle related the observations of a "Neapolitan lord" about the odd properties of the land surrounding Mount Vesuvius, he noted not only that this lord was "a person of high quality" but also that he was "very curious" and had lived "divers months at a country house but little distant" from the site and had examined the land there on numerous occasions.[92] In a similar manner, when he recorded observations made in Prussia about the odd phenomenon of swallows' being frozen alive in lakes, he noted that his informant was "an inquisitive person, that having gone through his studies in the university, travelled through divers countries, to make himself the more fit for the profession of physic."[93]

Merchants and government officials, if they were observant, could also make good witnesses, because they often spent a considerable amount of time in foreign countries. In *Experimenta et observationes physicae*, for example, Boyle reported an observation about an "extraordinary in-draught of the sea" on "the East coast of *Sumatra*" that he had received from a "merchant, rich and judicious, and more addicted to letters than is usual to men of his calling" who had traveled extensively in the "remoter parts of the *East-Indies*."[94] In "Cosmical Suspicions" he used the reports of "an inquisitive gentleman that had lived in *New England*" and "the governor of a colony" there, who had both observed that the climate was becoming milder. They based their observations not merely upon their personal sensory experiences but also upon what they had seen of "the remisser operations of the cold upon running and standing waters."[95]

Boyle preferred to speak directly with his informants or at least to corre-

spond with them so that he could ask additional questions concerning the circumstances surrounding their observations. When he gave the account of the
merchant's observations in Sumatra, for example, he noted that it contained a
part that was "manifestly fabulous." He included this part all the same, because
it was "easily distinguishable from the rest" and he did not wish to alter the
observer's words until "I can see him again, and propose my scruples to him."[96]
Sometimes, however, he had to depend upon the information in books whose
authors were no longer living. In his "Proëmial Essay" Boyle maintained that
these authors should not be treated as "judges" or absolute authorities, but
rather as additional witnesses "to attest matters of fact." Because they were witnesses, their credibility also had to be assessed. Boyle noted, for example, that
among the ancient writers he was more prone to accept the reliability of reports from "the ancient physicians" than from philosophers such as Aristotle,
because they were "more conversant with things" and thus "usually more credible." Once again he stressed that he wished to include in his works only those
narratives that writers "deliver upon their own particular knowledge, or with
peculiar circumstances," not because such personal experience is incorrigible
but because too many writers, such as "Pliny, Solinus, Aristotle, Theophrastus,
Ælian," and others often included accounts of "historical traditions" that Boyle
found upon examination to be "either certainly false, or not certainly true."[97]

While the literate members of society often made good witnesses, the
illiterate could be every bit as qualified, if not more so, to make accurate observations for inclusion in the history of nature. In the preamble to his second
volume of *The Usefulness of Experimental Philosophy*, Boyle noted that "those, that
are mere scholars, though never so learned and critical, are not wont to be
acquainted enough with nature and trades."[98] He was adamant about the value
of the knowledge possessed by tradesmen, and he showed his contempt for
scholars who did not wish to speak with them:

> It seems to me to be none of the least prejudices, that either the haughti-
> ness and negligence, which most men naturally are prone to, or, that
> wherewith they have been infected by the superciliousness and laziness,
> too frequent in schools, have done to the progress of natural philosophy,
> and the true interest of mankind, that learned and ingenious men have
> been kept such strangers to the shops and practices of tradesmen. For
> there are divers considerations that persuade me, that an inspection into
> these may not a little conduce, both to the increase of the naturalist's
> knowledge, and to the melioration of those mechanical arts.[99]

Because the "phaenomena afforded by trades" are "part of the history of nature,"
they may "both challenge the naturalist's curiosity, and add to his knowledge."
The excuses made by those who neglect this part of the history of nature because it must be learned from "illiterate mechanicks" are "indeed childish, and

too unworthy of a philosopher, to be worthy of a solemn answer." Boyle had no patience with the "learned men" among his contemporaries who showed their open "contempt for this part of natural history." He insisted that "he deserves not the knowledge of nature, that scorns to converse even with mean persons, that have the opportunity to be very conversant with her [nature]."[100]

Illiterate tradesmen, craftsmen, farmers, and other practitioners not only make excellent witnesses, but sometimes they are better than those who have learned from the schools, because they "are usually more diligent about the particular things they handle," since "their livelihood depends upon it." Because of their diligence, "tradesmen . . . do often observe in the things they deal about, divers circumstances unobserved by others, both relating to the nature of the things they manage, and to the operations performable upon them."[101] As we have seen, Boyle wished to learn about the qualities of natural things, those properties that bodies have that in turn define their essences. In his inquiry into stones, Boyle maintained that he had "learned more of the kinds, distinctions, properties, and consequently of the nature of stones, by conversing with two or three masons, and stone-cutters, than ever I did from *Pliny*, or *Aristotle* and his commentators." A knowledge of trades and crafts could contribute to an inquiry into the qualities and essences of bodies, because it was of fundamental importance to those who made their living working with natural bodies to understand what could and could not be done with them. From "carpenters, joiners, and turners," for example, "we may learn, that some woods, as oak, are fit to endure both wet and dry weather; others will endure well within doors, but not exposed to the weather; others will hold out well above ground, but not under water; and others on the contrary will last better under water, than in the air."[102]

The information that Boyle obtained from tradesmen often played a crucial role in his theoretical writings. In his essay on "the intestine Motions of Particles of quiescent Solids" he used observations about the disposition of glass to break, which he took as an indication of an imperceptible motion in its parts, made by "a famous and experienced maker of telescopes," "some other observing men, that deal in optical glasses," an "honest man, that furnishes the greater part of London with large looking glasses," and "an ingenious master of a glass-house."[103]

He was also interested in the practical knowledge possessed by the illiterate members of society because of its direct usefulness for human health and welfare. In his first volume of *The Usefulness of Natural Philosophy,* for example, he argued that

the knowledge of physicians might be not inconsiderably increased, if men were a little more curious to take notice of the observations and experiments, suggested partly by the practice of midwives, barbers, old women, empiricks, and the rest of that illiterate crew, that presume to meddle with physick among our selves; and partly by the Indians and other barbarous nations, without excepting the people of such part of *Europe* it self, where the generality of men are so illiterate and poor, as to live without physicians.[104]

The "dietetical part" of medicine could be improved by the information that travelers to foreign lands brought back to England concerning the food and drink that the natives of those countries found healthful. The introduction of new materials into the English diet, such as Chinese tea or American sugar, might involve the cultivation of previously unfamiliar plants, but other additions could be made simply by learning new methods of preparing familiar fruits and vegetables, such as "the drink they call *mobby*" in Barbados, "made of potatoes fermented with water, which . . . would be of excellent use, if it were but as wholesome as it is accounted pleasant."[105] In a similar manner, the therapeutic part of medicine could be improved if physicians would seriously investigate the practices employed by the "illiterate crew," because successful practices, such as those of the Chinese, "may afford very good hints to a learned and judicious observer."[106]

Boyle argued that European physicians were sometimes hampered in their practice because of their "too great reliance on the Galenical, or other ancient opinions" that led them to "neglect useful remedies, because presented by persons, that ignore them, and perhaps too hold opinions contrary to them." He cautioned his readers "to beware of relying so much upon the yet disputable opinions of physicians, as to despise all practices, though usually successful, that agree not with them." One could learn a good deal about the preparation and use of specific medicines, for example, that were often found "where the practitioners of physic are altogether illiterate."[107] One could also learn to use less radical forms of therapy from the Brazilian Indians, who used "gums and balsams" to cure "the limbs of soldiers wounded with gun-shot" that would otherwise have "been cut off by the advice of our European surgeons, both Dutch and Portuguese."[108] Boyle did not advocate an empiricist approach to medicine, however. He believed that theory was necessary for the improvement of medical practice, but he also believed that such theory would only be "clearly established" once "the *historia facti* shall be fully and indisputably made out." If a practice is successful, it should be included in the history of nature regardless of the educational, social, or ethnic class of the practitioner. It is prudent to place one's trust in those who have a practical working knowledge in a particular area. It is also quite common. Boyle related Linschoten's account of the successful practices of the "heathenish physicians" of Goa and noted that the Portuguese "Viceroy himself, the archbishop, and all the monks and friars, do put more trust in them [the natives] than in their own countrymen."[109]

In his search for a complete history of nature, Boyle maintained that one must seek to broaden the factual basis. Thus one must expand the pool of witnesses from whom observations are collected. Anyone who is competent to judge may be considered a credible witness, and Boyle also often included women in this category.[110] Scattered throughout his published works on medicine, for example, are numerous references to women's testimony concerning the course of their own illnesses and the successful use they made of remedies to cure illness in others.[111] Often the information provided by women had important theoretical implications. In his *Experimental History of Colours* Boyle used the testimony of a lady of "unquestionable veracity" to show how the perception of colors could sometimes be the result of a physiological condition and not a result of the presence of a colored object.[112] In addition to the testimony that he used from tradesmen in his essay on "the intestine Motions of Particles of quiescent Solids," he also used the report of a woman "who had observed more about gems than any lady I have yet met with" about the changing aspects of diamonds that seemed to indicate that even such hard substances possessed internal motion.[113] Finally, he used the testimony of the two ladies at court in his "Mechanical Origin of Electricity" as evidential support for his speculations about the existence of electrical effluvia. When he recounted the discussion he had with them, he noted that at first he was dubious when these ladies told him that sometimes their hair would be attracted to their faces in much the same way that hair was attracted to "amber or jet excited by rubbing," and he suspected that "there might be some trick in it." But one of the ladies, who was "no ordinary virtuosa . . . very ingeniously removed my suspicions."[114]

No prior conditions were set by Boyle to indicate a specific class of people that would qualify as witnesses. Rather, each person was to be assessed for credibility based upon individual experience and competence. The reports of different classes were mingled together in Boyle's essays and histories: in his work on "vitiated sight" he included, along with his own observations, those made by eight "gentlemen," seven women, some of whom were identified as "ladies," one mathematician, one farrier, and one "dextrous artificer."[115] If a witness had the requisite experience, Boyle would accept the observation reported, whether it came from an illiterate tradesman, an apothecary, a woman, a doctor, a lawyer, a mathematician, or a theologian. Such a tactic was of crucial methodological importance. Not only did an expanded witness pool increase the quantity of information in the factual basis, but it also served to improve the quality of the information included, because it helped to guard against the prejudiced observations that resulted from having a too-limited experience of the world. But sometimes, even when the observations were made by careful and experienced witnesses, conflicts might arise between their testimonies, or between what was related by them and what was taken to be true about nature in general.[116] In such a situation, independent criteria would be required to assess the credibility of the report itself.

In much the same way as Boyle used the comparison of passages to determine the accuracy of interpretations in his biblical hermeneutics, he used the comparison of observations to determine the accuracy of reports about nature. In the first instance, observation reports have to be judged by whether they are internally coherent and intelligible. This is in part accomplished by judging the likelihood of the occurrence of an event given what else is known about natural processes. The more background knowledge one has about the variety of phenomena in nature, the better one can judge the credibility of reports about new phenomena. While Boyle was sometimes too credulous and sometimes too skeptical, on the whole he had the requisite experience to be an excellent judge in this area. As Peter Shaw noted, to appreciate the fact that Boyle was well able to judge reports even of phenomena that would sound "strange and shocking to vulgar ears," we need only consider the enormous number of phenomena that "must have been presented him, by the process of his laboratory, the experiments of the pneumatic engine, his mechanical, optical, hydrostatical, and other kinds of trials; the correspondence he happily cultivated with the whole learned world, and particularly with the chemists of all nations."[117] Sometimes, however, the things reported to him were so strange that he could find no analogous instances in his own experience with which to compare them. Another criterion, corroboration, would be required for cases of this type.

As Boyle noted, some observations "are of that strangeness, and of that moment, that they need and deserve to be verified by more than a single attestation."[118] As in the legal realm, corroborating testimony provides additional evidence for the truth about a matter of fact. While corroboration could produce a consensus of opinion, the two notions should not be confused. Consensus concerns a decision procedure about the beliefs of individuals, whereas corroboration concerns an investigative procedure whereby one attempts to discover the quantity and quality of the evidence that may be available to attest to the truth of a particular claim about nature. When Boyle related the reports of the two gentlemen from New England who had observed that the climate there had become more temperate, for example, he stressed that their testimony had been corroborated by that of the natives.[119] In order to determine what actual change was taking place in the climate, that is, whether the milder weather was indeed unusual or merely marked a return to normal temperatures that had been interrupted by a few winters of excessive coldness, one had to consult with the natives, who had the length of experience and the knowledge of tradition by which to judge the situation.

Corroboration was not limited to a multiplication of witnesses, however.

Often a multiplication of the objects or events witnessed was required. In his first volume of *The Usefulness of Natural Philosophy*, for example, Boyle reported the "ingenious attempts of *Sanctorius*, in his *Medicina Statica*" to determine the wholesomeness of different types of foods by observing the amount of perspiration produced after their consumption. While he accepted the accuracy of Sanctorius's accounts, he did not believe that they provided sufficient evidence for the proof of general matters of fact concerning the value of these foods. He stated that he would withhold his assent until similar observations had been "carefully made in bodies of differing ages, sexes, and complexions, and with variety of circumstances."[120]

Sometimes corroboration might not be forthcoming. In situations like this, Boyle discussed how reason would have to be employed to judge the accuracy of an uncorroborated observation. If he was "satisfied of the abilities and circumspection" of the observer, he would "not presently reject his observation as untrue," even if he found "that it seems to be contradicted by a contrary and more undoubted observation, or to contradict a received and plausible either hypothesis or tradition." Unless he could "imagine something or other, which might probably lead him [the observer] to mistake," he would "try, if by fit distinction or limitation I can reconcile" the new observation with previously accepted background knowledge. Just as there can be no contradictions in Scripture, there can be no contradictions in the book of nature. The method of reconciliation, used in biblical hermeneutics and illustrated by the example of Pythagoras's reconciliation of the apparent contradiction in observations about Venus, was to be used in natural philosophy also. As Boyle noted, "The contradiction betwixt the observations may be but seeming (by reason of the want of some unheeded circumstance necessary to make them inconsistent) and so they may both be true." When an observation conflicts with theory, a number of options can be explored. The theory may have to be rejected in light of the new information. On the other hand, the observation may be rejected if the theory with which it conflicts is highly confirmed. One should attempt a reconciliation prior to taking these more drastic steps, however. Perhaps a slight emendation in the observation report or a simple alteration in the theory could produce a resolution of the conflict. "Sometimes the received hypothesis, though perhaps not to be rejected as to the main, will not hold so universally as men presume."[121]

The strategy of reconciliation, used to assess the accuracy of uncorroborated observations, together with the corrigibility of sense perceptions, shows how theory and observation, reason and sense, are inextricably involved in the construction of the factual foundation.[122] Observations must be made by people with experience and competence. In addition, the reports of observations should include as much detail as possible about the conditions under which the observations were made, so that an accurate comparison and reconciliation of the phenomena with other known truths can be produced, which in turn will lead to a higher level of understanding about nature's processes.[123] An excellent example of how Boyle used corroboration and reconciliation in an attempt to establish a matter of fact can be seen in "An Hydrostatical Discourse Occasioned by the Objections of the Learned Dr. Henry More, against some Explications of New Experiments Made by Mr. Boyle".[124]

In his *Enchiridium metaphysicum* More explained various hydrostatic phenomena by incorporeal principles, in opposition to Boyle's earlier explications, which employed "the mechanical affections of matter, without recourse to nature's abhorrence of a vacuum, to substantial forms, or to other incorporeal creatures."[125] According to Boyle, in the course of his argument More had included as a premise the "resolute denial, that there is any such gravitation as I pretend, of bodies, or their particles, in their proper places."[126] In support of this denial More appealed to what Boyle termed the "grand argument . . . employed by the schools, and others, for the vulgar opinion" that water does not weigh in water. This assertion was based upon the traditional testimony of pearl divers, who, it was claimed, did "not find themselves oppressed . . . by the pressure of the incumbent and ambient water" when diving.[127] In response to More, Boyle questioned the accuracy of the divers' observations and relied instead upon evidence he had obtained from experimental trials:

> The pressure of the water in our recited experiment having manifest effects upon inanimate bodies, which are not capable of prepossessions, or giving us partial informations, will have much more weight with unprejudiced persons, than the suspicious, and sometimes disagreeing accounts of ignorant divers, whom prejudicate opinions may much sway, and whose very sensations, as those of other vulgar men, may be influenced by predispositions, and so many other circumstances, that they may easily give occasion to mistakes.[128]

This may appear to be a surprising move on Boyle's part, given what we have seen about the importance of observations and the credibility of witnesses. Those who make their living by diving in the sea certainly should be assessed as competent to make accurate observations about the effects that they experience. Indeed, Shapin and Schaffer, in their attempt to display the social constitution of evidence, cite the above passage to support their claim that Boyle had a preference for the "authority of the experimental community" over the "unsupported testimony of nonmembers." They claim that in it he used the "status of witnesses" to discredit More's argument. In elaboration of this claim, Shapin elsewhere has referred to this case as "the rejection of testimony from persons of suspect perceptual and moral reliability."[129] Certainly factual claims are socially constituted, because they must be established upon the basis of information from a number of witnesses. But the further claim that

the social status of the witnesses will play a determinate role in the establishment of such facts does not follow. Despite its appearance in the "Hydrostatical
Discourse," the above passage represents the conclusion of a ten-page discussion in which the issue of the reliability of the pearl divers' testimony was part
of a larger discussion concerning how best to reconcile their testimony with
conflicting evidence from the testimony of other types of divers, from the results of experimental trials, and from the background knowledge contained in
newly established pneumatic and hydrostatic theories.

Boyle had a large and varied collection of observation reports from divers.
He was fully aware of the traditional testimony of the pearl divers, but he also
had reports from a man who salvaged shipwrecked goods and from a "mechanician." The salvager told him that he had "felt a great pain in both his ears"
when "he staid at a considerable depth . . . under the surface of the sea," and
the "mechanician" had found it necessary to construct armored clothing to
avoid having his breast and abdomen compressed while underwater.[130] In addition, Boyle referred his readers to the essay that followed, "New Experiments
about the Differing Pressures of Heavy Solids and Fluids," which contained
reports from other salvagers who had been "much incommodated" by the pressure they experienced while diving.[131] In the presence of conflicting testimony,
Boyle maintained that the circumstances surrounding the reported phenomena
"need to be more heedfully observed" in order to determine "what is true in
point of fact" before one makes the attempt to "indagate the reasons" for
them.[132]

Boyle did not reject the pearl divers' testimony but sought instead to reconcile it with the other observations that he had compiled. He suggested a
number of ways in which a preliminary reconciliation of the conflicting reports
could be made. First, he looked more closely at the circumstances surrounding
the observations made by the pearl divers and noted that "as far as I have yet
learned by pursuing voyages, and enquiring of travellers of my acquaintance,
the places, where they are wont to dive for pearl, are but moderately deep, and
indeed shallow, in comparison of the great depths of the sea."[133] If the pearl
divers did not descend to a great depth, then perhaps the pressure would not
be extreme. Since they usually dive "in such haste" with their "minds so intent"
upon their task, maybe they would not take notice of "such lesser alterations."[134]

Boyle did not question the perceptual reliability of the pearl divers. Instead, he offered an explanation for their observations. He went on to note
that even if their dives occurred at greater depths, they still might not feel any
pressure because it could be the case that no direct pressure was being exerted
upon them. The hydrostatic principle concerning the equilibrium of fluids,
which had been experimentally confirmed by both Boyle and Pascal, could be
used to reconcile the apparent inconsistency between the divers' observations

and the principle of the "gravitation of water in water," because it could explain why some divers "that are a hundred feet beneath the surface of the sea, are not crushed inwards, especially in their chests and abdomens, or at least so compressed as to endure a very great pain."[135] Boyle recounted the various experiments that he had performed on inanimate bodies, not because he wished to elevate experimental results over observation reports but because such experiments could be used to illustrate his points concerning the equilibrium of fluids. He discussed a series of experiments performed on bladders "so thin and delicate, that a piece of fine Venice paper is very thick in comparison" in order to show how these and other "such frail bodies as eggs and thin glasses" are able to sustain "great pressure" so long as "the pressure be exercised by the intervention of an ambient liquor, as water." Based upon these experimental results, Boyle speculated that an explanation for the divers' failure to feel any pain or pressure would "depend chiefly upon these two things, the uniform pressure of the fluid ambient, and the robust texture of a human body exposed to this pressure."[136]

Boyle's doubts concerning the credibility of the pearl divers' reports had nothing to do with their suspect moral reliability. At this point he did introduce some doubts concerning their perceptual reliability, but these doubts were also not associated with their status. In his work on cold, Boyle had expressed reservations about using the human body as an instrument for determining degrees of temperature, and here he noted much the same problem when he discussed how the "robust texture" of the human body could give misleading information regarding the pressure of water, especially if one took pain to be an indicator of the existence of pressure. To show how the human body was not a reliable instrument and to illustrate how the equilibrium of pressure together with the stronger "membranes and fibres" of a human frame could withstand a great pressure, Boyle drew a comparison with the more commonly experienced pressure of the air. He asked his readers to consider "what great effects gusts of wind have upon doors, trees, nay masts of ships, blowing them down, nay breaking them; and that yet a man, without being extraordinary strong, will stand against the impetuosity of such a strong wind . . . without so much as complaining that he feels any pain."[137] Boyle questioned "whether the common reports that are made concerning divers be fit to be relied on, without further examen and observation." He also certainly believed that the gravitation of water in water had been "directly proved by particular experiments."[138] Nevertheless, he took the observations of the divers seriously and sought a way to reconcile them in a coherent and adequate manner with other facts that he believed to be more firmly established. The pearl divers' observations could be accurate, and the absence of any experience of pressure or pain could be due to the shallow depth of their dives, to the equal pressure exerted by the water, or to the robust texture of the divers' bodies.

On the other hand, Boyle allowed that it was possible that their observations were faulty. Most of the common observations from the "less sober part of the illiterate vulgar" had been "procured by leading questions" and "proposed with the common opinion of the non-gravitation of water in its own place." Boyle doubted the reliability of the divers' testimony not because they were ignorant but because prejudice and partiality could have been introduced into their observations by their questioners. Any alterations that they experienced in diving, such as the feeling of cold reported by some of them, could be "a disaffection produced in the nervous and membraneous parts, occasioned by the compression of the ambient water," but they would be "prone to refer the inconvenient alteration they feel, to any other cause than the pressure of the water, which they are taught to be none at all."[139] Boyle noted that the case was not much different from the prejudice he had encountered among "learned and intelligent men" who "when prepossessed (as these common divers usually are) with the vulgar opinion about the non-gravitation of water and air in their natural places, do almost always refer an experiment of my engine to suction, which is indeed the effect of the pressure of the ambient [air]."[140]

Reconciliation is not an easy matter. Boyle's confidence in his explications of hydrostatic phenomena was mitigated by his critical and flexible attitude toward the factual foundation. The evidence was mostly on his side, and he had a choice of ways in which to reconcile observations that conflicted with his explications, but he was not fully satisfied about the resolution of the case. Thus his presentation remained somewhat tentative. The controversy about "what happens to men under water" could not be settled until "intelligent men, who mind more the bringing up out of the sea instructive observations, than shipwrecked goods" made the dives themselves and gave accurate accounts of all of the relevant circumstances.[141] As was typical in most of Boyle's works, here conflicts in observations at the factual level gave rise to ideas for experimental trials that could be designed to resolve the conflicts. Experiments would thus play a dual role. They would be used for the correction and verification of items included in the factual foundation and for the correction and justification of speculations that were still hypothetical.

Seven

EXPERIMENTING

EXPERIMENTAL PRACTICES, if developed upon rational grounds, can lead to a knowledge of the qualities of bodies by providing a more expedient and a more reliable means of discovering the "corporeal agents" responsible for such qualities. As Boyle noted in his *Origin of Forms and Qualities*, he often gave experimental evidence rather than simple observations

> (though without at all despising, or so much as strictly forbearing to imploy the latter) because the changes of qualities made by our experiments will, for the most part, be more quick and conspicuous; and the agents made use of to produce them being of our own applying, and oftentimes of our own preparation, we may be therefore assisted the better to judge of what they are, and to make an estimate of what it is they do.[1]

Because artifice allows the experimenter to control the production of effects by repeating them and varying the number of concurrent circumstances surrounding their production, such activity may "either hint to us the causes of them, or at least acquaint us with some of the properties or qualities of the things concurring to the production of such effects."[2] The artificial environment created by experiment provides one of the best means for the establishment of facts concerning causes and effects. Thus, while Boyle was acutely aware of a number of legitimate criticisms that could be made against his experimental practice, its artificiality was not one of them.

Creating an Artificial Environment

A number of critics within science studies today have developed accounts designed to display the amount of human agency involved in the production of observational and experimental evidence as part of a broader agenda to question the "objectivity" or legitimacy of such knowledge. What is often ignored in these studies is that Boyle and other early proponents of experimental science faced the same types of criticisms. They responded with reasoned arguments that did not attempt to hide the role of human activity in knowledge production but rather made successful activity one of the criteria by which to judge the soundness of experimental knowledge.[3] Some of Boyle's early critics, for example, charged that results obtained through the manipulation of nature were artificial and could not, therefore, provide reliable information about natural processes. According to Boyle, this criticism represented one more remnant of Aristotelianism. The Scholastic philosophers always attempted to elude criticism of their hypotheses by "some pitiful distinction or other," and the distinction they drew between the natural and the artificial was no better than their other distinctions between natural and violent motions or natural and preternatural states of bodies.[4] In some "factitious bodies," such as the artifacts produced by carpenters and statuaries, the "sensible (not insensible) parts of the matter" may be altered solely by the art of man. Yet there are "other factitious productions wherein the insensible parts of matter are altered by natural agents who perform the greatest part of the work among themselves, though the artificer be an assistant, by putting them together after a due manner."[5]

Among the latter type of effects, Boyle included "all the productions of the fire made by chymists," because "the fire, which is the grand agent in these changes, doth not, by being employed by the chymist, cease to be and to work as a natural agent." He compared these productions with those that were "of nature's own making." If, for example, the "ashes and metalline flowers . . . sublimed by the internal fire" of volcanoes, such as those observed by the Neopolitan lord "about the vents" of Vesuvius, could not "be denied to be natural bodies," then, Boyle wrote, "I see not why the like productions of the fire should be thought unworthy that name, only because the fire that made the former was kindled by chance in a hill, and that which produced the latter was kindled by a man in a furnace."[6]

Not all, but many so-called artificial productions could be categorized as natural. Indeed, because "there seems to be really nothing in the corporeal world but the bodies . . . acting upon one another according to settled laws," it follows that "natural and artificial are but Notional Attributes that the Mind affixes to Things according to its differing manners of considering them."[7] The distinction between natural and artificial rests purely upon the means by which the productions are initiated and does not apply to the productions themselves. As Boyle explained in his *Usefulness of Experimental Philosophy*, most of the things produced by tradesmen depend upon natural agents working in accordance with fixed physical laws governing the action of matter. Many of "those productions that are called artificial" by the majority of philosophers

> do differ from those that are confessedly natural, not in essence, but in efficients; there are very many things made by tradesmen, wherein nature

appears manifestly to do the main parts of the work: as in malting, brew-
ing, baking, making of raisins, currans, and other dried fruits; as also
hydromel, vinegar, lime, etc. and the tradesman does but bring visible
bodies together after a gross manner, and then leaves them to act one
upon another, according to their respective natures.[8]

That Boyle found artificiality to be a virtue of his practice is apparent, not
only from his programmatic statements but also from the numerous places in
his published works where he found it necessary to apologize for the lack of a
sufficient degree of artificiality in the experiments reported there.[9] The ability
to re-create natural processes provides the experimenter with a more expedient
means of making observations, and often there are processes, such as the formation of minerals in the earth or stones in a human body, that cannot be
investigated directly in nature but must be inferred from effects produced in
the artificial environment of the laboratory. In addition to the obvious advantage of expediency, however, experiments also produce "the most instructive
condition, wherein we can behold" nature.[10]

As we have seen, Boyle was opposed to the use of sense impressions that
were "only received and not improved," and he warned his readers against the
habit of "contenting ourselves with the superficial account given us of things
by their obvious appearances and qualities."[11] As the level of artificiality in an
experiment increases, the level of superficiality decreases. In his "Examen
of Mr. Hobbes's *Dialogus Physicus*" Boyle made this point when he noted that
Hobbes's doctrine of simple circular motion was based upon a few superficial
observations and experiments and thus could not account for all of the varied
and complex phenomena exhibited by the air-pump experiments. Boyle did
not discount the use of observations and simple experiments, but as he told
Hobbes, "it is not safe in all cases to content one's self with such; especially
when there is reason to suspect, that the phaenomenon they exhibit may proceed from more causes than one, and to expect, that a more artificial trial may
determine, which of them is the true."[12]

The differences between Hobbes and Boyle extended beyond their debate
about the roles of mathematics and experiment in the construction of hypotheses. A more fundamental disagreement existed between them concerning the
establishment of facts. While Hobbes argued for the superiority of mathematical reasoning at the theoretical level, at the factual level his epistemology was
strictly empirical.[13] Boyle's advocacy of experimentalism entailed a rejection of
empiricist assumptions. Not only did he reject the belief that existence claims
are to be established solely upon the basis of the ideas one has from previous
sense experience, but he also rejected the empiricist equation of ideas with
images in the mind. In a manner reminiscent of Descartes, Boyle maintained
that the intellect, a mental faculty distinct from imagination, is able to form

ideas about the existence of an object without the necessity of ever having had a sense experience of it.[14] This point was crucial not only for Boyle's defense of the knowledge-producing capability of experiment but also for his defense of the rationality of religious belief. In *Things Said to Transcend Reason* he discussed the theological dangers inherent in an empiricist epistemology. It is obvious that he had Hobbes's heretical beliefs in view when he wrote that

> those, that are wont to employ their imaginations about things, that are the proper objects of the intellect, are apt to pronounce things unconceivable, only because they find them unimaginable; as if the fancy and the intellect were faculties of the same extent: upon which account some have so grossly erred, as to deny all immaterial substances, and chose rather so far to degrade the deity itself, as to impute to it a corporeal nature, than to allow any thing to have a being, that is not comprehensible by their imagination, which themselves acknowledge to be but a corporeal faculty.[15]

For both philosophical and theological reasons, Boyle rejected the empiricist claim that knowledge of facts depends upon simple sense experiences. As we saw in chapters 4 and 5, Boyle's philosophical goal was to discover the invisible corporeal agents that give rise to the qualities of bodies, most of which depend upon the relations that obtain between bodies in the vast cosmic mechanism. The artificiality of experiment can help in the discovery of these qualities because, in addition to being less superficial, experiments reveal the effects that bodies are able to produce in combination with other bodies and thus provide the best means available for establishing the truth of causal inferences about the powers that bodies possess.

Boyle often used an analogy with the processes employed to test the genuineness of gold coins to illustrate these virtues of artificiality. In *The Christian Virtuoso*, for example, he described how different the situations would be "if a piece of coin, that men would have pass for true gold" were to "be offered to an ordinary man, and to a skilful refiner." The refiner

> will examine it more strictly, and not acquiesce in the stamp, the colour, the sound, and other obvious marks, that may satisfy a shop-keeper, or a merchant; yet when he has tried it by the severer ways of examining, such as the touch-stone, the cupel, aqua fortis, etc., and finds it to hold good in those proofs, he will readily and frankly acknowledge, that it is true gold, and will be more thoroughly convinced of it, than the other person, whose want of skill will make him still apt to retain a distrust, and render him indeed more easy to be persuaded, but more difficult to be fully satisfied.[16]

The experimenter in natural philosophy has the same advantage as the refiner. By going beyond the superficial appearances of things and combining skill with extensive background knowledge, the experimenter is able to achieve a degree of certainty at the factual level that would be lost to a more empirical investigator.

As the above example illustrates, however, the experimental production of effects that were to be used for the factual foundation upon which theory would be built could not be achieved without some prior theoretical knowledge about the bodies manipulated in such trials. This situation created a practical problem. It was not so much the confirmation of theoretical knowledge that was at issue but the establishment of matters of fact. In a classic statement of the experimenter's regress, Boyle wrote that "a solid and complete theory of nature is not to be had without such a history of nature, as is not to be made without the theory it should minister to."[17] The regress did not make the situation hopeless, but it certainly contributed to Boyle's critical and diffident attitude toward knowledge claims. Not only could new factual information lead to revisions in theory, but new theoretical information could also lead to revisions in the way that experiments were performed or interpreted. All levels of inquiry from the factual through the hypothetical remain open to revision, because the experimental process depends upon a gradual reconciliation of the claims in all levels. The factual level has to remain open to revision, for example, because some of the "things in Natural History cannot be well and certainly related" without a knowledge of "some theories which perhaps are yet unknown."[18]

Boyle's belief in the necessity of joining together the practical and the theoretical parts of knowledge can be seen from an examination of his published works. He always included some discussion about the theories that had been used in the design and execution of his experiments and speculations about what the experiments could in turn indicate about his own and others' theories. As early as *The Sceptical Chymist,* Boyle noted that the validity of inferences made from chemical experiments depended upon the truth of the theoretical suppositions employed in their performance.[19] Even those works that he hastily prepared for publication shortly before his death were not totally devoid of theoretical considerations. In *Experimenta et observationes physicae* (1691), for example, he apologized for the fact that the work was "for the most part plainly historical," yet he still managed to include some "scholia" and "preambles" designed to provide hints about the theoretical implications of the experiments and observations recorded there.[20]

The inextricable involvement of theory and practice was an issue that concerned Boyle throughout his career. Near the end of his life he mentioned a paper "never published" that he had written "to settle the bounds of reason and experience, in reference to natural philosophy."[21] The draft material to which

he referred is most likely that which is preserved in volume 9 of his unpublished papers and described there as a collection of "Observations and Reflections about the Bounds and Uses of Experience in natural philosophy, as some
years familiar converse with divers productions and Laws of nature has given
me opportunity to make."[22] Boyle's appreciation of the reciprocal relation between theory and practice, philosophy and experiment, is most perspicuous in
this draft material, where he left out discursive embellishments and merely
listed the number of ways in which the two processes interact:

"Of the Use of Experiments to Speculative Philosophy"

1. To supply and rectify our senses
2. To suggest Hypotheses both more general and particular
3. To illustrate Explications
4. To determine doubts
5. To confirm truths
6. To confute errors
7. To hint luciferous inquiries and experiments and contribute to the
   making them skilfully.

"Of the Use of Speculative Philosophy to Experiments"

1. To devise philosophical experiments which depend only, and
   mainly, upon Principles, notions, and Ratiocinations
2. To devise instruments both mechanical and others to make inquiries
   and tryalls with
3. To vary and otherwise to improve known experiments
4. To help a man to make estimates of what is physically possible and
   practicable
5. To foretell the events of untried experiments
6. To ascertain the limits and causes of doubtful and seemingly
   indefinite experiments
7. To determine accurately the circumstances and proportions, as
   weight, measures and duration etc. of experiments[23]

It is not possible to design or interpret an experiment without the aid of
theory, yet as we have seen, Boyle cautioned his readers not to let theory unduly influence their practice. Even though it had become popular to write
about "the necessity of experiments," too many of his contemporaries tended
to "omit experiments in particular cases" because they believed that their outcomes might be "foretold by meer ratiocination or rational inferences from
Truths, that are known already." Such an attitude was harmful, not only because the theories employed to make such inferences might be faulty but also
because a number of practical limitations are associated with the performance

of experiments and naturalists must learn to "contest with the difficulties that frequently occur" in their attempts to manipulate nature.[24]

The Contingencies of Experiments

At the most general level, numerous practical problems may arise concerning the correct determination of what has actually been produced by an experimental trial. Indeed, sometimes an experimenter can be misled into believing that artificial manipulation was responsible for an effect whose production was in fact entirely natural. As an example of this problem, Boyle discussed Bacon's experiments on the effects of pruning. According to Bacon, wrote Boyle, "If a rose-bush be carefully cut as soon as it has done bearing, it will again bear roses in the autumn." Boyle noted that he and others had failed in their attempts to replicate this experiment, but he was "very apt to think" that Bacon had successfully performed it. In order to clear up the difficulty, Boyle "made trial of it" by having a number of bushes, all growing in the same row and therefore under the same soil and light conditions, pruned at the same time. He observed that only one of the bushes, more "vigorous" than the rest, flowered the following autumn and concluded that the vigor of the particular bush and not the art of pruning had been responsible for this result. For Boyle, the lesson was clear. Experimenters must exercise extreme caution in the interpretation of the production of effects, since "that may be mis-ascribed to art, which is the bare production of nature."[25]

On other occasions, the effect produced may be a result of the experimental manipulation of nature, yet other experimenters may not be able to replicate the phenomenon successfully because of a failure to understand all of the circumstances necessary for its production. Experiments are "seldom solitary."[26] The complexity of experimental methods, which makes artificial manipulation especially suited for exploring the complexity of natural processes, also gives rise to a number of problems.

> He that knows not the nature or Properties of all the other bodies [in the experiment], wherewith that on which the Experiment proper is made, . . . can hardly discerne what Effects the experiment may possibly concurre to produce, for many Inventions and Experiments consist as it were of several parts or require distinct actions, to some one of which, though it should happen to be the Principall, a thing may not be useful, which by being requisite to an other is of use to the Experiment in general though not to each distinct part of it.[27]

According to Boyle, the number of "seemingly dispicable and unheeded circumstances" surrounding a trial often led experimenters to conclude, incorrectly, that their failures were due to a radical contingency in nature. He noted that in the practice of grafting, for example, it had been observed that some trees "have borne fruit the same year that they were grafted . . . and others not till the year after the insition" even though the goodness of the graft and the stock could not be questioned. The explanation of this "seeming contingency" depended upon a thorough knowledge of the properties of fruit trees that could only be gained by careful observation and "frequent dealing" with such bodies. Boyle therefore sought an answer to the problem by questioning "one of the most skilful and experienced grafters of these parts," who told him that fruit would be borne the first year if the graft had "blossom buds, as they are wont to be called, upon it; whereas if it were only leaf-buds, as they may be termed, it will not bear fruit till the second season. And this not being taken notice of by vulgar gardeners, makes them, as we have said, impute a needless contingency to the fruitfulness of such kind of grafts." In some experiments the seeming contingency may proceed from "easily discoverable causes," whereas in others "the cause of the contingency is very abstruse and difficult to be discerned." Boyle was confident, however, that, even in the latter case, causes would be discovered "by men's future skill and diligence in observation."[28]

Boyle's experimental interests extended beyond the laboratory, and he acquired significant factual information from his frequent visits to the shops and workplaces of tradesmen and artisans. He also gained important methodological insights from his observations of their practices.[29] In particular, he learned that natural philosophers were not the only ones who encountered problems in attempts to manipulate nature. Glassworking, for example, is a "trade that obliges the artificers to be assiduously conversant with the materials they employ," and yet "even to them, and in their most ordinary operations," there could occur "little accidents" that kept them "from doing sometimes what they have done a thousand times." Boyle discussed one of his visits to a "glasshouse," where he had gone to pick up some vessels that the "eminently skilful workman, whom I had purposely engaged," had not been able to produce, in order to make the complexity of experimental practice vivid for his readers. He advised them that "it need to be no such wonder, if philosophers and chymists do sometimes miss of the expected event of an experiment but once, or at least but seldom tried, since we see tradesmen themselves cannot do *always*, what, if they were not able to do *ordinarily*, they could not earn their bread."[30]

From his converse with tradesmen, Boyle found that "in general" there are "in divers cases such circumstances as are very difficult to be observed, or seem to be of no concernment to an experiment," that "may yet have a great influence on the event of it." One could learn from smiths and farriers, for example, that something as seemingly inconsequential as a few moments of time could have a significant effect on the outcome of a trial. He noted that he had often "observed, how artificers in the tempering of steel, by holding it but a minute or two longer or lesser in the flame," were able to "give it very differing tempers, as to brittleness or toughness, hardness or softness," and "none but an artist expert in tempering of iron would suspect that so small a difference of time of its stay in the flame could produce so great a difference in its tempers."[31] In order to acquaint his readers with the problems that might be encountered in the performance of experiments, Boyle devoted two essays in his *Certain Physiological Essays* of 1661 to a discussion of the various types of "contingencies" that could arise in the experimental investigation of nature.

The first essay, "Concerning the Unsuccessfulness of Experiments," contained Boyle's "divers Admonitions and Observations (chiefly Chymical)" about problems that could arise from the state of the materials used in experimental trials.[32] He advised his readers to be cautious in their use of substances obtained from the shops of chemists, for example, because they could often be "adulterated by the fraudulent avarice of the sellers." Such corrupt materials present a problem for the experimenter, because they "may acquire an aptitude to produce such effects, as, had they not been adulterated, they would not have been fit to do."[33] Therefore, when "experiments succeed not," that is, when an experimental result cannot be easily replicated, it could be that the experiment was "at one time tried with genuine materials, and at another time with sophisticated ones."[34]

The fraudulent addition of ingredients was not the only problem associated with the materials used in trials, however. All substances "prepared by art," even by trustworthy and expert chemists, may be subject to slight differences. It could be that only genuine ingredients were used, yet the resultant compound could differ significantly from a good sample because the ingredients had not been "purified and exalted enough" in the process. In addition, even the mere passage of time could affect the nature of an artificial substance. Using an example that would be familiar to most readers, Boyle noted that a significant alteration takes place when wine is turned into vinegar. Admittedly, no one "is like to lose an experiment by mistaking vinegar for wine" because the change can be easily ascertained by taste, but the example could serve as a reminder for the need for caution and circumspection in the laboratory. Boyle asked, "Who knows what changes there may be in other bodies, with whose alterations we are unacquainted, though the eye, which is oftentimes the only sense employed about judging of them, discern no change in them?"[35]

While the purity of artificial substances may be more easily questioned, at a very early stage of his practice Boyle learned that many natural products could also be deceptive at the level of appearances. The "individuals of the same ultimate subdivision of plants," for example, could possess different qualities as a result of differences in the "temperature of the air," the "nature of the soil" they are grown in, and "many other causes" that would lead to variations in their efficacy to produce consistent results. Metals and minerals, which "almost all men without hesitancy" look upon as being "of the same nature as well as denomination," could also differ between themselves as much as "vegetables and animals of the same species" do. The differences between animals of the same species are "wont to be more obvious to the eye," whereas minerals may appear to be "perfectly similar." Yet it may often be true that small quantities of other minerals "lie concealed" in a sample. The presence of such adulterating minerals "is hardly to be discerned, before experience have discovered it," either by "exquisite separations" or by the "unexpected operations" of the substances in experimental trials."[36]

The only way to detect significant differences in the materials used for trials is to observe contingent outcomes and avoid the "prepossessions" that make many experimenters "ascribe the variations they meet with in their experiments, rather to any other cause, than the unsuspected difference of the materials employed about them."[37] A naturalist must carefully observe and accurately attribute such differences in outcomes before attempting to use experimental trials in the construction of theory. The recognition of the differences in the properties of artificial and natural materials was also crucial because of the practical uses made of these materials. Physicians should be aware of all of the ingredients that go into the medicinal remedies they use and not "presume that drugs or other materials are more simple than indeed they are."[38] If a medical practitioner is not aware of the properties of particular substances and thus not able to determine whether a particular medicine is adulterated or pure, then serious practical consequences for human health and welfare could follow. Boyle discussed how, for example, anyone "acquainted with the violent emetic qualities of *Venus* [copper]" could "scarcely doubt" the "ill effects" that could result from "the mixture of copper . . . in such medicines, as ought to be of pure silver."[39]

Clearly these and the many other material contingencies identified by Boyle created numerous problems for the correct design, execution, interpretation, and application of experimental trials. But the purity of materials was only part of the problem. In Boyle's second essay, "Of Unsucceeding Experiments," he addressed all of the other "contingencies, to which experiments are obnoxious, upon the account of circumstances, which are either constantly unobvious, or at least are scarce discernible till the trial be past."[40] In addition to its purity, the suitability of a substance for a particular trial must be assessed. As we saw in chapter 3, for example, the use of a wax ball to test whether the specific gravity of water differed with a change of temperature was inappropriate because wax is itself a substance whose properties change with differing temperatures.[41] The quantity of material used could also produce misleading results. As Boyle noted, "Divers experiments succeed, when tried in small quantities of matter, which hold not in the great," because "oftentimes a greater and unwieldly quantity of matter cannot be exposed in all its parts to a just degree of fire, or otherwise so well managed, as a less quantity of matter may be ordered."[42]

Boyle included within his discussion of these other types of contingencies a number of examples drawn from medicine, admittedly an inexact science.[43] He went on to note, however, that "even mathematical writers themselves," who tend "to be more attentive and exact, than most other men in making almost any kind of philosophical observation," often "deliver such observations as do not regularly hold true." Because they focus upon the abstract quantitative properties of bodies, "we must not expect from mathematicians the same accurateness, when they deliver observations concerning such things, wherein it is not only quantity and figure, but matter, and its other affections, that must be considered." In addition to the contingencies that arise from the "nature of the material objects, wherewith the mathematician is conversant," however, the "imperfection of the instruments, which he must make use of in the sensible observations, whereon the mixed mathematics (as astronomy, geography, optics, etc.) are in great part built," may also "deceive the expectations grounded on what he delivers." Because such "material instruments" are "framed by the hands and tools of men," they may "in divers cases be subject to some, if not many, imperfections upon their account."[44] As Boyle described it, problems surrounding the use of instruments are

> but too manifest in the disagreeing supputations, that famous writers, as well modern as antient, have given us of the circuit of the terrestrial globe, of the distance and bigness of the fixed stars and some of the planets, nay, and of the height of mountains: which disagreement, as it may oftentimes proceed from the differing method and unequal skill of the several observers, so it may in divers cases be imputed to the greater or less exactness and manageableness of the instruments employed by them.[45]

From his own area of expertise, Boyle discussed similar problems that had arisen in hydrostatic and pneumatic trials, where the significantly different results obtained could not be attributed to a lack of skill on the part of those performing them, but rather must owe to some as yet unheeded contingencies surrounding them:

> I shall observe, that the observations even of skilful mathematicians may hold so little, or disagree so much, when they pretend to give us the determinate measures of things, that I remember of three very eminent modern mathematicians, who have taken upon them, by their experiments, to determine the proportion betwixt air and water, the one makes not the weight of water to exceed above 150 times that of air; the other reckons water to be between 13 and 14 hundred times; and the third no less than 10,000 times the heavier.[46]

Boyle admitted that he had encountered similar difficulties in his own attempts to determine such measures. He did not give the details of the "experiment partly statical, and relating to the weight of the air" that "we made divers times in an hour," but he did mention that in it "we missed of the like success twice as often in the same hour, without being able to know beforehand, whether the experiment would succeed within some pounds weight."[47]

The appreciation of the complex conditions surrounding experimental trials that Boyle gained from his observations of the practices of tradesmen and his own laboratory work led him to urge his readers to be cautious in their acceptance of his own and others' reports. Yet despite the vast catalog of failures that he detailed in *Certain Physiological Essays,* at the end of the second essay he maintained that all he had said should only be "fit to persuade" his readers of the need for "watchfulness in observing experiments, and wariness in relying on them." He did not wish to create in them "such a despondency of mind, as may make you forbear the prosecution" of experiments. The practice is admittedly fallible, but that does not mean that all of its products will be faulty. As he explained, a physician does not "renounce his profession, because divers of the patients he strives to cure are not freed from their diseases by his medicines, but by death; nor doth the painful husbandsman forsake his cultivating of the ground, though sometimes an unseasonable storm or flood spoils his harvest, and deprives him of the expected fruit of his long toils."[48]

The experimental philosophy is fundamentally a practice subject to human error, but one should not reject a practice because of its fallibility. While practitioners may "sometimes miss of their ends, yet they oftentimes attain them, and are by their successes requited not only for those endeavours that succeed, but for those that were lost."[49] The fallibility of the practice, however, did mean that naturalists would need to employ "reason in natural philosophy" in order "to devise apposite Experiments, and contrive the wayes of making them, and of examining whether they be well made."[50]

Making Experiments

For naturalists to "surmount" the many "difficulties" that arise "in the serious and effectual prosecution of experimental philosophy," they must acquire a sufficient degree of expertise, the lack of which is "the most obvious cause of error."[51] As a first step, inquirers should recognize that not all experiments are the same. According to Boyle, experiments comprise two distinct types of methodological activity: they can be either "probatory" or "exploratory."[52] Probatory experiments function at the practical level to produce a proof, not of a theory, but of the reliability of the experimental materials and instruments that are to be used in more complex and theoretically significant experiments. As Boyle described it, the "scope" of a probatory experiment "is to discover whether or no, some quality or attribute do belong to the subject to be examined, as we know, or believe, ought to be found in it, if it be genuine."[53]

Because a number of the contingencies of experiment result from practical problems, practical methods must be in place to detect them. Probatory trials are thus designed not to test theoretical knowledge but to use such knowledge to test the reliability of the conditions surrounding an experiment. The same type of reasoning process used for the construction and confirmation of theoretical claims is also employed at this level, however. In an unpublished note, Boyle used the coin analogy to explain how the process of testing coins provided an instance of a probatory experiment that could be used at the methodological level to produce a concurrence of probabilities for the establishment of a matter of fact. When "proving that a piece of yellow coin is a true guinea," it may

> be no concluding argument either that 'tis yellow, or that 'tis malleable, or that it touches well upon the stone or that 'tis heavier than Brass, or that it sustains the cupel or that 'tis not dissoluble in AF [aqua fortis] (though I say) to prove any one of these be not a sufficient argument that the proposed piece of coin is true gold, yet the evincement of each of these particulars, is of considerable use towards the compiling of a demonstration of the grand conclusion this being the result of all these proofs taken together.[54]

Boyle illustrated this type of trial in the context of the problem concerning the purification of experimental materials. In order to determine if spirit of wine had been "sufficiently dephlegmed," he advised his reader to dip a wick into the prepared substance. If the saturated wick could be set aflame, then the material could be considered good. If not, further rectification would be required.[55] The development of such indicator tests was one of Boyle's major contributions to the new philosophy, and he provided numerous practical instances of how they functioned throughout his published works. In "Concerning the Unsuccessfulness of Experiments," for example, Boyle recounted an attempt to perform an experiment with some "spirit of salt" (hydrochloric acid) that he had obtained from a chemist's shop. He thought that he had procured a good sample and believed at first that its apparent weakness to perform what he expected was due to the fact that it had not been purified sufficiently. He found after numerous rectifications of his own, however, that the sample could not be made "pure enough, to perform what we expected from it." This led him to suspect that the sample was adulterated. His suspicions were confirmed when, from the "peculiar and odious smell" produced during rectification, he discovered that the sample was indeed "sophisticated with either spirit of nitre, or aqua fortis [nitric acid]." Because spirit of salt and spirit of nitre are both colorless and corrosive liquids, he could not have discovered the adulteration from sight, but from his laboratory experience he had learned that, when rectified, spirit of nitre has an "odious smell; whereas spirit of salt skilfully and sincerely drawn" has "usually a peculiar, and sometimes . . . a not unpleasing smell."[56]

An experimenter has to be aware of such differences in the effects of various types of materials and to observe accurately all of the effects produced by a particular substance in order to determine if it can be used reliably in further trials upon other bodies. Boyle noted that there is a "great disparity in the operations of those two liquors [spirit of salt and spirit of nitre], whereof . . . the former will precipitate silver, when the latter has dissolved it."[57] Probatory experiments are also required when the materials are to be used for direct applications. Physicians, for example, should not assume that the drugs they use are pure. In a planned work, entitled "The First Sort of the Uses of Distillation, tends to the examen of the Genuineness or Adulterateness of Medicine," Boyle listed six "chief signs" of impurity that ought to be observed before a drug is administered:

1. If the adulterate medicine affords something it should not.
2. If the suspected medicine does not afford something that it should.
3. If the quantity of all or any of the Substances obtained by Distillation do too much vary from what might have been expected from a genuine drug.
4. If any manifest quality or operation of any of the obtained substances, differ too much from what might have been expected, if the drug had been genuine.
5. If there be a notable difference from what is wont to happen in a genuine drug (of the same name) in any other considerable circumstance, one or more, as the volatility or slugishness of the ascending matters, the greater or lesser heating of the vessels, the order or *series* wherein the differing substances ascend etc.
6. And lastly (when the elevated substances have been examined) if the *caput mortum* or Remains do too much differ from what the Genuine Drug is upon distillation wont to leave behind, as in colour, quantity, weight, consistence, disposition, or indisposition to be calcined, burnt, etc.[58]

The performance of a probatory experiment obviously depends upon theoretical knowledge and previous experience in working with familiar materials. Sometimes, however, an experimenter may be faced with a new substance or a new piece of apparatus and thus not possess the knowledge required for such trials. In these cases, a long series of "exploratory" experiments may have to be performed. As Boyle described them, "experiments of discovery, or as for brevity sake I sometimes call them, indagatory, detecting, or exploratory ways of trial," may often use methods "very near of kind and sometimes perhaps coincident" to probatory experiments, but the aim "is, to discover what quality or other attribute may be found in the subject, of which we have but a suspicion not knowledge, that it ought to belong to it, and perhaps wherein we endeavour to discover whether it have such an Attribute or one that is very remote from it or perchance even contrary to it."[59]

The type of experiment performed will be dependent upon the current state of knowledge and technological development in the environment within which the experimenter is working. It is this context that gives rise to the ideas upon which experiments are designed. In a list concerning the ideas upon which the two types of experiment are based, Boyle noted that probatory trials involve a more direct form of testing, whereas exploratory trials often arise from vague ideas or informed intuition:

"Probatory Experiments"

1. Immediate sensations
2. Statical
3. Hydrostaticall
4. Chymical strictly so called
5. Mathematical
6. Magnetical
7. Chym[ical]: less strictly so called
8. Medicinal
9. Anatomical
10. Technical
11. Mechanical
12. Compounded
13. Miscellaneous.

"Exploratory Experiments"

1. Almost all the wayes of making Probatory Experiments being a little vary'd and improved may be reduced to the above mentioned title.
2. Analogisms.
3. The framing of Hypotheses and then examining them by proper tryals.
4. The drawing consequences from vulgarly received opinions and examining them by proper tryals.
5. The devising of new and convenient tools or other instruments for altering the usual state or the common course of things, and for thereby reducing nature to vary her course and afford the enquirers some new phaenomena.
6. Composition of two or more of these wayes.
7. Uncertain sagacity.[60]

The two types of experiments differ primarily because of the different contexts within which they are designed and performed. Sometimes, therefore, the same experiment could be probatory in one context and exploratory in another. In his discussion of distillation, for example, Boyle noted that "the uses that may be made" of it were of two sorts. We may distill a substance "first to examine whether a drug proposed be genuine or counterfeit," but we may also distill a substance in order "to make some such discoveries about the more simple substances that it contains or their qualities as may be of use to the physician."[61]

The distinction between these two types of experiments had a practical significance for the way in which Boyle went about making experiments, and one can see it at work in the way he reported his results. In *New Experiments Physico-Mechanical,* for example, he began with a number of probatory experiments designed to show that his engine was capable of evacuating a large portion of atmospheric air from the sealed receiver. He then went on to show that the results obtained from Pascal's Puy-de-Dôme experiments could be replicated by his engine and thus could provide additional evidence for the establishment of factual claims concerning the pressure of the air.[62] In addition, he described a number of exploratory experiments that he had performed in a somewhat haphazard manner with his engine in order to determine more accurately the type of effects it was capable of producing, such as those performed with a "good loadstone" to discover "whether or not divers magnetical experiments would exhibit any unusual phaenomena, being made in our evacuated receiver instead of the open air."[63] Years later, in his essays on phosphorus, he would employ exploratory experiments in a similarly haphazard manner and justify the procedure by noting that "it is not easy to know, what phaenomena may, and what cannot, be useful, to frame or verify an hypothesis of a subject new and singular."[64]

In his *New Experiments* and elsewhere, Boyle also included a discussion of some exploratory experiments that belonged to a type that he called *"fiat experimentum,"* such as those that could not actually be performed until technical improvements had been made on his engine.[65] As he wrote to Oldenburg in 1666, if experimental ideas were properly identified as such, their discussion could serve as "the great bridge" that would lead to future research.[66] Among the types of reports that ought to be included in a natural history are those concerning "designed trials or fictitious Experiments, wherein processes and other ways of operating are proposed to supply the defect of real Experiments when we want them to determine doubts, to resolve questions, or for other purposes."[67] Boyle compiled vast lists of such experiments. Sometimes he made notes about ideas he had for new types of probatory experiments, but most of the items on his lists concerned ideas for new types of exploratory trials. As he noted, it is "of great use" to "have in readiness a good number of experiments for discovery."[68]

In his notes for a planned work entitled "Chromatic Examen," for example, he listed a number of exploratory experiments designed for the discovery of the unknown properties of bodies and recorded a number of methodological observations about how the experiments ought to be prosecuted and what they could reveal. Color changes produced in bodies by artificial means could provide helpful hints about the underlying configuration of their parts. In particular he noted that "it may be useful and instructive, to try and observe what difference time, especially a moderate space of it, will make in Chromatical Experiments." The "quickness of the operation" by which a change of color is produced in a substance sometimes "proves instructive, by showing a Body to be of a more loose or open texture, or to abound more with certain parts, than the experimenter knew before." At other times "observations no less instructive" could be made "by finding that some bodies will not have the expected change of color produced, or will have them produced but slowly, or sparingly, or both." In addition, he noted that not only the time taken to produce a change but also the permanence of the change ought to be observed. Boyle suggested that the experimenter should set the experimental substance aside for a while to see if the color altered. This could "give a heedful observer some information of the Texture or Qualities of one or more of the Ingredients of the mixture."[69]

Similar lists of experiments "to try" are scattered throughout Boyle's papers. He designed a number of experiments on the nerves of animals "to try what effect" would occur when the nerves were cut in various places, and he compiled a number of experiments to discover "the sphere of activity of electrical bodies," which included experiments designed "to try whether electrical bodies will at all attract under water, or other liquors" and "in what proportion if in any the attractive power in electrical bodies decreases according as the body to be drowned is remoter from it."[70] The most complete record of Boyle's daily experimental activity is preserved in a notebook that dates from 1684–89. Interspersed with the addresses of tradesmen and lists of apparatuses that he wished to have constructed by them are extensive lists of "Experiments to be Done" concerning chemical, pneumatic, hydrostatic, and botanical investigations. Each page contains at least fifteen entries, and all entries are prefixed by the date when they were to be performed. In separate lists, entitled "Experiments Done," results are recorded for each day of the year except Sundays. On average, three or four experiments are listed for each day, but sometimes as few as one or as many as seven are recorded.[71]

Probatory experiments should be performed to check the goodness of ingredients and the accuracy of instruments before the results of more complex

experimental trials that employ these materials are used for theoretical speculation. Exploratory experiments should be performed when one desires to learn
more about the properties of these materials. Both are useful as ways either to
guard against or to discover the latent qualities of things that may lead to
contingent outcomes. They are simpler than the complicated experiments designed to test hypotheses, yet even they are subject to contingencies, and thus
more general methodological strategies would be required to guide the practice of all experimentation. As Boyle explained, a naturalist "must often make
and vary experiments."[72] The two practical strategies of variation and repetition would provide the best means by which one could ensure the reliability
of the evidence that would compose the informational basis of the new philosophy.

Experimental Strategies

The materials and instruments used in the performance of an experiment must
be varied in order to eliminate the possibility that the outcome is merely an
artifact of the procedures employed. As we saw above, for example, a variety
of tests were used in one complete probatory experiment designed to test the
genuineness of gold coins. In a similar manner, exploratory experiments will
only yield the discovery of the properties of bodies if they are made up of a
variety of tests.[73] In order to determine the true nature of a substance, one
must test a variety of samples, since a particular experimental result may not
be generalizable because of some unsuspected peculiarity within the original
sample chosen for a trial.

In his essays, Boyle provided a number of practical examples of the ways
in which variation could be used to reconcile apparent inconsistencies in experimental reports. He had been unable to replicate in "several trials before
witnesses" a report made by Bacon of the "somewhat unlikely truth, that spirit
of wine will swim upon oil (of almonds)." Because of his "tenderness of the
reputation of so great and so candid a philosopher," however, he did not immediately reject Bacon's report. Instead, he speculated "that (though he mentions
it not, nor perhaps thought of any such thing, yet) possibly he may have used
spirit of wine more pure than ordinary." Upon further trial, using a sample
of spirit of wine that was "well rectified," Boyle found "that the oil . . . did
readily sink."[74]

Another seemingly irreproducible experimental report was that made by
"the learned Dr. *Brown*," who had claimed "that aqua fortis will quickly coagu-
late common oil." At first Boyle suspected that his failure to repeat the experiment could be due to the temperature at which the mixture was made, and so
he attempted the trial again, first keeping the mixture "in a cool place, and after
in a digesting furnace." When these variations also failed to produce the

desired result, he then questioned "whether or no the unsuccessfulness we have related might not proceed from some peculiar though latent quality, either in the aqua fortis or the oil by us formerly employed: whereupon changing those liquors, and repeating the experiment, we found after some hours the oil coagulated almost into the form of a whitish butter." Whenever he had reason to believe in the trustworthiness of an experimenter, Boyle maintained that he would "forebear to reject his experiments, till I have tried, whether or no by some change of circumstances they may not be brought to succeed."[75]

The instruments employed to take measurements of the properties of bodies must also be varied as much as possible because of the imperfections to which they are prone. In his *History of Air,* for example, Boyle complained of the inaccuracy of thermometers and suggested that, in addition to making them more exact, naturalists should use a number of different types and locate them in different places within a building to ensure that "the air of the chimney, cranny of a wall or door, breath of people, or other such accidents, do not interpose to deceive a man's observation, which must be circumspectly foreseen and considered."[76] Boyle expressed similar reservations about experiments performed in furnaces, because the degree of heat could vary significantly between them, and in air-pumps, because they could be subject to different degrees of leakage. The only results that would be completely reliable were those that were consistently obtained from a variety of instruments.

Often variation could be achieved simply by following Boyle's other practical strategy: repetition. As he advised his readers in "Of Unsucceeding Experiments," "The instances we have given you of the contingencies of experiments, may make you think yourself obliged to try those experiments very carefully, and more than once, upon which you mean to build considerable superstructures either theoretical or practical; and to think it unsafe to rely too much upon single experiments."[77] Repetition is absolutely necessary as a way in which to ensure that the experimental result is not a one-time chance occurrence. But in his draft for "Uses and Bounds of Experiments," Boyle made it clear that repetition involved a more complex methodological strategy that could be useful even for those "experiments of whose truth we are persuaded." He listed four types of "utility that the repeating of known experiments may afford the heedful naturalist."[78]

In his first point Boyle noted that repetition is not done solely to increase the quantity of information in the factual foundation. Rather, it is designed to assess the quality of the experimenter's result: "It may assure him of the truth of the matter of fact, and that it may be relied on for such as it has been represented, which, when the subject is important, or some circumstance of moment seems doubtful or lyable to have been mistaken is no small satisfaction to a strict searcher of truth." Repetition is often done in the presence of witnesses, and thus the strategy increases the number of people directly involved in the experimental process. But the reliability of the result is more likely not because of the number of witnesses but because of the role that the witnesses play in checking the accuracy of the conditions surrounding the experiment's performance. Sometimes, though, repetition is a private affair. Boyle often repeated his own experiments and reported those instances when he failed. He recalled, for example, that one time he had performed an experiment in which he changed the color of olive oil from "pale yellow to a deep red with a few drops of a liquor, that was not red, but almost colourless." But "because we do not willingly rely on a single trial of such things, as we know not to have been ever tried before, we thought it not amiss for greater security to make the experiment the second time, but could not then find it to succeed, nor even since upon a new trial (probably by reason of some peculiar quality in that particular parcel of liquor we first made use of)."[79]

Boyle's second point about the usefulness of repetition was directly related to the main concern that he had for naturalists to acquire the expertise needed to perform experiments well. Through the repetition of experiments, inquirers could learn not only about natural processes but also about the method itself and ways in which it could be improved: "The repeating of an Experiment confers on the maker of the tryal a facility of making it well, and oftentimes suggests to him expedients to compass his end more speedily or more easily, or more cheaply, or more certainly, or on some other account more advantageously than he could before." Repetition could also aid in the discovery of new information, because, as Boyle said in his third point, "Nature, and Attention almost always presents some novelty, either in the Event, or the intermediate circumstances" of a repeated experiment.[80] Repetition could function as a way to achieve a variation of experimental conditions. Even the simple fact that a repeated experiment is performed at a different time is significant. As Boyle had learned from tradesmen, gardeners, and bakers, the time of year is a frequently overlooked variable that may affect the successful production of artificial effects.[81]

Finally, in his fourth point Boyle returned to the issue of the complexity of experimental practice: "The most of capital Experiments, especially if they be somewhat long or compounded, are indeed made up of several partial experiments which as so many new phaenomena, are exhibited in the process or operation, though they were not from the beginning directly or particularly designed by the Experimenter."[82] Even when an experimenter was assured of the reliability of the main result, repetition of a complex experiment could still be of use. It would allow for a more detailed inspection of all of the subsidiary parts that contributed to the main experiment and thus could lead one to discover useful information about natural processes and about the ways to learn more about them.

The strategies outlined by Boyle were not strict rules but general guidelines for learning about the conditions that might be responsible for contingent outcomes. The above accounts mostly concern the practical failures that occur when one is unable to replicate previous results of one's own or those reported by others. Experiments are also subject to theoretical failures, however, when they fail to turn out as background knowledge would have led one to expect. Experiments are performed with expectations in mind, and when these results do not obtain, we must use reason to determine whether the failure is due to a problem with the theories upon which our expectations were grounded or to the practical circumstances surrounding the experiment itself. Once again, reason and practice must be employed in unison. While Boyle maintained that "the well circumstanced Testimony of Sense is to be preferred to any meer Hypothesis, or to Ratiocinations not grounded upon sense or either mathematical or metaphysical truths," yet "the negative Testimony of Sense ought not to be admitted without distinction and caution."[83]

Theoretical failure cannot be taken as an immediate refutation of the theory upon which the experiment is based, yet it may provide significant insights into the processes operative in nature.[84] As Boyle explained, "in experiments it not unfrequently happens, that even when we find not what we seek, we find something as well worth seeking as what we missed." In much the same way as merchants and navigators, when "having been by storms forced from their intended course," are led to make "discovery of new regions much more advantageous to them," so also "in philosophical trials, those unexpected accidents, that defeat our endeavours, do sometimes cast us upon new discoveries of much greater advantage, than the wonted and expected success of the attempted experiment would have proved to us."[85]

Failed experiments may also set inquiry on a new course by suggesting innovative lines of research or improved experimental techniques. By accurately recording and reviewing the details of failed experiments, one may learn about the types of previously unheeded circumstances that can lead to contingent outcomes and thus also learn how to correct for the many variables that must be taken into account in order to perform experiments successfully. Boyle was learning his practice as he went along. As he noted, an experimenter works "for the most part in the dark, or with but dim-lights," and therefore "must vary his methods and alter the structures from time to time, as he is prompted by cross or lucky incidents and further discoveries."[86]

Unlike some of his contemporaries, Boyle wrote in order to make the difficulty of experimental practice vivid for his readers and to advise them that "our way is neither short nor easy."[87] He often reported his failures so that others would appreciate that the experimental establishment of facts is not a simple matter of the inductive collection of data but also requires a sophisticated reasoning process whereby the data are interpreted and reconciled into a coherent narrative. In order to achieve such coherence, the third activity of the experimental philosophy is crucial. One must collate the data into written form and communicate it to others. Only in this way could one produce the collaborative effort with workers in all areas of inquiry that would be required for a solid grounding of the new philosophy.

Eight

WRITING

ONCE OBSERVATIONS AND EXPERIMENTS have been made and recorded, the "undigested heap" of data must be ordered according to topics so that one can begin the preliminary survey necessary to find the theoretically significant concurrences among facts required for the construction of a coherent interpretation of nature. The first part of this process takes place in private. The data compiled in notebooks must be reviewed, collated, and reconciled. The final and crucial stage of the process, however, is the communication of these compilations, and any speculations suggested by them, in published form. Boyle had definite ideas about the way to write for this purpose, yet little attention has been given to the details of his explicit precepts, despite the large number of recent studies designed to display the rhetorical strategies employed in his works.

Among the many rhetorical studies of past science, some have concentrated upon the influence that the traditional rhetoric of the schools had on a particular thinker. Others use the term *rhetoric* in a derivative sense to refer to any persuasive argument form and normally discuss its usage in connection with a writer's attempt to persuade readers about the truth of a particular scientific theory.[1] Contextualist historians have extended the latter type of study to include analyses of how the choice of a particular writing style could be used to enhance the readers' sense of the writer's trustworthiness or expertise and thus lend credibility to the data communicated. This type of analysis is prominent in rhetorical studies of Boyle's works. It has become accepted practice, for example, to talk of Boyle's vivid and circumstantial reporting style as a "literary technology" designed to turn his readers into "virtual witnesses" of the truth of the matters of fact he reported.[2] My intention in this chapter is different. For Boyle, the written report of experimental results was not merely a literary device used after reaching definitive conclusions concerning a particular area of inquiry. Writing was also part of the learning process and thus a crucial part of experimental activity itself.

As we have seen, Boyle favored an argument structure for natural philosophy based upon a concurrence of probabilities. Limited truths would be discovered by a gradual and dynamic process of inquiry. In order to achieve such a concurrence, as much information as possible would have to be collated and communicated to as many readers as possible. Writing in a clear and vivid style would help to ensure a large audience. It was also necessary to describe as many circumstances as possible surrounding the observations or experiments reported, because these readers were not witnesses but judges who would assess the reliability of the matters of fact and the intelligibility of any speculations based upon these facts. In addition, Boyle hoped that by writing in such a fashion, he would convey an excitement for experimental learning and encourage at least some of his readers to become experimentalists themselves. Thus the number of workers actively involved in natural inquiries would increase. He urged these nascent experimentalists to follow his writing style so that his vision of a truly collaborative research effort would be realized.

In his works Boyle stressed repeatedly the importance of an orderly presentation of data. In *The Sceptical Chymist,* for example, Eleutherius praises Carneades, Boyle's spokesman in the dialogue, not so much for the individual experiments that he reported as for the way that he "laid them together in such a way, and applied them to such purposes, and made such deductions from them."[3] Over twenty years later, in his *Natural History of Human Blood,* Boyle repeated this point when he noted the self-correcting nature of his surveys of various natural phenomena and observed that a "heedful collation" of data would suggest "new topics of inquiry and hints" for the composition of such histories.[4] In his collections Boyle did not merely compile a record of his own results, however; he also presented those reported by others. The work of all investigators into nature had to be collated, because

> divers particulars, which whilst they lay single and scattered among the writing of several authors were inconsiderable, when they come to be laid together in order to the same design, may oftentimes prove highly useful to physiology in their conjunction, wherein one of them may serve to prove one part or circumstance of an important truth and another to explicate another, and so all of them may conspire together to verify that saying, *Et quae non prosunt singula, multa juvant.*[5]

The production of such a collation, even for a single subject, requires a massive effort. Boyle noted that it is "a much easier task to censure experimental composures than to write such." We have seen how he emphasized the complexity and difficulty involved in the assessment of observation reports and the verification of experimental results. Concerning the final stage of writing, he once again noted the difficulty of the work and asked that his reader "believe, that sometimes a short essay of this nature, . . . may have cost me more than a whole treatise written on such a subject."[6] He would return to the topic of writing in a number of his works, but his most extended discussion took place in the "Proëmial Essay" in *Certain Physiological Essays* (1661), which he described as an "Introduction to all those [essays] written by the Author, as is necessary to be perused for the better understanding of them."[7]

Composing Experimental Essays

Boyle maintained that the "form of writing" that "we call essays" is the most appropriate style for scientific communication because it best mirrors experimental practice.[8] An essay, whose name has a dual meaning, either an active attempt to perform something or the report of that attempt, could be used as a means by which to convey the dynamic nature of his philosophy. It was a better form of communication than the systematic way of writing practiced by some of his predecessors and contemporaries, because systems tend to give the impression that the work has been completed. In addition, systems also tend to perpetuate errors. According to Boyle, those who construct systems may be tempted to fill their books with discussions of things that they have not themselves tried. The "great conveniency" of writing in experimental essays is that the writer "is not obliged to take upon him to teach others what himself does not understand, nor to write of anything but of what he thinks he can write well."[9]

The essay format also allows for a more straightforward style of reporting. In *New Experiments Physico-Mechanical* Boyle had maintained that the "chief requisites of historical composures" are "candor and truth."[10] In his "Proëmial Essay" he enlarged upon this idea. An experimenter should "write rather in a philosophical than a rhetorical strain," making the expressions "clear and significant," because the subject is of a "serious and important nature" and "our design is only to inform readers, not to delight or persuade them."[11] Boyle was moving away from earlier types of persuasion that were achieved by rhetorical argumentation or logical demonstration. Rather, he would marshal evidence to support his views and give full reports to his readers so that they could judge the acceptability of his facts and hypotheses for themselves.

Two things must be included in experimental essays. First, they "should be competently stocked with experiments," so that readers may learn about new processes.[12] Second, speculations concerning the circumstances surrounding experiments, the reasons for their performance, and their theoretical significance ought to be included. But such speculations have to be clearly labeled. Once again, candor and truth are essential. In order for natural philosophy to be "solidly established," writers should "more carefully distinguish those things, that they know, from those, that they ignore or do but think, and then explicate clearly the things they conceive they understand, acknowledge ingenuously what it is they ignore, and profess so candidly their doubts, that the

industry of intelligent persons might be set on work to make further inquiries." [13]

Because of his misgivings concerning mathematical modes of demonstration, Boyle did not use the type of presentation found in the works of Descartes
or Spinoza or that of his younger contemporary Isaac Newton, who brought
the style to its fullest expression in his *Principia mathematica*. In his optical works
Newton wrote in a more experimental style, but even though his *Opticks* has
been seen by some as the model for this way of writing, it would be misleading
to think that Boyle's writing was similar. [14] Before the publication of the *Opticks*,
Newton had presented his experimental results in a 1672 letter to the *Philosophical Transactions* "containing his New Theory about Light and Colours." He began with a simple temporal narrative of his experiments, presented in an empirical framework without the admission that any theoretical prepossessions had
influenced his work. The narrative led logically to an *"experimentum crucis"* from
which Newton apparently derived his grand conclusion that *"Light* consists of
Rays differently refrangible" in a direct and unproblematic fashion. [15] Boyle's experimental essays differed significantly from Newton's form of presentation. He
did not write his works as simple temporal narratives. Rather, he grouped phenomena together under various headings, and he explicitly stated the theories
that he had used in the design and performance of his experiments. He made
no pretense that theories would follow simply from the data presented, and as
we have seen, he was careful to include all of his results whether favorable or
not to his preferred hypotheses.

All of these details were present in his first major work, *New Experiments
Physico-Mechanical*. He began with experiments 1 and 2, which concerned the
phenomena exhibited by his engine, and then proceeded in an orderly progres-
sion to those experiments performed in order to prove the weight of the air.
He also included a number of inconclusive exploratory trials such as those
grouped together under experiment 37, concerning "a kind of light" that sometimes would appear, "almost like a faint flash of lightning," in the receiver
"when the sucker was drawn down, immediately upon the turning of the
key." He noted that these experiments had been performed "among the first,"
but he had waited until this point in his report to present them in order to
"observe as many circumstances of it as we could." [16] He also freely admitted
that he was unable to identify the cause of the phenomenon because of
the "unsuccessfulness" he experienced in his attempts to repeat it in any
consistent manner. [17]

In *Experimental History of Colours* Boyle followed a similar format. He began,
in the preface, by writing that in the following he would give "speculative
considerations and hints which perhaps may afford no despicable assistance
towards the framing of a solid and comprehensive hypothesis." [18] In part 1 he
went on to discuss some "general considerations" about color and light. He

clearly favored the notion that color is produced by a modification of light, although he also believed that the texture of the colored body was in part causally responsible for the sensation of color in humans.[19] Only after the theoretical discussion in part 1 did he proceed to a discussion of the production of black and white in part 2. Following Bacon's usage, he presented fifteen "experiments in consort" designed to show how the texture of bodies is responsible for the different capacities they have for absorbing or reflecting light, which in turn produce in us the perceptions of black and white.[20] In Boyle's last work, *Experimenta et observationes physicae*, he explicitly discussed the role of these "experiments in consort" once again and described them as those "wherein divers experiments and observations, all of them relating to the same subject or purpose, are set down together."[21] In accordance with his programmatic statements concerning the necessity for a concurrence of probabilities, Boyle did not write his essays as simple temporal narratives, nor did he present any single experimental result as crucial.

His diffidence also led him to present his results in a tentative manner, which has led one historian to speak of his "curious inability" to present a complete and systematic treatment of any one subject.[22] But it was not so much an inability as a conscious decision about style. Because of the interrelations of nature, the contingencies of experimental practice, and the poverty of the informational basis, a systematic way of writing would be premature at best. Yet Boyle did have a prescribed way of writing, and it was systematic in the sense that he sought an orderly presentation of data. Unlike a number of his contemporaries, he simply left the process open-ended. He was not seeking a reputation as an ingenious theorist but was trying to lay the foundation for a new way of doing science. He would be content if others were enabled by his efforts to complete his work. As he said, although the number of experiments, observations, and inferences that he reported were often "not conclusive, being yet experimental, the mention of them, which in a strictly logical way of reasoning must have been forborn, might well make you amends for the exercise, to which I intended they should put your reason."[23]

The inconclusiveness of an experimental essay is not a philosophical vice but a virtue. Boyle sought to popularize the style, believing that "I should thereby do real learning no trifling service, by bringing so useful a way of writing into the request it deserves."[24] The dynamic, open-ended structure of his experimental essays reflected well the nature of experimental practice in general. In addition, Boyle had at least two other reasons for defending the usefulness of such essays. First, because of their vivid and informal style, they could be used to excite the curiosity of readers and thus lead to greater involvement in experimental practice. Second, they provided the best means for both displaying and achieving the collaborative effort that Boyle maintained was crucial for the advancement of learning.

Exciting Curiosity

Given the tentative manner in which Boyle presented most of his conclusions, his statement concerning his desire to inform but not persuade his readers would seem to reflect accurately the scientific content of his works. His writing does have a persuasive force, however. At the methodological level Boyle was attempting, as Galileo had done before him, to open up the investigation of nature to a new audience. He clearly recognized the need to "rouse up" and "loudly excite, and somewhat assist, the curiosity of mankind; from which alone may be expected greater progress."[25] In his "Proëmial Essay" he discussed a number of ways in which the essay format could generate the type of excitement that he himself felt for experimental practice.

First, Boyle noted that the practice of giving full reports of experimental trials in his work would enable others to repeat the experiments, which would be crucial for the verification of their reliability. Providing his readers with copious details surrounding the design and performance of his trials would also help those just beginning their study of nature to learn how to do experiments and thus encourage them to become more deeply involved in the practice. He explained that it was for the latter reason that he would often report the details of experiments "which may seem slight and easy, and some of them obvious also." These experiments "are more easily and cheaply tried," and it would be "somewhat unkind . . . to refer you most commonly for proof of what I delivered, to such tedious, such difficult, or such intricate processes, as either you can scarce well make, unless you be already, what I desire by experiments should help to make you, a skilful chymist; or else are as difficult to be well judged, as the truth they should discover is to be discerned."[26]

Boyle's concern for teaching others the details of the experimental method is evident throughout his published works. In *The Sceptical Chymist*, for example, he noted that he wrote his experimental reports in such a way as to make them understandable to an "ordinary reader."[27] In *Experimental History of Colours* he explained that he would provide details of even his simplest experiments because his aim "was not barely to relate them, but . . . to teach a young gentleman to make them." Sometimes he would include more difficult experiments, however, as he did in his work on colors. Because "so many persons of differing conditions, and even sexes, have been curious to see" such experiments, they could serve to "gratify and excite the curious and lay perhaps a foundation" for further work.[28] The "difficult and elaborate processes" are more intriguing and challenging, and as he observed in *Experimenta et observationes physicae*, not all of his readers were novices. He thus thought that the inclusion of complex experiments in his reports was "necessary to excite and maintain the curiosity and sustain the attention of the reader" who was either advanced in study or would be interested only in such intricate processes.[29]

Boyle was not always able to give as full a report as his methodological precepts and altruistic tendencies made him desire. Sometimes he had to leave the details vague because he had learned about an experimental procedure from a tradesman whose livelihood depended upon the secrecy of the process. At other times he found it necessary to withhold details of his own processes so that he would have a store of secrets that he could trade with alchemists.[30] Even these incomplete reports could be useful for exciting his readers' curiosity. As Boyle explained, by indicating that a particular process was possible, the reports could challenge the reader to discover how to perform it. There would be "a kind of necessity laid on you to exercise your own industry, and thereby increase your experience."[31] Failed experiments could serve a similar function, because the reader would be challenged by them to discover the circumstances responsible for the contingent outcomes reported.[32]

Full reports of successful experiments remained the most useful, however. Boyle wanted to teach his readers the entire process, which included not only the design and performance of experiments but also the interpretation of results. Because experiments are costly and difficult to perform, not all readers would be in a position to repeat Boyle's experiments, but they could still contribute to the new philosophy by assessing the reliability of the circumstances surrounding his trials and by judging the coherence and adequacy of the inferences that he drew from them. As long as a full account is given of how the experiments were performed, then, even if the experimenter's opinions are "erroneously superstructed upon his experiments, yet the foundation [provided by the experiments] being solid, a more wary builder may be very much furthered by it in the erection of more judicious and consistent fabrics."[33] Indeed, it could "oftentimes" be the case that "the very experiments, that he delivers, besides that they may be applicable to many other purposes unthought of by him, may be either sufficient, or at least helpful to the very discovery of the erroneousness of the opinions they are alleged to countenance."[34]

To excite the curiosity of readers more interested in theoretical speculations, Boyle did not hesitate to include "inferences, which I discerned well enough not to be cogent," because he "was willing to exercise thereby your reasoning faculty, and try how far you would discern the tendency of several things, all of them pertinent enough to the subject in hand, but not all of them concluding to the main design, in order whereunto they were alleged."[35] Through his writing style, Boyle attempted to engage the attention of both practical experimenters and speculative theorists. He began the project at the start of his career in his "Proëmial Essay." He continued this attempt up to his last published work, *Experimenta et observationes physicae,* in which he included "conjectures and opinions, whose proofs I do not insist on," merely as "occasional things, mentioned principally to excite, and give hints to the inquisitive and sagacious."[36] Once again in his last work he reported a number of simple experiments, because they were easier for others to repeat and also because they were "more easy to be judged of as to their causes, phaenomena and effects, and consequently more fit to ground notions and reasonings upon." [37]

It is important to note that the type of "curiosity" Boyle wished to generate went beyond a mere inquisitiveness. He also used the term to connote the care that must guide such inquisitiveness. He invited his audience to read his works critically and be as diffident toward his reports as he was toward others. True diffidence requires not only that one refuse to accept opinions about natural phenomena on the basis of faith or authority but also that one not be hasty in rejecting such opinions. In *The Sceptical Chymist* Boyle discussed the type of diffidence that was appropriate and once again used the coin analogy to illustrate his meaning. [38] He asked his readers to consider the case in which "a man, more rich than skilful, should bequeath me a purse of guineas." It would be "very imprudent" to "take them all for good," since there would be "strong presumptions" that some were likely to be counterfeit. On the other hand, it would be "downright folly" to "throw them all away." The critical naturalist would proceed by the directives of "common prudence" and

> take them all out, and examine them one by one, first with the touchstone, and then, if need be, by the cupel, and by aqua fortis too: and this I should do with desire to find all the pieces true, having also care not only to preserve and put back into the purse those that prove right; but if any be but partly adulterated, to preserve the good portion by purifying it (by the cupel or some other fit way) from the falcifying alloy, by whose admixture it had been imbased. [39]

In the "Proëmial Essay" Boyle explained that diffidence was a crucial attitude for an experimental philosopher because hypotheses could be like these "counterfeit pieces of money." Some counterfeit coins may "endure . . . one proof, as the touch stone, others another, as aqua fortis, some a third, as the hammer or the scales, but none of them will endure all proofs." So also hypotheses "may agree very fairly with this or that or the other experiment; but being made too hastily, and without consulting a competent number of them, it is to be feared, that there may still after a while be found one or other (if not many) [experiments] their inconstancy with which will betray and discredit them." [40] Boyle used his works to test the acceptability of his own and others' opinions. He was critical toward all, yet he would only reject them when he was confident that he "could bring experimental objections against them." He invited his readers to share in his diffidence and "embrace or refuse opinions, as they are consonant to experiments, or clear reasons deduced thence, or at least analogous thereunto; without thinking it yet seasonable to contend very earnestly for those other opinions, which seem not yet determinable by such experiments or reasons." [41] The inculcation of an inquisitive and critical attitude in his readers could enlarge the pool of workers in natural philosophy and thus set up the conditions for a collaborative effort. But more would be required to achieve an effective collaboration.

Collaborating

Experimentalists must be critical, but they must also remain civil. Collaboration can only be achieved if certain social precepts are followed. When Boyle discussed the contingencies of experiments that could lead to failure, he noted that one of his purposes was "that they may serve for a kind of apology for sober experimental writers, in case you should not always upon trial find the experiments or observations by them delivered answer your expectations." It would be a hindrance to the growth of knowledge if naturalists were afraid that "they must expose their reputation to all the uncertainties, to which any of their experiments may chance to prove obnoxious." It would be "but a piece of equity" on the part of the reader to look for "some latent reason" why the experiment reported "does not constantly hold."[42]

Controversies must be conducted with "gentlemanly decorum." In his "Proëmial Essay" Boyle noted that when one was opposed to the report of another, the argument should be directed only to that person's opinion. The "(very much too common) practice of many, who write, as if they thought railing at a man's person, or wrangling about his words, necessary to the confutation of his opinions" is not only un-Christian but also unphilosophical. If writers think that they will be disgraced by the publication of an error, they will be discouraged from communicating their ideas. In addition, the practice is also as "unwise, as it is provoking" because it will make an enemy of the person and thus make it that much harder "to convince him of his error."[43] To achieve effective collaboration, one cannot engender ill feelings within the community of investigators. Boyle sought to promote collaboration not only in his programmatic statements concerning civility and communication but also by the numerous examples that he gave in his works of the actual collaboration that went on in his own laboratory. Although he did not always name his assistants, he made no attempt to hide the fact that a number of his works had depended upon their contributions.

Boyle's intense curiosity led him to design a large number and variety of experiments, but his time was not completely his own. His work was often interrupted by his social and business obligations and also by a constant flow of foreign visitors. As John Evelyn wrote, he "had so universal an esteeme in foraine parts" that "Travellers come over to see this Kingdom" would not "think they had seen any-thing, til they had visited Mr. Boyle."[44] He was clearly frustrated at times by these interruptions, but in part because he wished to repay the kindness that had been shown him while he was a traveler on the Continent, he remained gracious.[45] As Thomas Dent, a witness to some of these visits, wrote of Boyle, "I have known him severall mornings (when I had ye honour to wait upon him) entertain persons of severall nations viz—french,—spaniards, Germans & English & yt in different dialects, & most readily with a most pleasing aire and genious; wch was in discourse alwaies agreable & instructive to ye whole company."[46]

Boyle received his visitors, said Dent, "with a certain openness and humanity which were peculiar to him," and the accounts recorded by foreigners about meetings with him clearly support such testimony.[47] He did not merely grant them short interviews but would often spend a day or more with them discussing philosophical topics in depth and satisfying their curiosity by performing intricate pneumatic and chromatic experiments in his laboratory. His civility and graciousness reflected his desire to promote collaboration, yet given such lengthy interruptions, it is clear that while Boyle planned a vast amount of work, he would not have had the time to perform it all himself. Indeed, the actual performance of some of his "experiments to be done" can be traced to his assistants. Those recorded in a notebook concerning "things to be sealed up" in Boyle's air-pump, for instance, were done by Denis Papin between 1676 and 1679 and reported in the second continuation of *New Experiments*.[48]

Among his many assistants, Papin is one of the better known, in part because Boyle, in his preface to the second continuation, credited Papin not only with the performance of the experiments reported there but also with the actual writing of the treatise. Papin, in his advertisement to the reader, wrote that Boyle was too generous in this attribution and maintained that he was deeply indebted to Boyle, whose patronage had "afforded me both occasion of meditation, and also leisure to operate."[49] Boyle had begun the practice of crediting his assistants for their contributions years earlier, in his first edition of *New Experiments*, when he attributed the successful manufacture of his air-pump to his best-known assistant, Robert Hooke.[50] While some of his assistants may have been mere technicians or household servants employed for menial laboratory tasks, others were clearly more like apprentices or associates.

Peter Staehl, the chemist that Boyle brought to Oxford, served for a time as his assistant and, like Hooke, went on to become an operator for the Royal Society.[51] Frederick Slare, who assisted Boyle in his investigations on phosphorus, became a fellow of the Royal Society and published a number of papers on the substance in the *Philosophical Transactions*.[52] Another assistant who worked with Boyle on the investigation of phosphorus, Ambrose Godfrey Hanckwitz, also became a fellow of the Royal Society. Godfrey, who dropped his German surname once he had settled in England, either out of gratitude or in hope of future patronage, christened his firstborn son Boyle Godfrey.[53] He also set up a successful chemical firm for the commercial manufacture of phosphorus and years after his patron's death, wrote: "I am indebted for the first Hints of the matter whence it [phosphorus] is made, to that Ornament of the *English* Nation the great Mr. Boyle, my kind Master, and the generous Promoter of my Fortune, who's Memory shall ever be dear to me."[54] In Boyle's published works on phosphorus in both its liquid (aerial noctiluca) and consistent (icy noctiluca) forms, one can gain the most insight into the type of relationship that he had with his assistants and how the experimental strategies he suggested in his programmatic works were employed in his laboratory.

As he reported in his essay "The Aerial Noctiluca," Boyle first learned about the possibility of creating phosphorescent materials from a foreign visitor, Johann Daniel Kraft, who in 1677 had shown some samples of it to Charles II in Boyle's presence.[55] Before Kraft left England, he visited Boyle's home and gave him a hint about the material that was used in the preparation of the samples. Boyle immediately set out to have some made in his own laboratory. After a period of trial and error, he hit upon a successful recipe for liquid phosphorus. At this point it became the task of his assistant to prepare good samples that could be used for exploratory trials. Although Boyle did not give the name of his "venturous laborant," it was most likely either Slare, who had been present at Kraft's last visit to Boyle, or Godfrey, who would later be instrumental in perfecting the production of consistent phosphorus.[56]

Boyle was confident that his assistant possessed the competence not only to produce good samples of the substance but also to bring any unusual circumstances to his attention. He reported, for example, that once after the "distillation of our luciferous matter had been freshly made," his assistant, "being in some haste" to wipe the phosphorus off a knife that he had used to place the substance in a vial, drew "the blade nimbly" through his apron, whereupon "he was much surprized to feel a smart heat; and presently looking upon that part of the apron, where it had been produced, perceived, that it had in it two holes of some bigness, which he concluded must have been produced there by burning, both because of the intense heat he had felt before, and because it was a new apron; which when I had called for, and heedfully inspected, I did, with him, impute those holes to the action of the fire."[57] This observation was significant because originally Boyle had believed that phosphorus, as a "light without heat," was nonflammable and thus could be used as a safe source of illumination in places, such as gunpowder rooms, where candles and torches often proved dangerous.[58] But he was led to doubt this assumption by observations occasioned by the accidents that frequently befell his assistant.

A number of these accidents occurred while the assistant was handling the "icy noctiluca," so called because it was normally cold to the touch.[59] As Boyle reported, his "venturous laborant found several times, to his no small pain," that his fingers would be "almost covered with blisters raised on them, by handling our shining matter with too bold a curiosity; and he complained to me, that, though according to the usual fate of chymists, he had been often burned on other occasions, yet he found blisters, excited by the phosphorus, more painful than others." On another occasion, his assistant performed a probatory experiment in order to determine if a newly prepared sample of phosphorus "was good and fit to be brought to" Boyle for his inspection. The assistant wrote some letters with it "upon a piece of plank" and "found to his surprise, that he had not only shining but burning letters." Many other instances of burned hands and clothing led Boyle to speculate about the flammable properties of the substance. He thus designed experiments to be made directly on gunpowder to see if it would ignite. His assistant's bold, almost reckless curiosity once again put him in danger. After the first experiment with gunpowder failed, Boyle "left the care of reiterating the experiment in my presence" to his assistant, who, "presuming it would not succeed, scrupled not to hold his head over it, that he might the better see what change was made in the mixture; but then upon a sudden the powder took fire, and the flame shooting up, caught hold of his hair, which made a blaze, that proving innocent enough, became more diverting than the smell of the smoke, that succeeded it, was delightful."[60]

Despite these numerous accidents, his assistant was never seriously harmed, and because of Boyle's understanding of the contingencies of experiments, the accidents did not diminish his confidence in his assistant's competence. Because his assistant had the skill to prepare phosphorus and took care to report any unusual occurrences to him, Boyle was free to concentrate on his primary interest: making observations and experiments designed to discover the unknown properties of the newly invented substance. He often carried a vial of liquid phosphorus with him during the day in order to make observations about the length of time that the substance would emit light on its own, and because he reasoned that the light would be more easily seen when his eyes had been "long kept in the dark (whereby the pupil uses to be much opened, and consequently capable of admitting more numerous beams)," he often brought the vial into his bedchamber. On one such occasion he noted that when he "was awake some time before break of day, I enclosed both the glass and my head between the sheets" and observed that "the light seemed to me to be very considerable, and to enlighten the compass of a foot or more in diameter, and probably would have diffused itself further, if it had not been bounded by the sheets."[61]

Boyle designed other trials, "to try by taste," for example, "whether our shining liquor" was composed of acidic or alkaline particles and "to try . . . what the air and agitation would do, towards the kindling or exciting" of the light.[62] In some of these trials, he found himself to be in as much danger as his assistant had been. He recorded one instance of an accident he had while performing an exploratory experiment concerning the flammability of consistent phosphorus. He put some of the substance on a piece of paper and found, to his surprise, that after "rubbing the paper between my hands . . . it did on a sudden take fire, and blazed out, so that it would have burned my hand, if this had not been kept from receiving much harm (for all it did not escape) by a thin glove, that was thereby scorched, and in part shrivelled up."[63]

Although Boyle often entrusted probatory trials to his assistants, his laboratory was not set up according to any strict division of labor. He sought to involve the assistants in more interesting work: "to incourage" his "industrious laborant," for example, he gave him "for his own use, some fragments of our icy noctiluca."[64] Boyle recorded one of his assistant's exploratory trials, in which he had "mingled a portion of this shining substance with a spirituous medicinal liquor, that he had prepared, by extracting several drugs with it," and observed "a phaenomenon, at which being surprized, he came to acquaint me with it, bringing me withal some of the liquor." According to Boyle, it was through his assistant's ingenuity that he first "had a hint" for the performance of a series of experiments, reported immediately after this account, designed to discover the extent to which the phosphorescent properties of his substance could be communicated "to a contiguous liquor."[65]

It was only from his own and his assistants' sometimes reckless curiosity that Boyle was able to observe the properties of phosphorus and gain a better understanding of the nature of this unfamiliar substance.[66] He was seldom alone in his laboratory. In his published reports the work of his servants, "laborants," and assistants is constantly visible.[67] Through his writing he sought to encourage the involvement of others in the experimental enterprise and to convey the collaborative effort that would be required to overcome the paucity of the informational basis and thus lay the foundation for an improved understanding of all aspects of nature.

We have seen Boyle's advice on how experimentalists ought to proceed in their study of nature along with some practical examples of the ways in which he conducted his own study. Observations must be collected and assessed, experiments must be performed and their reliability validated, and finally results must be collated, with any concurrences noted and variations reconciled, and then communicated in published form with as many details as possible. To illustrate how these three experimental activities were incorporated into Boyle's works, I will examine one of his most extensive, but rarely discussed, studies: his investigations into the nature of cold, which, unlike phosphorus, was an exceedingly familiar phenomenon.[68]

An Experimental History of Cold

The discovery of the cause of and the various effects produced by cold was one of Boyle's lifelong areas of inquiry. His *New Experiments and Observations*

Touching Cold; or, An Experimental History of Cold Begun (1665) was the longest work published by him at one time.[69] It was followed, almost ten years later, by "A Sceptical Dialogue about the Positive or Privative Nature of Cold," wherein he examined the adequacy of the various theories that had been proposed concerning cold in light of the experiments and observations he had compiled in his *History*.[70] In addition, Boyle made many references to cold throughout his other works, beginning with *New Experiments Physico-Mechanical*, continuing through his investigations of the causal powers of bodies in such works as *The Origin of Forms and Qualities* and *The Mechanical Origin of Qualities*, and finally ending in his posthumously published *History of Air*. Boyle considered the discovery of the causal properties of cold an important undertaking, and in his preface to the *History* he discussed the many motives that had led him to pursue his investigation.

Despite the familiarity of cold, Boyle noted that the study of it had "hitherto been almost totally neglected." His first task was to compile a history of cold in order to "repair the omissions of mankind's curiosity towards a subject so considerable." Not only was the inquiry into the causes and effects of cold a study important in its own right, but because of the interrelation of natural processes, it would also contribute to an increase of knowledge about other aspects of nature. Because cold is one of the most "eminent and diffused" attributes of the air, a better understanding of its nature could aid pneumatic investigations, for example, and our understanding of heat might be improved, "since . . . confronted opposites give themselves mutual illustration."[71] Experimental methodology itself could also be improved. It was possible that chemical separations might be able to be made by cold as well as by heat.[72] And cold might also provide a means to compress air without the use of an engine.[73]

In his preface Boyle also discussed a number of things about his manner of presentation. He admitted that his "method" was "not exact" in the main body of the work, where he listed experiments and observations under twenty-one different categories by title, because of the comprehensiveness of the subject and his belief that "future discoveries and hints might make it" even "more comprehensive." As he explained, he did not wish "by a confinement to a strict method" to "discourage others from continuing the history, by adding new titles to those twenty-one, I have treated of." The titles were also "unequal" in length because he had intended not to "write a complete and regular treatise" but rather to set down "matters of fact." Thus "the length of each section was to be determined, not by its proportion to that which went before it, or followed after it, but by the number and condition of the particulars, that were to compose it." While his primary concern was the communication of factual information, he once again noted that he would also mention those "conjectures" that occurred to him along the way because they "might more conduce" to "promote the history of cold."[74]

Because of the massiveness and comprehensiveness of this work, it will not be possible for us to explore all of the issues raised by Boyle. Although somewhat tedious, the following short synopsis of the history will provide a helpful illustration of the character of the work, particularly the scope of its contents and the scrupulous manner of its presentation. After the author's preface, Boyle inserted an essay entitled "New Thermometrical Experiments and Thoughts," about the problems encountered in attempts to make accurate measurements of the degree of cold.[75] This was followed by the main history, in which numbered groups of observations, experiments, and considerations were ranged under the twenty-one titles. The numbers in brackets after each entry below indicate the number of groups recorded there.

   I. Experiments touching bodies capable of freezing others. [23]
   II. Experiments and observations touching bodies disposed to be frozen. [5]
   III. Experiments touching bodies indisposed to be frozen. [6]
   IV. Experiments and observations touching the degrees of cold in several bodies. [13]
   V. Experiments touching the tendency of cold upwards or downwards. [8]
   VI. Experiments and observations touching the preservation and destruction of bodies . . . by cold. [28, plus appendix]
   VII. Experiments touching the expansion of water, and aqueous liquors, by freezing. [10]
   VIII. Experiments touching the contraction of liquors by cold. [7]
   IX. Experiments in consort, touching the bubbles, from which the levity of ice is supposed to proceed. [20]
   X. Experiments about the measure of the expansion and the contraction of liquors by cold. [12]
   XI. Experiments touching the expansive force of freezing water. [12]
   XII. Experiments touching a new way of estimating the expansive force of congelation, and of highly compressing air without engines. [11]
   XIII. Experiments and observations touching the sphere of activity of cold. [9]
   XIV. Experiments touching the differing Mediums, through which cold may be diffused. [11]
   XV. Experiments and observations touching ice. [11, plus 4 passages appended from travelers' journals]
   XVI. Experiments and observations touching the duration of ice and snow, and the destroying them by the air and several liquors. [7, plus appendix]

XVII. Considerations and experiments touching the *Primum Frigidum*. [32]
XVIII. Experiments and observations touching the coldness and temperature of the air. [24]
XIX. Of the strange effects of cold. [30, including quoted passages from books and letters of travelers]
XX. Experiments touching the weight of bodies frozen and unfrozen. [14, plus appendix]
XXI. Promiscuous experiments and observations concerning cold. [7, plus appendix][76]

The reports presented under these titles comprise more than one hundred pages in Thomas Birch's edition of Boyle's works. They are followed by an appendix, belonging to title 17, which includes speculations about whether a particular frigorific agent is responsible for cold, and a "postscript," which raises further theoretical speculations concerning the reports made in the other sections.[77] After these, an essay entitled "An Examen of Antiperistasis" is included, along with another "postscript," called "A Sceptical Consideration of the Heat of Cellars in Winter and Their Coldness in Summer," and "An Examen of Mr. Hobbes's Doctrine Touching Cold."[78] The final piece is "An Account of Freezing," which was composed by Christopher Merret concerning observations that he made in December 1662 and January 1662/63.[79] Throughout this work, observations, experiments, and speculations are intertwined in a complex manner.

Boyle said that his design was to present the factual element of his history in a "plain and simple way," yet it is clear that there was nothing simple about the way in which the observations were gathered and collated.[80] Some of the observations about cold were Boyle's own. In order to have enough information upon which to theorize, however, he required an account of the effects that cold produces in all circumstances, and London was not the ideal location from which to compile a full survey. Boyle was particularly interested in the unusual effects that extreme degrees of cold could produce, since, as he had learned in his chemical work, the degree of heat was often an important factor in the effects produced. Because of this, he actively sought "reports from travellers and navigators, into those cold regions, that afford much considerabler, or at least much stranger observations concerning ice, than are to be met with in so temperate climate as ours."[81]

He grouped a number of reports about oddities under title 19, but a number of others were scattered throughout the work, because such reports "may afford the ingenious such strange phaenomena that the explication of them may serve both to exercise their wits, and try their hypotheses."[82] He had to rely upon the observations made by others for this part of the historical work, but he wished to include only "credible testimony," which meant that the witnesses had to have the requisite skill and experience to make accurate observations.[83] Among the witnesses that Boyle found credible were Dr. Fletcher, a physician and ambassador to Moscow, upon whom he relied for reports on frostbite; Captain James, a well-known navigator, from whom Boyle learned about the activity of icebergs; and a "most experienced mason," who gave Boyle accounts about frozen stone.[84]

Also, the reports themselves had to be assessed for credibility on the basis of their concurrence with other reports. In his examination of the sphere of activity of cold in title 13, Boyle recorded a report from "a very learned friend" who claimed to have felt the cold from an iceberg at a distance, but he noted that the report was unlikely to be trustworthy because of a vast amount of evidence to the contrary from navigators who had been taken by surprise by icebergs hidden in fogs.[85] Boyle expressed his reservations about the reports of another odd phenomenon that he had received from more than one witness concerning the crumbling of bricks during a Moscow winter. He did not doubt the truthfulness of these reports, but he did question whether cold was the true cause of the phenomenon and suggested that it might rather be the effect of a fault in the manufacture of the bricks. Yet he did not entirely reject the idea that cold was the cause, since he had observed that other substances, such as marble, would sometimes crack in cold weather.[86] As he explained elsewhere, such odd reports could be "consistent with the laws of nature" and "appear improbable, only, because of the intense degree of cold, that they suppose."[87]

Most of the observations that Boyle reported were those that were made about cold's effects on inanimate bodies and not about the observers' felt perceptions of cold, because from the outset he had realized that cold could be considered either sensibly or "philosophically."[88] The "unheeded dispositions" that "our own bodies may have in the estimates we make of the degree of cold" meant that our senses could not be relied upon as a standard for determining its presence.[89] Oftentimes things that feel cold to the human sense, such as warm air forced through a bellows, or tepid water to a heated hand, are not strictly cold in a philosophical sense. He admitted that it "may to most men appear a work of needless curiosity, or superfluous diligence, to examine sollicitously, by what criterion or way of estimate the coldness of bodies, and the degrees of it are to be judged." He argued in response that "we can scarce imploy too much care and diligence in the examining of those touchstones, which we are to examine many other things by."[90] Because of the problems associated with our tactile sense, Boyle wrote that

> though it be true, that cold in its primary and most obvious notion be a
> thing relative to our organs of feeling, yet since it has also notable operations on divers other bodies besides ours; and since some of them seem

more sensible of its changes, and others are less uncertainly affected by them, it would be expedient to take in the effects of cold upon other bodies, in the estimates we make of the degrees of it.[91]

In the investigation of cold the "testimony of our senses" may "easily and much delude us."[92] Observing the effects of cold on other bodies could act as a check on human perceptions of cold, but even then problems could arise, because our senses only receive impressions. They do "not perceive what is the cause or manner of these impressions."[93] Boyle wanted to know what effect cold had on metal, for example, because it was a well-known fact that its contrary, heat, had an appreciable effect. Some observations, such as those concerning how the iron hoops around water barrels would sometimes break in very cold weather, were not reliable indicators about the power of cold, however, because one could question whether this phenomenon was a direct effect of the cold or merely concurrent with the presence of cold.[94] Indeed, even the cause of the formation of ice in a vessel of water surrounded by snow was open to question. All that "is clear in matter of fact" is that once the vessel has been placed in such a position, "within a while after, this water begins to be turned into ice," but the cause of this effect "is not perceived by sense but concluded by ratiocination."[95]

In order to discover which effects are actually produced by cold, artificial trials would have to be performed, and the best way to avoid errors from our tactile sense would be to use instruments such as thermometers.[96] But man-made instruments are themselves fallible. In "New Thermometrical Experiments and Thoughts" Boyle began with the "paradox" that "not only our Senses, but common Weather-glasses, may misinform us about Cold."[97] He discussed seven different types of weather-glasses (thermometers) in all and noted how many circumstances, such as the temperature of the liquid enclosed, its disposition to freeze, and the pressure of the air, could affect the accuracy of readings.[98] Because of these problems, experimenters "are not rashly to rely upon the information" of these instruments. As Boyle explained, "Though they be an excellent invention, and their informations, in many cases, preferable to those of our senses . . . yet I fear they have too much ascribed to them, when they are looked upon as such exact instruments to measure heat and cold by, that we neither can have, nor need desire any better." To show how contingencies could arise from the use of these instruments, Boyle discussed how the "differing gravities of the atmosphere may, as well as heat and cold, have an interest in the rising and falling of the liquor in common weather-glasses." He reported a series of experiments that he had performed with two weather-glasses, one sealed and one open. He noted that they did not always give the same reading and that he could "by comparing the two weather-glasses together . . . usually foretell, whether the mercury in the Torricellian tube . . .

were risen or fallen, and consequently whether the external air was heavier or lighter than before."[99]

In his second discourse of this essay, entitled "Observations about the Deficiencies of Weather-Glasses, Together with Some Considerations Touching the New or Hermetical Thermometers," Boyle argued that there was a need to have "a standard, or certain measure of cold, as we have settled standards for weight, and magnitude, and time." He suggested that the use of sealed thermometers would provide a better means for achieving such a standard, because the liquid within them would not spill or evaporate over time and would be protected from the effects of the pressure of the air.[100] Despite these advantages, however, he doubted whether an accurate measuring instrument could be constructed until the nature of cold was more fully explicated. "Even the particular nature of the liquors, employed in weather-glasses is not altogether to be neglected, till we have a better and more determinate theory of the causes of cold."[101]

Although the instruments were more reliable than human sense, Boyle noted that in the end he "would have a philosopher look upon both these and our sensories, but as instruments to be employed by his reason, when he makes estimates of the coldness of bodies." He had discussed these problems not to discourage his readers but to emphasize the recurrent theme in all his writings about the need for a number and variety of experimental results. As he said, these difficulties were presented so that the reader would "consider, whether it would be amiss to take in . . . as many collateral experiments and observations, as conveniently we can, to be made use of, as well as our sensories and weather-glasses in the dijudications of cold." He repeated this point in his summary of the discourse, warning his readers that "it may be prudent . . . to make use also of other ways of examining the coldness of bodies, that the concurrence or variance to be met with in such ways of examination, may either confirm the testimony of the weather-glass, or excite or assist us to a further and severer inquiry."[102]

Because there was as yet no standard for judging the presence and degrees of cold, Boyle had to perform many probatory experiments in his investigation. As one can see in his twenty-one titles, Boyle also performed a large number of exploratory experiments. Both types of experiments were necessary because, while familiar, the phenomenon of cold had not previously been studied in any serious, organized manner. There were, however, numerous notions about the subject that had come to be accepted as fact by the authority of tradition or as conclusions drawn from theoretical beliefs about other natural processes. The general and unquestioning acceptance of this traditional knowledge hindered the study of cold. Boyle noted, for example, that if he had believed the Aristotelian arguments against the possibility of a vacuum, he would not have attempted the design, or sought the manufacture, of a sealed thermometer.[103]

It was also an accepted "fact" in Boyle's day that hot water would freeze sooner than cold, and the Aristotelians had constructed an elaborate theory of antiperistasis to explain it.[104] In "An Examen of Antiperistasis" Boyle showed that the testing of so-called matters of fact upon which theories were based was a crucial stage in the elimination of faulty theories. Through a series of simple experiments, in which he set out vessels filled with water of varying temperatures—cold, tepid, and hot—he was able to show that the temperature of the water did not affect the rate at which it would freeze. Because the theory of antiperistasis explained an effect that did not actually obtain in nature, the theory itself was therefore worthless. As Boyle said, "We may see, how well bestowed their labour hath been, that have puzzled themselves and others, to give the reason of a phaenomenon, which perhaps with half the pains they might have found to be but chimerical."[105] Some other "facts" that had received general acceptance among various segments of Boyle's audience had more basis in experience, yet Boyle was often able to show that they also were chimerical because they had been based upon too few observations.

If experiments are not repeated, there is a danger that a mere chance occurrence will be taken as confirmation of a favored theory. This had apparently been the case with some of Boyle's contemporaries, who affirmed that if a solution "made of the ashes . . . of a burned plant be frozen, there will appear in the ice the idea [image] of the same plant."[106] Boyle noted that it is "not impossible" that "among the many figures" found in frozen liquors "there may appear something so like this or that plant, that, being looked upon with the favorable eye of a prepossessed beholder, it may seem to exhibit the picture of the calcined vegetable." When he tried the experiment himself with a solution made from grapes, he was "surprized" to find in the ice "divers little figures, which were indeed so like to vines." But not resting content with this result, he thawed the ice "for further trial sake" and then refroze it "in the same phial, and after the same manner." He "could not discern, in the second ice, anything like that, which we had admired in the first."[107]

Boyle continually emphasized the need for repetition to validate the results of experimental trials. Because some of his experiments were so strange, he often reported "several trials of the same thing, that they might naturally support and confirm one another."[108] He preferred those cases where trials had been repeated by independent experimenters, however, and he welcomed the opportunity to append those made by Christopher Merret to his *History of Cold*.[109] Merret's observations on "freezing bodies" were significant. Before one can have an adequate account of the nature of cold, one must first determine, among other things, which bodies are capable of having the quality produced in them. Repetitions carried out independently by others can serve as corroboration, because when the results "are the same, it is like the truth will be more confirmed." Even if the results are not the same, at the very least reports of repeated trials might increase the readers' curiosity. As Boyle noted, "In cases where the successes are very differing, the reader will be excited to make further trials" in order to judge which results may be trusted.[110]

Some popular theories about cold and its productions were eliminated when Boyle's experiments showed that they were not founded upon a reliable factual basis. Theoretical and factual work has to be done in unison. One must critically compare both types of knowledge and through a gradual process of reconciliation eliminate those elements found to be faulty and retain those that have passed the tests. This was the type of work that Boyle performed when he attempted to discover the cause for the breakage of the iron hoops around water barrels. Boyle thought it likely that the hoops broke from the expansive force that the enclosed water exerted on the sides of the barrel when it was frozen into ice. But as he noted, a number of preliminary inquiries would have to be made before one could conclude that this alternative account was acceptable.

To begin, it was necessary to show, in opposition to the predominant Aristotelian theory, that water expands when frozen.[111] The obvious thing to do would be to place the water to be frozen in a measured container, and then observe whether its volume increased or decreased when it turned to ice. Even this simple case would require a number of experiments skillfully designed and carefully performed. Boyle noted that glass vessels would have to be used, because if any other material, such as metal or pottery, was used, then an opponent of expansion could argue that any increase of volume was actually caused by the addition of particles that passed from the vessel to the water. Also, it would be best if the glass vessels were sealed, so that an opponent could not argue that the addition of air had been the cause of the increased volume.[112]

To perform the experiment, Boyle had glass vials constructed in such a way that they could be sealed after the water was added. But then, he noted, if he allowed the water in the glass vessel to be frozen in the common way by exposing it to the air, "it is odds" that "the freezing water would break the glass and so spoil the experiment." To proceed rationally, a more artificial manner of investigation had to be devised. In Boyle's analysis, the glass broke because in "inartificial glaciation" the ice formed first at the top of the vessel, and as the water further down was frozen it met resistance from the top layer of ice and thus expanded in the direction of least resistance, which was into the glass. Boyle altered "the common way of freezing" by immersing the bottom of the vial in a mixture of snow and salt. By having the ice form from the bottom up, he was able to prevent the glass from breaking and to show that ice was indeed an expanded state of water.[113]

Because of the many control conditions in place, Boyle concluded that it was "abundantly confirmed by divers of the experiments" that water expands when frozen.[114] But his work was not complete. After proving expansion, he then had to measure its quantity in order to determine whether the pressure exerted by the expansion would be sufficient to break not only glass vessels but also the much stronger iron hoops. He did so by reverting to the common way of freezing. Using strong vessels of metal and pottery filled with water, he placed weights on the top surfaces as they became frozen and thus calculated the pressure exerted. From this series of experiments, he concluded that the breakage of the iron hoops was a result of the expansive force of the ice and not a direct effect of cold, although the brittleness produced in the metal would be a concurring condition of the effect.[115]

Most of Boyle's work was designed to establish a factual foundation for the phenomenon of cold, but he did not discount the use of theory in this project. As he said, "Experiments and notions will reciprocally direct to one another."[116] His ultimate goal was to find the theory that could best account for all of the information that he had compiled, and his speculations filled a large part of the *History*, especially the appendixes and postscripts, and were central to his later "Sceptical Dialogue." He was able to eliminate a number of the prevalent theories of his day by a strict adherence to the requirement that a theory be consistent with all relevant matters of fact.[117] In "An Examen of Mr. Hobbes's Doctrine Touching Cold," for example, Boyle proceeded in this way when he noted that "this learned author [Hobbes] has past by far the greatest part even of the more obvious phaenomena of cold, without attempting to explicate them, or so much as shewing in a general way, that he had considered them, and thought them explicable by his hypothesis."[118]

In *De corpore* Hobbes had presented a mechanical explanation of the production of cold and ice, both of which he attributed to a "constant wind" that pressed upon bodies.[119] Boyle found his account to be "partly precarious, partly insufficient, and partly scarce intelligible." Among the numerous criticisms to which it was subject, Boyle noted that not only was it internally inconsistent with some of Hobbes's other mechanical explications, especially those in which he employed his plenist theory, but it was also based upon superficial observations concerning the felt perceptions of cold.[120] As Boyle explained, Hobbes maintained that his theory could explain "why a man can blow hot or cold with the same breath" depending upon the motion that the breath is put into, apparently without realizing that "in estimating the temper of the produced wind, our senses may impose upon us."[121] To show the precariousness of Hobbes's use of such superficial observations, Boyle recorded a simple experiment that he had performed before witnesses:

> Having taken a very good and tender sealed weather-glass, and blown upon it through a glass pipe . . . that was chosen slender, to be sure, that my breath should issue out in a small stream; by this wind beating upon the ball of the weather-glass, I could not make the included spirit of wine

subside, but manifestly, though not much ascend, though the wind, that I presently blew through the same pipe, seemed sensibly cold, both to the hands of by-standers, and to mine own.[122]

Boyle also found Hobbes's account "of turning water into ice . . . the most unsatisfactory I have ever yet met with." According to Hobbes, when wind blows upon a fluid body, it raises the parts of it in such a way that the uppermost parts become pressed together and thus "coagulated."[123] This account was unintelligible, since, as Boyle pointed out, Hobbes had given no reason to suppose that a wind could cause such a coagulation and Boyle had found that "though you long blow upon a glass of water with a pair of bellows, where there is not an imaginary wind, as Mr. *Hobbes's*, but a real and manifest one; yet the water will be so far from being frozen, that our formerly mentioned experiments (of blowing upon thermometers) make it probable, that it will scarce be cooled."[124] In addition, the experiments in which Boyle had produced ice in "hermetically sealed" glass vessels could not be explicated by Hobbes's doctrine unless he assumed, which he would not, that glass is pervious to the air.[125]

Among the other theories that Boyle examined, the most common were those that postulated a *primum frigidum*, some physical agent that by its natural coldness is able to produce cold in other bodies.[126] Because ice takes up a greater volume than the water from which it is formed, *primum frigidum* theories gained credence. It seemed likely that something had been added to the water in its frozen state. Yet when Boyle looked at the particular candidates that had been proposed for the *primum frigidum*, he found that the evidence for them was not strong enough to warrant his assent.

Gassendi, for example, had postulated nitre as the *primum frigidum*. In response, Boyle maintained that Gassendi had failed to deliver an account of his experiments in enough detail to allow others to judge the nature of his support, and he also suspected that Gassendi may have been "imposed upon by others to write that, as matter of fact, which he never tried." In addition, the details that he did provide were not sufficient to confirm his theory. Boyle noted that Gassendi presented the case of ice forming in a vessel surrounded by snow mixed with nitre as confirmation that nitre was the cause of cold. Gassendi had been misled in this conclusion, however, because he had failed to vary the experimental circumstances. Boyle had found that other substances, such as "sea-salt" or the "inflammable part of wine," would produce the same result when mixed with snow, which showed that nitre could not be the causal variable responsible for the experimental effect.[127]

General objections could also be made to such theories. The experiments using sealed glasses, for example, made it doubtful that anything had been added to the water, and most *primum frigidum* theorists assumed that the increased volume of ice indicated a corresponding increase in weight. In his experimental measurements, however, Boyle was not able to detect any such weight increase.[128] Because of these general objections, Boyle stated his preference for the Baconian notion that cold resulted from a privation of motion, and he argued for the superiority of this theory in his "Sceptical Dialogue." Boyle preferred mechanical accounts for the origin of qualities. The privation theory did not require the postulation of any mysterious element; it was in agreement with the mechanical account of heat, its contrary; and it could account for most of Boyle's experimental results.[129] But some oddities in Boyle's experimental results could not be readily reconciled, such as the result that he recorded describing how two liquids when mixed together would boil and yet produce cold.[130] In addition, an alternative *primum frigidum* theory could not be as easily eliminated as others had been because it did not equate the agent with a particular material substance but instead with "frigorifick corpuscles."[131]

In "A Sceptical Dialogue" Boyle's spokesman, Carneades, noted that he would not be able to prove that "cold must be a privation," but he could "shew that the arguments brought to evince it to be a positive one, are not concluding."[132] Because ice does not weigh more than the water from which it is made, for example, defenders of frigorific corpuscles would have to postulate that their particles lacked specific gravity. It would also have to be assumed that the particles had the capacity to permeate the glass that was used in the hermetically sealed trials. But other acceptable entities, such as the rays of light and the effluvia of lodestones, seemed to possess such properties.[133] Therefore, while Boyle clearly stated his preference, he had to admit that, because there was a viable alternative to his account, the privation theory had not been demonstrated beyond a reasonable doubt. His research had come full circle. He had failed to reach a definitive conclusion, but he had narrowed the range of possibilities. It would be left to "men's future industry" to build an investigation upon his initial round of inquiries.[134] As always, Boyle's work was only a beginning. Before a full understanding of the nature of cold and its effects could be achieved, more data would be required, and the procurement of such data would depend upon future improvements in technology and upon a better theoretical understanding of all natural processes.[135]

CONCLUSION:
THE EXPERIMENTAL PROCESS

BOYLE'S *History of Cold* provides an excellent illustration of his comprehensive approach toward the study of nature. Within his history he included accounts of the different effects that cold may have on numerous types of bodies, the various natural and artificial ways by which cold may be produced, and the many extant theoretical explanations concerning the production of cold. Yet at the end of his exhaustive history, as well as in his later "Sceptical Dialogue," Boyle reached no definitive conclusions. Indeed, the best he could offer was a discussion of the ways in which various theories, his own included, were faulty. As we have seen, Boyle's extreme eclecticism led him to publish works in many different areas, while his extreme caution led him to leave most of these studies in an unfinished state. It is not surprising that a number of his contemporaries, as well as later commentators, reacted negatively to his approach.

Hobbes, for example, was highly critical of what he perceived to be Boyle's failure to use the principles of the mechanical philosophy to provide a fundamental explanation of his pneumatic trials.[1] In correspondence with Huygens around the time of Boyle's death, Leibniz expressed his disappointment concerning Boyle's refusal to construct a comprehensive chemical theory. He attributed this fault to Boyle's overly cautious attitude when he noted that Boyle "est peut-estre trop reservé."[2] A century later William Brande, in his article "The Progress of Chemical Philosophy" for the *Encyclopaedia Britannica*, also found reason to criticize Boyle. Unlike Leibniz, Brande attributed Boyle's failure to provide a new foundation for chemistry to his overly eclectic approach. He wrote that Boyle's "experiments are too miscellaneous and desultory to have afforded either brilliant or useful results" and explained that Boyle had been "too fond of mechanical philosophy to shine in Chemistry, and gave too much time and attention to theological and metaphysical controversy to attain any excellence in either of the former studies."[3]

From the standpoint of modern science, Boyle's eclecticism and caution do seem a bit out of place. It would be accurate to say that Boyle was not a great scientist in the tradition of Newton, because he never provided a systematic theoretical account for any of the areas that he studied. Boyle's methodological achievements were significant, however, and they had lasting influence on the generations of scientists that followed.[4] Although his cautious and eclectic approach to the study of nature was viewed by some as a reason for his failure at the theoretical level, others saw these same traits as reason for his success at the methodological level. In 1696 John Evelyn praised Boyle for being a "great and happy analyzer" and a "generous and free philosopher" who was "addicted to no particular sect."[5] A few years later Eustace Budgell repeated Peter Shaw's assessment of Boyle's achievements when he wrote that Boyle had *"animated* philosophy; and put in *Action* what before was little better than a *speculative* Science."[6]

Boyle sought to establish the experimental philosophy as a free and active mode of inquiry. He made a conscious decision to avoid what he perceived to be the premature theoretical speculations of a number of his contemporaries because he believed that such systems would leave readers with the impression that there was no more work left to be done in natural philosophy and thus would close the door to future lines of inquiry. Boyle's eclecticism was more than an idiosyncratic personality trait.[7] Because he wished to encourage as many members of society as possible to become actively involved in experimental practice, he resisted advocating any particular philosophical system. In order to keep inquiry free and open, he also rejected the practice of beginning the study of nature with a set of first principles. Instead, he followed the broadly Baconian practice of beginning with the establishment of facts and then rising gradually to intermediate causal explanations, without requiring an ultimate theoretical description of the mechanisms involved in these causal processes. Boyle believed that his account of the springlike quality of the air, for example, provided a causal explanation for the effects produced in his air-pump trials. The spring, as an intermediate causal process, had explanatory power, even though Boyle was unable to provide a theoretical account of the mechanisms that caused the air to have such a quality.

The way in which Boyle established the spring of the air provides an example of his active conception of knowledge. Boyle insisted that one can learn about the way in which natural effects are produced by consistently reproducing these effects in the controlled conditions of the laboratory. As Antonio Pérez-Ramos has argued, this aspect of Boyle's philosophy reflects the makers' knowledge tradition employed by Bacon for the justification of experimental practice. The usefulness of the knowledge gained through experimental manipulation is not limited solely to utilitarian concerns or to the ways in which experimental results may be used to test theoretical claims. Numerous properties of bodies can be discovered directly through the active manipulation of them, and such properties can therefore be known with a type of "operational certainty."[8]

The origins of Boyle's free and active conception of knowledge acquisition can thus be traced to the influence of Bacon. The way in which Boyle built upon Bacon's precepts, however, resulted in an experimental philosophy unique for its cautious and eclectic approach. Boyle's eclecticism and caution were products of two elements integral to his practice: his development of and strict adherence to a strong standard of proof for theoretical and factual claims and his prolonged experience of the numerous problems involved in experimental activity.

Boyle's Philosophy of Experiment

Boyle described his notion of experimental proof as a concurrence of probabilities that occurs only when a "judgment of reason" based upon "the most information procurable, that is pertinent to the things under consideration" has been achieved.[9] We saw in chapter 3 that Boyle believed that the justification of Harvey's theory of the circulation of the blood provided a successful instance of this mode of proof. In his account, Boyle stressed the point that Harvey's theory could not be proved true by one or two crucial experiments. Rather, a number of "particulars" had to be gathered from a variety of trials. When considered singly, these particulars would be merely probable signs of the truth of the theory, but when combined, they could be "attributed to the truth of what they jointly tend to prove."[10] Boyle did not believe that concurrence would be so easily achieved in other areas, however. His theological beliefs led him to have a conception of the world as a vast interrelated mechanism, the parts of which are related to the complex whole in much the same way as individual passages are related in a coherent text. In order to achieve a successful concurrence of probabilities, then, the experimentalist must employ a method of interpretation whereby different areas of learning are compared and reconciled into a coherent account of natural processes.

Because Boyle's experimental philosophy was designed to produce a "confederacy, and an union between parts of learning," it had the eclecticism that such a method of interpretation required.[11] A cautious attitude toward the generation and justification of knowledge claims would also be required, however, because a method of interpretation ultimately involves a type of circularity.[12] Theoretical interpretations must adequately account for the facts of nature, but the accuracy and relevance of the facts is at least partially determined upon the basis of accepted theoretical knowledge. Boyle's mode of proof did not remove circularity from the process, but it did mitigate its negative aspects. It is relatively easy to support a given theory by a careful selection of data that agree with it. It is much more difficult to find one theory that not only accounts for the full spectrum of data concerning a particular subject but is also consistent with other pieces of theoretical knowledge. Although a concurrence of probabilities lacks the logical compulsion of a deductive system, it does produce a strong argument that is rationally compelling. As Boyle explained, "Rational assent may be founded upon proofs, that reach not to rigid demonstrations, it being sufficient, that they are strong enough to deserve a wise man's acquiescence in them."[13]

Boyle's concurrence of probabilities is a decidedly vague epistemological notion, in large part because it is based upon a consideration of relevant information, and the determination of relevance clearly depends upon the current state of knowledge concerning a particular subject. Such context dependence requires that scientific investigation must remain an open-ended process. Boyle said that he would not "debar myself of the liberty" to change opinions in the event that "further progress in history shall suggest better hypotheses or explications."[14] This does not mean that knowledge is forever out of our reach, however. Harvey's theory had been proved conclusively, in part because identifying the relevant information that would be required for proof was somewhat easy, and in part because the technology was available by which the experiments could be performed in order to obtain the relevant information. Clearly the procurability of evidence is also context dependent, and developments in experimental technique may significantly affect the amount and kind of information upon which theories are to be justified. Another reason for Boyle's cautious attitude toward most of the theories of his day, therefore, resulted from his awareness of the difficulty in procuring reliable evidence. More than most of his contemporaries, he appreciated the problems involved in compiling observations, performing experiments, and making rational judgments concerning which of these items could be used as reliable indicators about processes operative in nature.

Observing is a rational activity that involves much more than the simple passive reception of sense perceptions. The classification of observation reports for inclusion in natural histories also involves the use of rational judgments. Boyle noted that the observations he classified as oddities, for example, might be so only from a limited European perspective and might represent the normal course of nature from a different experiential standpoint. In order to acquire a full understanding of the ways in which nature is able to produce effects, Boyle warned his readers to guard against those prejudices that accrue from learning and custom that could adversely affect the ability to make accurate observations. The best way to overcome such a limited outlook was to broaden the witness basis by actively seeking observations from all sectors of society and from all parts of the world.[15] This eclectic approach to the collection of observation reports had to be balanced by a cautious approach in assessing their suitability for use in the establishment of the factual foundation.

The credibility of observation reports must be assessed by determining whether or not the witnesses have the requisite experience to make an accurate observation. The likelihood of the purported fact of the observation must also be determined. Gathering corroborating testimony from independent witnesses was one way in which this likelihood could be ascertained. It is also necessary to assess credibility by examining how well the purported fact fits with other things believed to be known about natural processes. When Boyle questioned the reliability of the pearl divers' reports concerning the pressure of water, for example, he did so because their reports conflicted not only with the reports of other types of divers but also with other facts known about nature, such as the pressure of the air. In addition, he also devised hydrostatic experiments to test the effects that could be produced by the weight of water on inanimate bodies. Clearly, Boyle believed that these experiments showed that the divers' testimony was faulty, yet his study remained inconclusive because he knew that experimental results are even more difficult to assess than observation reports.

Boyle recognized that experiments are subject to a number of systematic errors. Experiments are actively constructed by us. We decide what to test and how we will test it. We choose certain materials rather than others, we construct instruments of varying degrees of complexity, and we perform experiments with varying degrees of skill. At every stage we are taking something for granted, and if we are mistaken at any one stage, we will also be mistaken in our final interpretation of the experimental result. In chapter 3 we saw that Boyle criticized the chemists for their failure to develop reliable indicator tests, and in chapter 7 we saw how Boyle attempted to rectify this situation by advising experimentalists to perform exploratory and probatory trials on the materials and instruments that they proposed to use in their more complicated experiments. Such preliminary trials must be done in order for one to understand more fully the causal properties possessed by experimental materials and to ensure that various pieces of apparatus are in proper working order.

The performance of these preliminary trials removes only some of the contingencies from experimental practice. Even when one is assured that the materials and instruments used in a complex experiment are suitable for the purpose, many things may still go wrong. Because of this, Boyle suggested that experimentalists ought to repeat their complex trials and vary the conditions of them, even when they appeared to have produced successful results. Despite our best efforts, however, some contingencies could still remain because some might result from the use of faulty theoretical presuppositions that would not be identifiable given the present state of theoretical or technological knowledge. The contingencies surrounding experimental practice require a cautious and critical approach to the establishment of knowledge claims. In addition, the methodological strategies for dealing with such contingencies demand an eclectic approach. Aside from the fact that a large number of workers would be needed to perform the tedious tasks of repetition and variation, workers from many different disciplinary backgrounds would also be needed in order to ensure that all of the factors involved in the performance of complex experimental trials were adequately understood. Boyle's experimental philosophy was thus a complex social activity that would require organization and collaboration.

The complexity of Boyle's philosophy has yet to be appreciated because previous studies by intellectualist historians have tended to use modern categories such as empiricism, inductivism, pragmatism, and hypothetico-deductivism to describe his methodology. All of these historical interpretations have been based upon references to passages in Boyle's texts, and thus none are entirely "wrong." The resultant reductions of Boyle's thought to one particular epistemological category are misleading, though, because each fails to capture the uniqueness and significance of his synthetic account wherein he incorporated elements from all of the above perspectives.

Although Boyle believed that experimental reasoning must begin with a vast collection of factual information, we have seen that he consistently rejected empiricist and inductivist assumptions concerning the manner by which matters of fact were to be justified and the way in which they could in turn be used for the construction of knowledge claims concerning natural processes. In addition, we have seen that Boyle rejected the empiricist goal of a phenomenal ordering of appearances described by J. E. McGuire and Margaret J. Osler, among others. Instead, Boyle designed his experimental trials in order "to discover deep and unobvious truths" concerning nature's causal powers.[16] In a similar manner, Boyle's discussions of the role of activity in knowledge acquisition certainly contained pragmatist elements, but his notion of usefulness went beyond that described by Philip Wiener. Boyle did not justify his acceptance of the corpuscular philosophy simply on the grounds that it was a useful hypothesis. He maintained that the way in which corpuscular principles provided suggestions for the successful manipulation of nature could serve as a sign of the likelihood that the causal properties of bodies did result from their corpuscular composition.[17] In this instance Boyle employed a version of the hypothetico-deductive method as described by James B. Conant, Larry Laudan, and Maurice Mandelbaum. We have also seen, however, that many of his experiments were not done to test theoretical claims. The hypothetico-deductive account of his method is therefore only partially accurate, because it leaves out the notion that some knowledge comes directly from the active manipulation of nature.[18]

Boyle's philosophy is difficult to characterize because in the interest of free inquiry he rejected all attempts to place general restrictions on knowledge acquisition. Each of the epistemological alternatives listed above is foundational, and thus each would preclude certain lines of investigation. Boyle sought instead to make the experimental philosophy a dynamic learning process. No one warrant is universally applicable. At different stages of inquiry, different warrants will be appropriate. Boyle's conception of proof as that which is based upon a concurrence of probabilities implied that the experimentalist must use whatever information or technique is required at a given time. In some contexts rational inference and the testing of theoretical consequences may be required. At other times, it may be necessary to devote more effort to the collection of observations or to the performance of exploratory experiments. It is not possible to decide which methodological strategy will work best without taking account of the theoretical and technological context surrounding a particular area of study.

Despite the fact that contextualist historians claim to study past science without the imposition of modern preconceptions, most of those who have worked on Boyle have perpetuated the earlier intellectualist interpretations of his method. Shapin and Schaffer, for example, described Boyle as an "empiricist" who, because he regarded "the man-made component of knowledge as a distortion of the mind's mirroring of reality," could not acknowledge the credibility of Hobbes's criticisms of the experimental method.[19] Their Boyle is not only an empiricist, however. He is also a positivist who supposedly presented his "experimentally produced facts as self-evident and self-explanatory."[20] It should be clear by now that Boyle's philosophy was much more complex and sophisticated than Shapin and Schaffer have allowed.

There is more than a problem of historical accuracy in Shapin and Schaffer's study. Like numerous other contextualist works, *Leviathan and the Air-Pump* has been used as "empirical" support for a radical relativist thesis concerning the status of scientific knowledge claims. That Shapin and Schaffer designed their study so that it would serve such a purpose can be seen in the way that they devoted most of their discussion of Boyle's experimentalism to a "display of its conventional basis: showing the work of production involved and exhibiting the lack of obligation to credit experimental knowledge." Like H. M. Collins, David Gooding, and Karin D. Knorr-Cetina, they apparently believe that the presence of conventions and human agency in the construction of scientific knowledge provides a reason to doubt its "authenticity."[21] What is missing from Shapin and Schaffer's superficial account of Boyle's methodology is a recognition of the significance of his extended discussions concerning how the presence of conventions and agency could be used as support for the knowledge-producing capability of experimental practice. My purpose in the following section is not to go over the criticisms of Shapin and Schaffer's historical analysis already made in previous chapters. Rather, I intend to suggest ways in which the relativist conclusions of their study are not warranted.

The Significance of Boyle's Philosophy

It is true that most, if not all, scientific knowledge has a conventional basis. Boyle argued that these conventions would be socially constituted because experimental practice is ultimately a social activity that will involve the interaction of a large number of workers. But the social negotiations and conventions embedded within experimentalism also have an epistemic dimension that has been totally dismissed by contextualists such as Shapin and Schaffer. Simply because experimental science is a social process, that does not mean that social analysis is the only appropriate means by which to understand the performances involved in the practice. Actors' reasons for what they do are often drawn from the internal dynamics of an activity itself, distinct from any ulterior goals or purposes. Thus a philosophical analysis of such practices is also required.

The presence of social conventions in the construction of knowledge does not warrant the conclusion that knowledge is nothing but a social construct established by "the aggregation of individuals' *beliefs*."[22] Boyle's notion of a concurrence of probabilities is not the same as the notion of consensus widely discussed in today's sociological literature. Although it is likely that a concurrence of probabilities will produce a consensus of opinion, the two notions should not be conflated. Consensus concerns a decision procedure about the beliefs of individuals, whereas concurrence concerns an investigative procedure whereby one attempts to discover the quantity and quality of the evidence that may be available to attest to the truth of a particular claim about nature. The purpose of concurrence, as both an epistemological and a methodological strategy, is not the simple quantitative multiplication of witnesses. Rather, it is required as a way by which to evaluate the reliability of the information attested to by such witnesses. Because of the inherent fallibility of human senses and reasoning, such a qualitative check is absolutely necessary.

The circularity and context dependence that characterize concurrence do not entail a logical compulsion to accept knowledge claims justified by this mode of proof. But radical relativism would follow from this circumstance only if one accepts the logicist equation of rationality with logicality—an equation that has been repeatedly called into question in this century by philosophers such as Ernan McMullin, Dudley Shapere, and Stephen Toulmin.[23] Boyle also rejected the idea that a "rigid demonstration" was necessary in order to obtain a "wise man's acquiescence." William Whewell made much the same point a century ago concerning his similar notion of a "consilience of inductions."[24] As Thomas Nickles has pointed out, the circularity involved in this type of "processual" confirmation is not vicious but "virtuous." When a "claim is backed by multiple arguments from different sets of premises rather than by just one," the degree of support for that claim increases appreciably. Such a "multiple determination of results" provides a strong argument in favor of the acceptance of the theoretical claim based upon it.[25]

The presence of conventional elements can be seen most perspicuously in the way in which theoretical knowledge must be used in our assessments of the reliability of observational and experimental reports, which in turn may give rise to what H. M. Collins has termed the "experimenters' regress."[26] The so-called theory-ladenness of experimental practice is also an issue that has been widely discussed in the philosophical literature by N. R. Hanson, Mary Hesse, Thomas S. Kuhn, and Michael Polyani, among others.[27] The significance of these discussions rests not upon the simple recognition of the role of theory in observation and experiment but upon the way in which such "ladening" is actually a positive virtue of the scientific process. As Boyle advised his readers, the use of background knowledge is necessary because the senses are inherently corrigible and thus must be corrected by the use of reason and judgment. The fact that our experiments employ theoretical knowledge does not mean that the "regress" is infinite, only that one must guard against taking any one experiment as conclusive. Pierre Duhem reached a similar conclusion when he noted that although the "realization and interpretation" of experiments "imply adherence to a whole set of theoretical propositions," any one of which might be faulty, it is not the case that experiments can never uniquely determine the acceptance of a particular theory. As Duhem explained, such a "state of indecision does not last forever. The day arrives when good sense comes out so clearly in favor of one of the two sides that the other side gives up the struggle, even though pure logic would not forbid its continuation."[28]

The presence of conventions, social and otherwise, within experimental practice does not imply that skepticism regarding the knowledge claims produced by such measures is the appropriate epistemic response. Boyle readily acknowledged the inherently fallible nature of experimental practice, and he continuously urged his readers to exercise caution in their acceptance, or rejection, of the knowledge claims presented for their consideration. Yet he remained confident that the experimental method provided the best means for discovering the processes operative in nature. Boyle's experimental philosophy was designed as a dynamic learning process. Experimentalists must remain open to the possibility that the knowledge claims proved beyond a reasonable doubt at one point may have to be revised at a later date. But the bare possibility that some of our knowledge claims may be found to be mistaken in the future does not mean that all of them will be. It is not appropriate to regard all of our knowledge as epistemically suspect unless we have a more specific reason to do so.

The other factor that relativists have used as support for their epistemological thesis is the role of human agency in the construction of experimental knowledge. Throughout their study, Shapin and Schaffer, for example, devote considerable time to displaying the amount of activity involved in the construction of experimental results, because they believe that "in common speech, as in the philosophy of science, the solidity and permanence of matters of fact reside in the absence of human agency in their coming to be." The normative character and intent of their study is revealed in their concluding passage, where they proclaim: "Knowledge, as much as the state, is a product of human actions. Hobbes was right."[29] This conclusion is striking, not because of the rather trivial claim that knowledge is produced by human activity, but because of the implication that neither Boyle nor the majority of today's philosophers of science have recognized the role of human agency in the construction of knowledge.

What is most striking, however, is the failure of Shapin and Schaffer even to acknowledge Boyle's extended discussions concerning the epistemic value of agency. As we have seen, Boyle designed his experimental philosophy as an active way of knowing. Philosophers today have also noted that the activity of the experimenter, not a set of static test results, often plays the crucial role in justifying our belief in the existence of theoretical entities. Ian Hacking, for example, has argued that "practice, not theory," provides "the 'direct' proof" for the existence of electrons.[30] In his account of "constructive realism," Ronald Giere has made a similar point in his discussion of how direct evidence for the existence of protons can be obtained from those practices in which scientists "are 'producing' and 'using' protons."[31] The experimental production of effects can be used as a sign by which to assess the accuracy of purported knowledge claims concerning the causal powers of bodies. When we are consistently successful in our attempts to use a certain type of body in the production of an effect, then it is rational for us to believe that bodies of that type have the causal powers to produce such effects.

Clearly, "constructivism" need not entail radical relativism. It is as compatible, if not more so, with a limited, localized version of scientific realism. It is not necessary to claim that all of our theoretical knowledge is accurate in order to say that some of our constructed entities actually exist and possess at least some of the properties that we have attributed to them. It may be true that scientists today tend to write agency out of their reports, as David Gooding has claimed, but Boyle certainly did not do so.[32] The active construction of artificial effects was part of Boyle's warrant for his justification of experimental practice as a privileged way of knowing. Activity is not merely a part of the experimental process; it is also the goal. If one desires to know the causal properties of bodies and to use such knowledge to manipulate and control them in order to produce beneficial effects, then the experimental method will provide the best means for achieving such an active understanding.

I certainly do not mean to imply that Boyle's philosophy of experiment is identical to any of the philosophical positions discussed above. What is significant, however, is the way his own extensive practice led him to appreciate the complex ways in which conventional background knowledge and human agency contribute to scientific investigation. The general conception of the dynamic nature of the facts, theories, and methods of experimental science is something that Boyle shares with numerous philosophers, past and present. There is also a growing awareness today of the dynamic nature of scientific explanation. Unlike earlier positivists, who attempted to describe all explanations as logical arguments produced by the deductive subsumption of facts under the laws to which they refer, today's philosophers discuss how the interests of the persons seeking an explanation must be taken into account in order to determine whether or not a good explanation has been given.[33] This does not make explanation epistemically suspect, however. There is no need to remove the subject from the process of knowing in order to achieve some type of mythic "objectivity." Knowledge is for us. As Boyle explained in his justification of the mechanical philosophy, for example, it could be the case that a natural process is actually produced by "an immaterial principle or agent," but an appeal to such an entity would "not enable *us* to explain the phaenomena," because "*we cannot conceive,* how it should produce changes in a body, without the help of mechanical principles."[34]

I also do not mean to imply that Boyle's motivations for his study of nature are identical with those of today's scientists. He did share with us a search for truth, but he also had a strong theological motivation that is not present in modern science. He believed that he had a religious duty to study nature. Although his experimental program was designed in part to yield useful knowledge, Boyle also maintained that "it is erroneous to say in the strictest sense that everything in the visible world was made for the use of man." He added, however, that "many parts of the material world" could be considered "highly useful" to man, "as he is a rational creature, that is capable, by contemplating the great and admirable works of God, to raise his mind to the acknowledgement of the divine Architect's power, wisdom, and beneficience, and thereby find produced in him due sentiments of veneration, gratitude, and love."[35]

Boyle's conception of the world as a vast interrelated mechanism was in part a product of his theological beliefs concerning the manner by which God created the universe. His experimental investigations reinforced this conception. He believed that all those who actively pursued experimental natural inquiries would come to the same realization. In *The Christian Virtuoso* he noted that an experimentalist would become accustomed to discovering "the weakness of those solutions, that superficial wits are wont to make and acquiesce in," which in turn would "insensibly work in him a great and ingenuous modesty of mind."[36] A cautious, eclectic, flexible, and critical approach was intrinsic to Boyle's experimental philosophy. As he explained to his readers, "It ought to be esteemed much less disgraceful to quit an error for a truth, than to be guilty of the vanity and perverseness of believing a thing still, because we once believed it. And certainly, till a man is sure he is infallible, it is not fit for him to be unalterable."[37]

Given the historical development of modern science, Boyle was wrong in his belief that experimental inquiry would lead to a modesty of mind and a reverence for the world. Indeed, as its critics have been quick to point out, science in the Western world has been characterized by a theoretical confidence that has resulted in the diminishing of alternative models of thought, and a technological hubris that has resulted in the near devastation of crucial parts of the earth's delicate ecosystem.[38] A retrieval of Boyle's attitude could provide a valuable resource for addressing such critical issues today. Most important, the ill effects of modern science need not be thought of as the necessary consequences of experimental practice, but merely as historically contingent outcomes that need not be accepted. As Boyle noted, knowledge in itself may be a good thing, even when the uses to which it is put are not. It is, he insisted, "not through the fault of the learning, but by the fault of the man, who being vitiously disposed himself, makes use of a thing, in its own nature good, to a bad end. An inconvenience incident to the most excellent things since the worst (and more pernitious) abuse is that of the best things."[39]

Boyle appreciated the complexity of the interrelated mechanisms operative in nature and what such complexity entailed for human understanding. He urged his readers to remember that "to discern particular things is one thing," but "to discover the intercourse and harmony between all truths, is another thing, and a far more difficult one."[40] Human knowledge is a "progressive, discursive thing."[41] The experimentalist must remain open to the possibility that new theories or improved technologies may radically alter the way in which inquiry proceeds. Boyle was keenly aware of the real limits, both theoretical and technological, that constrained his investigations. He believed that an "enlightened posterity" would much surpass his accomplishments, but he did not think that they would achieve total success.[42] Diffidence and caution must be retained. Because of "the connection of physical truths, and the relations that material bodies have to one another," an "inquisitive naturalist" in any age will find "his work to encrease daily," because even his successful trials will "but engage him into new ones."[43] Boyle illustrated this last point with a metaphor that provides an apt characterization of his unassuming and diffident attitude toward the never-ending process of experimental inquiry:

> If knowledge be, as some philosophers have stiled it, the aliment of the rational soul, I fear I may too truly say, that the naturalist is usually fain to live upon sallads and sauces, which, though they yield some nourishment, excite more appetite than they satisfy, and . . . reduce us to an unwelcome necessity of always rising hungry from the table.[44]

NOTES

Introduction

1. Boyle most frequently referred to himself as an "experimental philosopher" or as a "Christian virtuoso." Virtuosity was a product of extensive experimental practice. See *The Christian Virtuoso*, pt. 1, in *The Works of the Honourable Robert Boyle*, 6 vols., ed. Thomas Birch (1772; reprint, Hildesheim: Georg Olms, 1965–66), 5: 524. My citation practices are discussed at note 72 below.
2. The three best-known biographies include that by Thomas Birch, who published Boyle's autobiographical account of his youth, together with the details of his mature life, in his preface to Boyle, *Works*; R. E. W. Maddison, *The Life of the Honourable Robert Boyle, F.R.S.* (New York: Barnes and Noble, 1969); and Louis Trenchard More, *The Life and Works of the Honourable Robert Boyle* (London: Oxford University Press, 1944).
3. Aside from her numerous articles on Boyle, see particularly Marie Boas Hall, *Robert Boyle and Seventeenth-Century Chemistry* (Cambridge: Cambridge University Press, 1958), and *Robert Boyle on Natural Philosophy* (Bloomington: Indiana University Press, 1965).
4. Peter Alexander, *Ideas, Qualities, and Corpuscles: Locke and Boyle on the External World* (Cambridge: Cambridge University Press, 1985); James B. Conant, "Robert Boyle's Experiments in Pneumatics," in *Harvard Case Histories in Experimental Science*, vol. 1, edited by James B. Conant (Cambridge: Harvard University Press, 1970), 1–63; Norma E. Emerton, *The Scientific Reinterpretation of Form* (Ithaca, N.Y.: Cornell University Press, 1984); Thomas S. Kuhn, "Robert Boyle and Structural Chemistry in the Seventeenth Century," *Isis* 43 (1952): 12–36; James G. Lennox, "Boyle's Defense of Teleological Inference," *Isis* 74 (1983): 38–52; F. J. O'Toole, "Qualities and Powers in the Corpuscular Philosophy of Robert Boyle," *Journal of the History of Philosophy* 12 (1974): 295–315; Richard S. Westfall, "Unpublished Boyle Papers Relating to Scientific Method," *Annals of Science* 12 (1956): 63–73, 103–17.
5. Alexander, *Ideas, Qualities, and Corpuscles*, chap. 2; Marie Boas Hall, "Boyle as a Theoretical Scientist," *Isis* 46 (1950): 261–68; Westfall, "Unpublished Boyle Papers," 63, 108.
6. *Empiricism* is a term that tends to be used rather loosely, but from ancient times to the present it has normally been identified with the position stated in the text.

For the history of empiricism, see G. E. R. Lloyd, *Science, Folklore, and Ideology* (Cambridge: Cambridge University Press, 1983); and Richard H. Popkin, *The History of Skepticism from Erasmus to Spinoza* (Berkeley and Los Angeles: University of California Press, 1979). See Bas C. van Fraassen, *The Scientific Image* (Oxford: Clarendon Press, 1980), for the latest defense of an empiricist philosophy of science. See Paul M. Churchland and Clifford A. Hooker, eds., *Images of Science* (Chicago: University of Chicago Press, 1985); and Joseph J. Kockelmans, "On the Problem of Truth in the Sciences," *Proceedings and Addresses of the American Philosophical Association* 61 (1987): 5–26, for criticisms of empiricist assumptions.

7. Philip P. Wiener, "The Experimental Philosophy of Robert Boyle," *Philosophical Review* 41 (1932): 594–609. Wiener maintains that Boyle had a "pragmatic conception of scientific theory" because, while he was a realist about the qualities of bodies, he treated the corpuscular theory as a mechanical hypothesis (601) and appealed to the nonempirical criterion of simplicity to justify its usefulness (609).

8. Some predecessor studies retained a focus upon the content of Boyle's theoretical beliefs. See, for example, Robert Hugh Kargon, "Walter Charleton, Robert Boyle, and the Acceptance of Epicurean Atomism in England," *Isis* 55 (1964): 184–92; Leroy E. Loemker, "Boyle and Leibniz," *Journal of the History of Ideas* 16 (1955): 22–43; and John J. O'Brien, "Samuel Hartlib's Influence on Robert Boyle's Scientific Development," *Annals of Science* 21 (1965): 1–14, 257–76.

9. Larry Laudan, "The Clock Metaphor and Probabilism: The Impact of Descartes on English Methodological Thought, 1650–65," *Annals of Science* 22 (1966): 73–104. Similar studies, designed to show the hypothetico-deductive nature of Boyle's method, are Conant, "Boyle's Experiments"; and Maurice Mandelbaum, *Philosophy, Science, and Sense Perception: Historical and Critical Studies* (Baltimore: Johns Hopkins University Press, 1964).

10. G. A. J. Rogers, "Descartes and the Method of English Science," *Annals of Science* 29 (1972): 237–55. Rogers's article was a response to the first appearance of Laudan's article. When Laudan reprinted his text as Larry Laudan, "The Clock Metaphor and Hypotheses: The Impact of Descartes on English Methodological Thought," in *Science and Hypothesis: Historical Essays on Scientific Methodology* (Dordrecht: D. Reidel, 1981), 27–58, I responded with "Robert Boyle's Baconian Inheritance: A Response to Laudan's Cartesian Thesis," *Studies in History and Philosophy of Science* 17 (1986): 469–86.

11. Gary B. Deason, "Reformation Theology and the Mechanistic Conception of Nature," in *God and Nature,* ed. David C. Lindberg and Ronald L. Numbers (Berkeley and Los Angeles: University of California Press, 1986), 167–91; Keith Hutchison, "Supernaturalism and the Mechanical Philosophy," *History of Science* 21 (1983): 297–333; Eugene M. Klaaren, *Religious Origins of Modern Science* (Grand Rapids, Mich.: Eerdmans, 1977); Barbara J. Shapiro, *Probability and Certainty in Seventeenth-Century England* (Princeton, N.J.: Princeton University Press, 1983); idem, "Early Modern Intellectual Life: Humanism, Religion, and Science in Seventeenth-Century England," *History of Science* 29 (1991): 45–71; Henry G. van Leeuwen, *The Problem of Certainty in English Thought, 1630–1690* (The Hague: Martinus Nijhoff, 1963); Richard S. Westfall, *Science and Religion in Seventeenth-Century England* (Ann Arbor: University of Michigan Press, 1973).

12. See, for example, Peter Dear, "Miracles, Experiments, and the Ordinary Course of Nature," *Isis* 81 (1990): 663–83; idem, "Narratives, Anecdotes, and Experiments: Turning Experience into Science in the Seventeenth Century," in *The Literary Structure of Scientific Argument*, ed. Peter Dear (Philadelphia: University of Pennsylvania Press, 1991), 135–63; Margaret J. Osler, "John Locke and the Changing Ideal of Scientific Knowledge," *Journal of the History of Ideas* 31 (1970): 3–16; Richard Rorty, *Philosophy and the Mirror of Nature* (Princeton, N.J.: Princeton University Press, 1979); and Steven Shapin and Simon Schaffer, *Leviathan and the Air-Pump* (Princeton, N.J.: Princeton University Press, 1985).

13. J. E. McGuire, "Boyle's Conception of Nature," *Journal of the History of Ideas* 33 (1972): 523–43. Osler and Shapin and Schaffer have made extensive use of McGuire's analysis. See Margaret J. Osler, "The Intellectual Sources of Robert Boyle's Philosophy of Nature: Gassendi's Voluntarism and Boyle's Physico-Theological Project," in *Philosophy, Science, and Religion in England, 1640–1700*, ed. Richard W. F. Kroll, Richard Ashcraft, and Perez Zagorin (Cambridge: Cambridge University Press, 1992), 178–98; and idem, "Providence and Divine Will in Gassendi's Views on Scientific Knowledge," *Journal of the History of Ideas* 44 (1983): 549–60.

14. Boyle, *The Usefulness of Natural Philosophy*, pt. 2, essay 2, in *Works* 2: 85.

15. The two most popular manifestos written by defenders of the "strong programme" in SSK are David Bloor, *Knowledge and Social Imagery* (London: Routledge and Kegan Paul, 1976); and Steven Shapin, "History of Science and Its Sociological Reconstructions," *History of Science* 20 (1982): 157–211. For helpful discussions of the various types of sociological projects, albeit from quite different perspectives, see Ronald N. Giere, *Explaining Science: A Cognitive Approach* (Chicago: University of Chicago Press, 1988), 50–61; and Andrew Pickering, "From Science as Knowledge to Science as Practice," in *Science as Practice and Culture*, ed. Andrew Pickering (Chicago: University of Chicago Press, 1992), 1–26.

16. David Bloor, "The Sociology of Reasons; or, Why 'Epistemic Factors' Are Really 'Social Factors,'" in *Scientific Rationality: The Sociological Turn*, ed. J. R. Brown (Dordrecht: D. Reidel, 1984), 295–324.

17. David Bloor, "Durkheim and Mauss Revisited: Classification and the Sociology of Knowledge," *Studies in History and Philosophy of Science* 13 (1982): 285.

18. See, for example, Christopher Hill, *Change and Continuity in Seventeenth-Century England* (Cambridge: Harvard University Press, 1975); idem, *The World Turned Upside Down: Radical Ideas during the Scientific Revolution* (Harmondsworth, Eng.: Penguin, 1975); James R. Jacob, *Robert Boyle and the English Revolution* (New York: Burt Franklin, 1977); Margaret C. Jacob, *The Newtonians and the English Revolution* (Ithaca, N.Y.: Cornell University Press, 1976); James R. Jacob and Margaret C. Jacob, "The Anglican Origins of Modern Science: The Metaphysical Foundations of the Whig Constitution," *Isis* 71 (1980): 251–67; Charles Webster, *The Great Instauration: Science, Medicine, and Reform, 1626–1660* (New York: Holmes and Meier, 1975); and idem, *The Intellectual Revolution of the Seventeenth Century* (London: Routledge and Kegan Paul, 1974). Shapin, in "History of Science," also maintains that these works support the strong programme.

19. Jacob and Jacob, "Anglican Origins of Modern Science," 257. The Jacobs main-

tain that Boyle's "metaphysics of God and matter" authorized a conservative interpretation of the social hierarchy" (253). They further claim that this metaphysics
was used to support "the clergy's authority as interpreters of God's ways and
will" (257).

20. Bloor, "Durkheim and Mauss Revisited," 287–88. In "Sociology of Reasons," (301)
Bloor provides a functionalist explanation of Boyle's corpuscularianism when he
argues that because Boyle insisted "that matter was inert and inanimate and inca-
pable of moving itself," nonmaterial active principles were introduced in a way
that "reinforced the social patterns" of Restoration England. He makes a similar
point in Bloor, *Wittgenstein: A Social Theory of Knowledge* (New York: Columbia University Press, 1983), 152–55. Recently, Elizabeth Potter, in "Modeling the Gender
Politics in Science," *Hypatia* 3 (1988): 19–33, has extended the work done by
the Jacobs to argue that the implications of the passivity of matter thesis went
beyond class politics. Because matter was a "female principle," theories about
its activity or lack of activity could be used to advance particular views in
"gender politics." According to Potter, those who worked for progressive
change of the traditional gender roles advocated the "principle that matter is
alive and self-moving," while those who, like Boyle, advocated the passivity of
matter, did so in part to assure the continued male dominance of seventeenth-
century women.

21. H. M. Collins, *Changing Order: Replication and Induction in Scientific Practice* (Beverly
Hills, Calif.: Sage, 1985); David Gooding, *Experiment and the Making of Meaning* (Boston: Kluwer, 1990); idem, "Putting Agency Back into Experiment," in *Science as
Practice and Culture*, ed. Andrew Pickering (Chicago: University of Chicago Press,
1992), 65–112; Bruno Latour and Steve Woolgar, *Laboratory Life: The Social Construction of Scientific Facts* Princeton, N.J.: Princeton University Press, 1979); Bruno
Latour, "One More Turn after the Social Turn. . .," in *The Social Dimensions of Science*,
ed. Ernan McMullin (Notre Dame, Ind.: University of Notre Dame Press, 1992),
272–94; Michael Lynch, *Art and Artifact in Laboratory Science: Shop Work and Shop Talk
in a Research Laboratory* (London: Routledge and Kegan Paul, 1985); and Andrew
Pickering, *Constructing Quarks: A Sociological History of Particle Physics* (Chicago: University of Chicago Press, 1984).

22. The division between conservative and radical sociology can be seen in the ways
in which Bloor and Collins have recently defended themselves against their sociological critics in Andrew Pickering, ed., *Science as Practice and Culture* (Chicago: University of Chicago Press, 1992). See David Bloor, "Left and Right Wittgensteinians," 266–82, which is a response to Michael Lynch, "Extending Wittgenstein: A
Pivotal Move from Epistemology to the Sociology of Science," 215–65; and
H. M. Collins and Steven Yearley, "Epistemological Chicken," 301–26, which is
primarily a criticism of the work done by Callon, Latour, and Woolgar. See the
responses by Michel Callon and Bruno Latour, "Don't Throw the Baby Out with
the Bath School! A Reply to Collins and Yearley," 343–68; and by Steve Woolgar,
"Some Remarks about Positionism: A Reply to Collins and Yearley," 327–43. Collins and Yearley respond to these responses in "Journey into Space," 369–89. It is
interesting that in this debate Callon and Latour say that their "network building"
model is similar to Shapin and Schaffer's concept of a "form of life" (348), while

Collins and Yearley maintain that they "think of Shapin and Schaffer as on our side of the debate" (378).

23. Shapin and Schaffer, *Leviathan and the Air-Pump*, 342. They wrote: "We have set ourselves the historical task of inquiring into *why* experimental practices were accounted proper and how such practices were considered to yield reliable knowledge" (13).

24. Ibid., 341. Prior to this quotation, they maintain that this explanatory goal was not central to their study. Most reviewers of their work, however, whether critical or complimentary, have tended to view this explanation as a major conclusion of *Leviathan and the Air-Pump*. See Christopher Hill, "'A New Kind of Clergy': Ideology and the Experimental Method," *Social Studies of Science* 16 (1986): 726–35; Bruno Latour, "Postmodern? No, Simply Amodern! Steps Towards an Anthropology of Science," *Studies in History and Philosophy of Science* 21 (1990): 145–71; and Richard S. Westfall, Review of *Leviathan and the Air-Pump*, by Steven Shapin and Simon Schaffer, *Philosophy of Science* 54 (1987): 128–30.

25. See Andrew Pickering, "Against Putting the Phenomena First: The Discovery of the Weak Neutral Current," *Studies in History and Philosophy of Science* 15 (1984): 85–117.

26. Steve Fuller, "David Bloor's *Knowledge and Social Imagery* (2nd Edition)," *Philosophy of Science* 60 (1993): 158; Pickering, "From Science as Knowledge to Science as Practice," 4. See also Bloor, "Left and Right Wittgensteinians," 270–76, for a defense of interest theory. One can see the sociological concept of interest at work in *Leviathan and the Air-Pump*, in the way in which Shapin and Schaffer describe Hobbes's belief that the "experimenters were just another conspiratorial group whose interests were in obtaining power over citizens, and whose devious confederacy sought an illegitimate autonomy from the state" (320).

27. Michael Lynch, "From the 'Will to Theory' to the Discursive Collage: A Reply to Bloor's 'Left and Right Wittgensteinians,'" in *Science as Practice and Culture*, ed. Andrew Pickering (Chicago: University of Chicago Press, 1992), 299.

28. Latour ("One More Turn," 276) accuses traditional SSK theorists of playing "a game in one dimension," and he notes that social explanation is impossible simply because it is a modern notion (284). See Callon and Latour, "Don't Throw the Baby Out," 347, for a discussion of how the dichotomy is a modern scientific distinction; 349, for a reactionary characterization of Collins and Yearley; and 364, where Collins's "social realism" is described as the "classic view of society as what abruptly puts a stop to the indefinite negotiability of scientists." See Woolgar, "Some Remarks about Positionism," 337–38, for a similar evaluation.

29. Latour, "Postmodern?"

30. Callon and Latour, "Don't Throw the Baby Out," 343–44. Collins and Yearley ("Journey into Space," 382) do appear to be dogmatic when they provide the following "prescription: stand on social things—be social realists—in order to explain natural things. . . . others will be standing on natural things to explain social things. That is all there is to it. There is nowhere else to go that is significantly more interesting." Latour ("One More Turn," 274) notes that SSK has "won the battle." See Woolgar, "Some Remarks about Positionism," 332–34, 329, and 339, for similar concerns about the entrenchment of SSK.

31. Steven Shapin, "Social Uses of Science," in *Ferment of Knowledge*, ed. G. S. Rousseau and R. Porter (Cambridge: Cambridge University Press, 1980), 111. See also Shapin, "History of Science"; and H. M. Collins, "Knowledge, Norms, and Rules in the Sociology of Science," *Social Studies of Science* 12 (1982): 299–309, for similar views of the project.

32. Fuller ("David Bloor's *Knowledge and Social Imagery*") and Gooding ("Putting Agency Back into Experiment") both write as if the battle had only just begun. They appear to ignore the difference in the newer criticisms and assume that any criticism of SSK is a call for a return to earlier rational reconstructions.

33. Michael Hunter, "The Conscience of Robert Boyle: Functionalism, 'Dysfunctionalism,' and the Task of Historical Understanding," in *Renaissance and Revolution: Humanists, Scholars, Craftsmen, and Natural Philosophers in Early Modern Europe*, ed. J. V. Field and Frank A. J. L. James (Cambridge: Cambridge University Press, 1993), 147–59; idem, introduction to *Robert Boyle Reconsidered*, ed. Michael Hunter (Cambridge: Cambridge University Press, 1994), 1–18.

34. Thomas Söderqvist, "Existential Projects and Existential Choice in Science: Scientific Biography as an Edifying Genre," in *Telling Lives: Studies of Scientific Biography*, ed. Richard Yeo and Michael Shortland (Cambridge: Cambridge University Press, forthcoming).

35. The need for greater localization is also expressed by Lynch ("From the 'Will to Theory' to the Discursive Collage," 299), who maintains that scientific practice is "more thoroughly and locally social than sociology is prepared to handle."

36. Antonio Pérez-Ramos, *Francis Bacon's Idea of Science and the Makers' Knowledge Tradition* (Oxford: Clarendon Press, 1988), 40–41. Although he describes his project as an "epistemological reconstruction," it does differ from earlier rational reconstructions in significant ways.

37. Ibid., 5–6 and throughout. In his conclusion, Pérez-Ramos emphasizes that Bacon's work must be seen in its entirety (289).

38. Daniel Garber, *Descartes' Metaphysical Physics* (Chicago: University of Chicago Press, 1992), 2–3. Garber also acknowledges that the social context of past science is an equally important area of study.

39. Ibid., 299–305. Garber notes that while an occasionalist interpretation of Descartes's theological beliefs may be plausible, there is no basis in his works for accepting such an interpretation. See also 136–43 for a detailed discussion of the controversy between Descartes and Pascal over the proper interpretation of the Puy-de-Dôme experiment. Garber shows the reasonableness of both sides.

40. Ibid., 2. Garber refers to Descartes's program as a "dead end." Pérez-Ramos (*Bacon's Idea of Science*) notes that few of Bacon's successors understood the complexity of his thought.

41. Michael Hunter, "Alchemy, Magic, and Moralism in the Thought of Robert Boyle," *British Journal for the History of Science* 23 (1990): 387–410; William Newman, "Boyle's Debt to Corpuscular Alchemy," in *Robert Boyle Reconsidered*, ed. Michael Hunter (Cambridge: Cambridge University Press, 1994), 107–18; Lawrence M. Principe, "Boyle's Alchemical Pursuits," in *Robert Boyle Reconsidered*, 91–105; idem, "Robert Boyle's Alchemical Secrecy: Codes, Ciphers, and Concealments," *Ambix*

39 (1992): 63–74. See also Antonio Clericuzio, "A Redefinition of Boyle's Chemistry and Corpuscular Philosophy," *Annals of Science* 47 (1990): 561–80.

42. John Henry, "Occult Qualities and the Experimental Philosophy: Active Principles in Pre-Newtonian Matter Theory," *History of Science* 24 (1986): 335–81; idem, "Boyle and Cosmical Qualities," in *Robert Boyle Reconsidered*, ed. Michael Hunter (Cambridge: Cambridge University Press, 1994), 119–38.

43. John Harwood, introduction to *The Early Essays and Ethics of Robert Boyle*, edited by John Harwood, xv–lxix (Carbondale: Southern Illinois University Press, 1991); idem, "Science Writing and Writing Science: Boyle and Rhetorical Theory," in *Robert Boyle Reconsidered*, ed. Michael Hunter (Cambridge: Cambridge University Press, 1994), 37–56; Michael Hunter, "Casuistry in Action: Robert Boyle's Confessional Interviews with Gilbert Burnet and Edward Stillingfleet, 1691," *Journal of Ecclesiastical History* 44 (1993): 80–98; idem, "Conscience of Robert Boyle"; Malcolm Oster, "Virtue, Providence, and Political Neutralism: Boyle and Interregnum Politics," in *Robert Boyle Reconsidered*, 19–36. See also Jan Wojcik, "Robert Boyle and the Limits of Reason: A Study in the Relationship between Science and Religion in Seventeenth-Century England," Ph.D. diss., University of Kentucky, 1992.

44. Hunter, "Conscience of Robert Boyle," 157.

45. While Boyle's experimentalism was a major component of his life, he was not as one-dimensional as it may appear if the present study is mistakenly taken to be a biographical account of his entire life.

46. See Söderqvist, "Existential Projects"; and Rose-Mary Sargent, "Explaining the Success of Science," in *PSA 1988*, ed. Arthur Fine and Jarrett Leplin (East Lansing, Mich.: Philosophy of Science Association, 1988), 1: 55–63, for more detailed discussions of this point.

47. Timothy Lenoir, "Practical Reason and the Construction of Knowledge," in *The Social Dimensions of Science*, ed. Ernan McMullin (Notre Dame, Ind.: University of Notre Dame Press, 1992), 168. Lenoir maintains that Shapin and Schaffer use the notion of a "form of life" to "denote a consciously deployed set of strategies to promote a constellation of interests" (168). See Steven Shapin, "Discipline and Bounding: The History and Sociology of Science as Seen through the Externalism-Internalism Debate," *History of Science* 30 (1992): 354–55, for Shapin's discussion of sociological collectivism that he sees as a corrective for the historian's tendency toward "atomizing particularism."

48. Shapin and Schaffer, *Leviathan and the Air-Pump*, esp. chap. 7, pp. 305–11. When they attempt to link their interpretation of Boyle's pneumatic experiments with sociopolitical factors, the quotations from Sprat and Hobbes far outnumber those from Boyle.

49. See, for example, Michael Hunter and Paul B. Wood, "Towards Solomon's House: Rival Strategies for Reforming the Early Royal Society," *History of Science* 24 (1986): 49–108.

50. For this reason, although my study of Boyle may bear a resemblance to history of philosophy, it will not be possible to provide a tight analytic structure for his thought, since such an approach would negate that which was unique to Boyle.

51. See Carl G. Hempel, *Aspects of Scientific Explanation* (New York: Free Press, 1965), 22–24, 102–4; and Karl R. Popper, *The Logic of Scientific Discovery* (New York: Harper and Row, 1968), 59–61, 98–103, for characterizations of the positivist conception of a fact.

52. Lorraine Daston, "The Factual Sensibility," *Isis* 79 (1988): 452–67.

53. As I will discuss briefly in chapter 1 and in more detail in chapter 6, while Boyle's conception is not identical, it is quite similar to the way in which Pérez-Ramos has described Bacon's nonempirical conception of the factual in *Bacon's Idea of Science*, esp. chaps. 5 and 12.

54. See especially Steven Shapin, "Pump and Circumstance: Robert Boyle's Literary Technology," *Social Studies of Science* 14 (1984): 481–520; and Shapin and Schaffer, *Leviathan and the Air-Pump*, chap. 2. Among the many other studies devoted to a rhetorical analysis of Boyle's works, see Dear, "Narratives, Anecdotes, and Experiments"; Jan V. Golinski, "Chemistry in the Scientific Revolution: Problems of Language and Communication," in *Reappraisals of the Scientific Revolution*, ed. David C. Lindberg and Robert S. Westman (Cambridge: Cambridge University Press, 1990), 367–96; James Paradis, "Montaigne, Boyle, and the Essay of Experience," in *One Culture*, ed. George Levine (Madison: University of Wisconsin Press, 1987), 59–91; and Steven Shapin, "Robert Boyle and Mathematics: Reality, Representation, and Experimental Practice," *Science in Context* 2 (1988): 23–58.

55. Andrew Pickering, for example, sets up this distinction in the title of his introductory essay to *Science as Practice and Culture*: "From Science as Knowledge to Science as Practice."

56. See David L. Hull, "In Defense of Presentism," *History and Theory* 18 (1979): 13–14; Söderqvist, "Existential Projects"; and Thomas Nickles, "Good Science as Bad History: From Order of Knowing to Order of Being," in *The Social Dimensions of Science*, ed. Ernan McMullin (Notre Dame, Ind.: University of Notre Dame Press, 1992), 85–129. The idea that some presentism not only is unavoidable but may also be beneficial can be found in the historiographical writings of R. G. Collingwood. For the fullest development of this idea of "effective-historical consciousness," and especially how some prejudgments may aid our understanding, see Hans-Georg Gadamer, *Truth and Method*, translation of *Wahrheit und Methode* (New York: Crossroads Publishing, 1975), pt. 2, sec. 2.

57. Hull, "In Defense of Presentism"; Giere, *Explaining Science*, 18–19. Giere also notes that one may draw generalizations about science only from a large sample of past scientific activity.

58. Nickles, "Good Science as Bad History," 115. See also Söderqvist, "Existential Projects," on how scientific biography can be "edifying" for the working scientist.

59. Nancy Cartwright, *How the Laws of Physics Lie* (Oxford: Clarendon Press, 1983); Allan Franklin, *The Neglect of Experiment* (Cambridge: Cambridge University Press, 1986); idem, *Experiment, Right or Wrong* (Cambridge: Cambridge University Press, 1990); Peter Galison, *How Experiments End* (Chicago: University of Chicago Press, 1987); and Ian Hacking, *Representing and Intervening* (Cambridge: Cambridge University Press, 1983). For a helpful review essay of recent trends in the philosophy and sociology of experiment, see Allan Franklin, "Experimental Questions," *Perspectives on Science* 1 (1993): 127–46.

60. Ian Hacking, "Statistical Language, Statistical Truth, and Statistical Reason," in *The Social Dimensions of Science*, ed. Ernan McMullin (Notre Dame, Ind.: University of Notre Dame Press, 1992), 138. As he explains, Hacking is attempting to find "something in between timeless metaphysics and the momentary social conjunctures" by which to characterize scientific practice (131). In "The Self-Vindication of the Laboratory Sciences," in *Science as Practice and Culture*, ed. Andrew Pickering (Chicago: University of Chicago Press, 1992), Hacking states that "the philosopher of experiment must descend from semantics and think about things and actions instead of ideas and expectations" (61).

61. Hacking, for example, developed his notion of "entity realism" in *Representing and Intervening*. Galison notes in the conclusion to *How Experiments End* that although "experimental physics cannot be rewritten as a logical fantasy in which all theorizing is forbidden until 'facts' clinch the argument," it cannot simply "be parodied as if it were no more grounded in reason than negotiations over the price of a street fair antique" (277). Among those who find the artifactual character of knowledge an epistemic problem are Karin D. Knorr-Cetina (*The Manufacture of Knowledge* [Oxford: Pergamon Press, 1981]) and Shapin and Schaffer (*Leviathan and the Air-Pump*).

62. Arthur Fine, *The Shaky Game: Einstein, Realism, and the Quantum Theory* (Chicago: University of Chicago Press, 1986); Giere, *Explaining Science*; David L. Hull, *Science as Process: An Evolutionary Account of the Social and Conceptual Development of Science* (Chicago: University of Chicago Press, 1988).

63. Fine, *Shaky Game*, 171.

64. Both accounts are developmental, but Giere follows a more cognitive approach, while Hull focuses more upon the social aspects of competition.

65. Giere, *Explaining Science*, 227. See also Fine, *Shaky Game*, 171. Fine discusses how we have reason to believe in the existence of some theoretical entities, even though they are embedded in "ever-open practices that constitute the judgment of those claims by the community of concerned scientists." Fuller ("David Bloor's *Knowledge and Social Imagery*") attempts to assimilate the work of Giere, Hacking, and others to what he terms Bloor's "naturalized" approach, but clearly Bloor's program is much more global than that of these philosophers.

66. Franklin, "Experimental Questions," 143; Joseph Rouse, "The Politics of Postmodern Philosophy of Science," *Philosophy of Science* 58 (1991): 607; Pickering, "From Science as Knowledge to Science as Practice," 8 (see also 22).

67. Stephen Toulmin, *Cosmopolis: The Hidden Agenda of Modernity* (Chicago: University of Chicago Press, 1990), 24. Toulmin discusses the trend away from the "charms of logical rigor" with its formal conception of "rationality" and the emergence of "informal procedures," similar to those at work in "the common law when it is functioning at its best" (193). As we will see in chapter 2, Boyle illustrated his own informal conception of rationality by reference to the procedures employed in the common law. See also Harvey Wheeler, "Francis Bacon's *New Atlantis:* The 'Mould' of a Lawfinding Commonwealth," in *Francis Bacon's Legacy of Texts*, ed. William A. Sessions (New York: AMS Press, 1990), 291–310, esp. 301 and 308, for Wheeler's discussion of how Bacon's philosophy of science was similar to "the hermeneutic process of law-finding that Bacon had developed for the

common law case method" and thus is similar to "postmodern" discussions of method.

68. Toulmin, *Cosmopolis*, 193, 188.

69. Ibid., 200–201, 204.

70. Hunter, "Casuistry in Action" and "Conscience of Robert Boyle."

71. Some discussions may appear redundant, but there are normally slightly different connotations that result from the context in which particular themes are discussed. In an undated letter to "M[y] S[ister] R[anelagh]," Boyle wrote that some redundancies in his works were not out of place because "some things are of that importance to my design, and yet not so easy to be taken notice of as such, it will not perhaps be amiss, that they be both proposed in various manners and by that means diligently inculcated." Royal Society, *Boyle Letters* 1: fol. 154r (see the following note).

72. Thomas Birch's 1772 edition of Boyle's *Works* is the most accessible version of Boyle's publications because of the 1965 facsimile edition. A new twelve-volume edition of his works, edited and annotated by Michael Hunter and Edward B. Davis, is due out in 1997–98 as part of the Pickering Masters collection. To make cross-reference as easy as possible, I will cite passages from Boyle's works by short title, and section heading where applicable, as well as volume and page numbers from the Birch edition. Boyle's unpublished writings at the Royal Society of London include forty-six volumes of manuscript material and notebooks and seven volumes of correspondence. I will refer to these as Royal Society, *Boyle Papers*, and Royal Society, *Boyle Letters*, with volume and page or folio numbers. This material has recently become available on microfilm as *Letters and Papers of Robert Boyle* (Bethesda, Md.: University Publications of America, 1991). In quotations from Boyle's unpublished papers, I have retained his punctuation and spelling except for abbreviations and contractions, which I have taken the liberty to spell out. The reader should also be aware that for stylistic purposes the first letter of a quoted passage may be capitalized even though it is not so in the original.

The Birch edition is not totally reliable because of the imposition of eighteenth-century editorial practices. Boyle's unpublished manuscripts are more trustworthy indicators of his style, but their arrangement into volumes in the nineteenth century, in categories such as "Theology," "Philosophy," "Science," and "Miscellaneous," is somewhat misleading. See Michael Hunter, *Letters and Papers of Robert Boyle: A Guide to the Manuscripts and Microfilm* (Bethesda, Md.: University Publications of America, 1991), for a detailed discussion of the history of Boyle's papers. Whenever possible in this book, I will compare passages cited from Boyle's unpublished manuscripts with similar passages from his published works.

73. It will become apparent that Boyle carefully defined his terms and used them in a consistent manner throughout his works. In his unpublished manuscripts one can also see the care he took to rework passages until he found the wording that best expressed his thought. Royal Society, *Boyle Papers*, volume 35, contains extensive notes on a type of dictionary compiled by Boyle in which words are listed together with their usages and compared with other words of similar meaning. See also Royal Society, *Boyle Letters* 4: fols. 23r–24v, for Boyle's editorial comments on a manuscript sent to him by an unnamed author. He suggested a number of lin-

guistic changes in the English words chosen by the author based upon a consideration of their Latin roots.

74. Royal Society, *Boyle Papers* 36: fol. 13v. Hunter (*Letters and Papers*) has determined that this outline is in the handwriting of an unidentified amanuensis employed by Boyle during the 1660s.

75. Hunter's discussion of Boyle's "dysfunctionalism" in "Conscience of Robert Boyle" is somewhat similar, in that he focuses in part upon Boyle's cautious, and at times overly "scrupulous," attitude. I prefer Boyle's term, *diffidence*, because of the way in which it provides a helpful contrast to the widespread "confidence" expressed by many modern thinkers, such as Descartes and Newton. Söderqvist ("Existential Projects") discusses how passions may have cognitive implications. While little is known about Boyle's passions, his diffidence appears to be at least partly a personal and emotional trait that is like a passion, and it did contribute to what Söderqvist has termed an "ethical life of reflection."

76. Boyle, *The Natural History of Human Blood*, postscript to "An Appendix to the Memoirs," in *Works* 4: 758–59.

Chapter One

1. The following account of Boyle's youth is based upon Boyle's third-person autobiography of his early years, an "Account of Philaretus in His Minority," in *Works*, vol. 1; his correspondence published in *Works*, vols. 1 and 6; Maddison, *Life*; and an account of Boyle's first opinions about Aristotle in Royal Society, *Boyle Papers* 38: fols. 80r–81r, printed in M. B. Hall, *Boyle on Natural Philosophy*, 177–79.

2. As attested to by John Carew in a letter to Boyle's father, the earl of Cork, 17 October 1635, and a 1636 letter from Wotton to the earl, both printed in Maddison, *Life*, 10, 13. It should be noted that Carew's praise seems to be more than that of a servant attempting to please his employer, since Carew did not have the same high praise for Francis, Boyle's older brother, who was enrolled at Eton at the same time and was also under the charge of Carew. The boys' later tutor, Isaac Marcombes, had the same assessment. He felt that Francis had received little profit from his stay at Eton but that Robert was "ye finest gentleman of ye worlde." Letter from Marcombes to the earl of Cork, n.d., printed in Maddison, *Life*, 14.

3. Boyle, "Account of Philaretus," in *Works* 1: xv.

4. Ibid., xvii.

5. Ibid., xviii.

6. Royal Society, *Boyle Papers* 38: fol. 80r, in M. B. Hall, *Boyle on Natural Philosophy*, 177.

7. Ibid. See Maddison, *Life*, 30–31, and idem, "Studies in the Life of Robert Boyle, F.R.S., Part VII: The Grand Tour," *Notes and Records of the Royal Society* 20 (1965): 67, on Boyle's reading of Diogenes Laertius; and M. B. Hall, *Boyle on Natural Philosophy*, 61, on his study of Epicurus during his stay in Italy.

8. Royal Society, *Boyle Papers* 38: fol. 80r.

9. Ibid., fol. 80v. In his later published works Boyle often referred to these early observations, such as the display of veins and arteries that he saw at Padua. See Maddison, "Grand Tour," 58, 63, 65–66, 70, for further references.

10. Boyle, "Account of Philaretus," in *Works* 1: xxiv.

11. Royal Society, *Boyle Papers* 38: fol. 80v.

12. See Richard S. Westfall, "Galileo and Newton: Different Rhetorical Strategies," in *Persuading Science: The Art of Scientific Rhetoric,* ed. Marcello Pera and William R. Shea (Canton, Mass.: Science History Publications, 1991), 110–12, for a discussion about how Galileo had framed his rhetorical strategy in order to gain the attention of this young aristocratic audience. It could be said that Boyle was a member of this intended audience and that he was thus an early convert to Galileo's program. I will say more about Galileo's influence on Boyle in chapter 3 and about the ways in which Boyle's rhetorical strategies were similar to Galileo's in chapter 8.

13. Much has been written about the anti-Scholasticism of the early founders of the Royal Society. See, for example, Mordechai Feingold, "The Universities and the Scientific Revolution: The Case of England," in *New Trends in the History of Science,* ed. R. P. W. Visser, H. J. M. Bos, L. C. Palm, and H. A. M. Snelders (Amsterdam: Rodopi, 1989), 29–52. For Boyle's account of his good fortune at having found his sister, see Maddison, *Life,* 53–54, from the British Library (hereinafter BL), Sloane Manuscript 4229, fol. 68r. See More, *Life and Works,* on how this meeting could have been an "accident." See also Flora Masson, *Robert Boyle: A Biography* (London: Constable and Co., 1914); and Antonia Fraser, *The Weaker Vessel* (New York: Knopf, 1984), on Katherine's close relationship with Milton.

14. Letter from Boyle to Marcombes, 22 October 1646, in *Works,* 1: xxxiv. After much speculation, the "philosophical college" has been identified by most as the Hartlib circle that was meeting in London at the time. Boyle's studies also included theology and ethics. See Robert Boyle, *The Early Essays and Ethics of Robert Boyle,* ed. John Harwood (Carbondale: Southern Illinois University Press, 1991). Harwood maintains in his introduction to this work that Boyle "began his career as a moralist, not as a natural philosopher" (xviii). It is true that the earliest samples of Boyle's writing are concerned with such issues, but as this letter and many other surviving documents attest, Boyle was clearly interested in natural philosophy at the same time and had begun to compile notes and observations for its furtherance. He did not become actively involved in experimenting, however, until 1649.

15. *Works* 1: xxxiii. James R. Jacob, in "Boyle's Atomism and the Restoration Assault on Pagan Naturalism," *Social Studies of Science* 8 (1978): 211–33, notes that Boyle's long absence from England made it easier for him to remain uninvolved in English politics. It is difficult to determine Boyle's political allegiances because of his silence. In 1654 Boyle spoke disapprovingly of the Royalists when he wrote to John Mallet from Youghall that the "Torys have of late been very destructively active, & particularly to the prejudice of diverse of my tenants" (BL, Add. MS 32093, fol. 318v). His sister Katherine was involved with the Parliamentarian Party, but other members of his family favored the Royalists, and Boyle apparently remained on good terms with all of them. See Oster, "Virtue, Providence, and Political Neutralism," for a discussion of Boyle's neutrality and his reasons for it.

16. Royal Society, *Boyle Papers* 38: fol. 80v. Boyle's attitude toward Aristotle was similar to that of other modern philosophers: sometimes his acceptance of Aristotle was explicit, and at other times it was revealed implicitly in his practice. See, for example, Desmond Clarke's discussion of Descartes's Aristotelianism in *Descartes' Phi-*

losophy of Science (University Park: Pennsylvania State University Press, 1982). For more detailed accounts of Boyle's Aristotelianism, see Alexander, *Ideas, Qualities, and Corpuscles*, esp. chap. 2; and Emerton, *Scientific Reinterpretation of Form.*

17. Boyle, *The Origin of Forms and Qualities*, "Author's Discourse," in *Works* 3: 9.

18. Boyle, *A Free Inquiry into the Vulgarly Received Notion of Nature*, sec. 1, in *Works* 5: 166.

19. Boyle, *Origin of Forms and Qualities*, "Author's Discourse," in *Works* 3: 10.

20. Boyle, *Free Inquiry*, sec. 1, in *Works* 5: 166. Because of the obscurity of Aristotle's writing, Boyle maintained that attempts to determine what the ancient philosopher actually meant were futile. Boyle clearly saw the epistemological tension between Aristotle's discussions of induction in his *Organon*, where universal generalizations could be constructed based upon a few particular instances, and Aristotle's careful catalogs of particulars compiled in his histories of animals. There is still considerable controversy surrounding Aristotle's theory of induction. See, for example, Jonathan Barnes, "Aristotle's Theory of Demonstration," *Phronesis* 14 (1969): 123–52; D. W. Hamlyn, "Aristotelian Epagoge," *Phronesis* 21 (1976): 167–84; and Richard D. McKirahan, "Aristotelian Epagoge in *Prior Analytics* 2.21 and *Posterior Analytics* 1.1," *Journal of the History of Philosophy* 21 (1983): 1–13.

21. Boyle, *Origin of Forms and Qualities*, advertisements to sec. 2, in *Works* 3: 75.

22. Ibid., "Author's Discourse," in *Works* 3: 4–5.

23. Ibid., "Of Substantial Forms," in *Works* 3: 40.

24. Ibid., "Author's Discourse," in *Works* 3: 8.

25. Ibid., advertisements to sec. 2, in *Works* 3: 75.

26. Ibid., "Nature of Physical Qualities," in *Works* 3: 18.

27. Ibid., advertisements to sec. 2, in *Works* 3: 75.

28. Ibid., "About Subordinate Forms," in *Works* 3: 136.

29. Ibid., "Nature of Physical Qualities," in *Works* 3: 17. Boyle's nominalist critique of Aristotle in this and other passages should not be taken to indicate that Boyle was himself a radical nominalist. The extent of his nominalism will be discussed in chapter 4 below.

30. Ibid., "Of Substantial Forms," in *Works* 3: 41. When Boyle said that the words were "probably devised," he was not speaking about the likelihood of the truth of his historical account, but rather he was saying that the words had been originally chosen for good reasons. His use of the term *probable* will be discussed in more detail in the next chapter.

31. Boyle, *The Excellency of Theology*, sec. 5, in *Works* 4: 58. Boyle's interpretation of the history of atomism may seem a bit unusual from today's perspective, but it was not too different from that which Bacon described in *The New Organon*, bk. 1, aphorism li, in *The Works of Francis Bacon*, 15 vols., ed. James Spedding, Robert Leslie Ellis, and Douglas Denon Heath (Boston: Taggard and Thompson, 1860–64), vol. 8.

It is interesting that in his account of history, Boyle showed an appreciation for the social factors that can partially determine the fate of philosophical schools and doctrines. For example, he attributed Aristotle's dominance in part to the prestige he had from being Alexander's teacher and to the "lucky accident" that his works survived while some of his predecessors' had not.

32. Boyle, *Excellency of Theology*, sec. 5, in *Works* 4: 57. The four qualities were hot,

cold, wet, and dry; the four elements were earth, air, fire, and water. Boyle's criticisms of the premature construction of systems of philosophy will be discussed in more detail in chapters 3 and 8.

33. Ibid., 58.

34. Ibid., 57. These are the three general objections mentioned in his earlier manuscript.

35. Ibid., 59. Boyle grouped the work by Harvey and Gilbert with that of Bacon, while he linked Gassendi with Descartes. For that reason, Gassendi's work is not treated here as a separate philosophical alternative. Although Gassendi might have been more of an "empiricist" than Descartes, Boyle always referred to him as a mathematician or, what is almost the same, as an astronomer. See Lynn Joy, *Gassendi the Atomist* (Cambridge: Cambridge University Press, 1987), for a detailed account of Gassendi's philosophy.

Scholars have speculated about Gassendi's influence on Boyle. Kargon ("Charleton, Boyle, and the Acceptance of Epicurean Atomism") argues that Boyle's brand of corpuscularianism was influenced by Gassendi through the writing of Charleton. But Charleton's *Physiologia Epicuro-Gassendo-Charltoniana* was not published until 1654, rather late to have influenced Boyle's views. In 1647 Boyle wrote to Hartlib that Gassendi was "a great favorite" (*Works* 1: xli). Osler ("Locke and the Changing Ideal") argues that Boyle fashioned his model of empirical inquiry after Gassendi. But Boyle did not merely rely upon the accounts of ancient atomism by Gassendi and Charleton; he read many of the original works for himself. Furthermore, the extent of Boyle's "empiricism" is questionable. While Boyle admired the work of Gassendi, his method is more akin to Bacon's "marriage" of the empirical and the rational.

36. Boyle, *Excellency of Theology*, sec. 5, in *Works* 4: 59.

37. René Descartes, *Discourse on the Method*, in *The Philosophical Writings of Descartes*, 3 vols., trans. John Cottingham, Robert Stoothoff, and Dugald Murdoch (vol. 3 only: also Anthony Kenny) (Cambridge: Cambridge University Press, 1985–91), 1: 147, 115, 146. The editions of Bacon and Descartes cited here are, of course, not those that Boyle read. It is not possible to determine with accuracy which editions Boyle did use, but the general themes developed within these two philosophical positions, particularly those to which Boyle referred, are evident in modern texts.

38. Bacon, *New Organon*, bk. 1, aph. xxxvii, in *Works of Bacon* 8: 76.

39. Bacon, *Advancement of Learning*, bk. 2, in *Works of Bacon* 6: 266.

40. Descartes, *Discourse on the Method*, in *Philosophical Writings* 1: 115.

41. Bacon, *Advancement of Learning*, bk. 2, in *Works of Bacon* 6: 266–67.

42. Bacon, *New Organon*, preface, in *Works of Bacon* 8: 29. Shapin and Schaffer (*Leviathan and the Air-Pump*, 305) note that finding a space between dogmatism and skepticism was an issue of particular concern to Restoration politics. Clearly, it was also an issue for Bacon almost forty years earlier.

43. Boyle, *The Usefulness of Experimental Philosophy*, pt. 2, essay 1, sec. 7, in *Works* 3: 423.

44. Boyle, *The Sceptical Chymist*, "Author's Preface" to the appendix, in *Works* 1: 591.

45. Boyle, *Origin of Forms and Qualities*, advertisements to sec. 2, in *Works* 3: 75.

46. Laudan ("Clock Metaphor and Hypotheses," 34) argues that Boyle's corpuscularianism "inclined him to adopt Descartes' method of hypothesis, while simultaneously taking his experimentalism from Bacon." Westfall ("Unpublished Boyle Papers") also finds a tension between Boyle's empirical and hypothetical approaches. Following chapters will elaborate upon the numerous ways in which Boyle mixed elements of the two methods and his reasons for doing so.

47. On the nonempirical elements in Bacon's philosophy, see Curt J. Ducasse, "Francis Bacon's Philosophy of Science," in *Theories of Scientific Method*, ed. Ralph M. Blake, Curt J. Ducasse, and Edward H. Madden (Seattle: University of Washington Press, 1960), 50–74; Mary Hesse, "Francis Bacon," in *A Critical History of Western Philosophy*, ed. D. J. O'Connor (New York: Free Press, 1964), 141–52; Mary Horton, "In Defence of Francis Bacon," *Studies in History and Philosophy of Science* 4 (1973): 241–78; Lisa Jardine, "*Experientia literata* or *Novum organum?* The Dilemma of Bacon's Scientific Method," in *Francis Bacon's Legacy of Texts*, ed. William A. Sessions (New York: AMS Press, 1990), 47–67; Pérez-Ramos, *Bacon's Idea of Science*; and Peter Urbach, *Francis Bacon's Philosophy of Science* (La Salle, Ill.: Open Court, 1987). On what could be called the "empiricist" elements in Descartes's philosophy, see Ralph M. Blake, "The Role of Experience in Descartes' Theory of Method," in *Theories of Scientific Method*, 75–103; Desmond Clarke, "Descartes' Philosophy of Science and the Scientific Revolution," in *The Cambridge Companion to Descartes*, ed. John Cottingham (Cambridge: Cambridge University Press, 1992), 258–85; idem, *Descartes' Philosophy of Science*; Daniel Garber, "Science and Certainty in Descartes," in *Descartes: Critical and Interpretive Essays*, ed. Michael Hooker (Baltimore: Johns Hopkins University Press, 1978), 114–51; and idem, *Descartes' Metaphysical Physics*.

48. Bacon, *New Organon*, preface, in *Works of Bacon* 8: 34.

49. Ibid., bk. 1, aph. xcviii, in *Works of Bacon* 8: 134.

50. Ibid., preface, in *Works of Bacon* 8: 33. As such, experiment is a third methodological category that mixes elements from classical empiricism and rationalism. Sometimes it is said that Bacon was not a true experimentalist because he did not himself practice in accordance with his rhetoric. But Bacon did report a number of experimental trials—in his *New Organon*, bk. 2, for example—and Boyle clearly saw Bacon as an experimentalist. We will see Boyle's assessment of Bacon's experimental activity in chapter 7.

51. Descartes, *Discourse on the Method*, in *Philosophical Writings* 1: 143. The term that Descartes uses here is *expériences*. In the Cottingham, Stoothoff, and Murdoch edition, this term is consistently translated as "observations." In Elizabeth S. Haldane and G. R. T. Ross, trans., *The Philosophical Works of Descartes*, vol. 1 (Cambridge: Cambridge University Press, 1931), it is usually translated as "experiments." In this particular passage Descartes spoke of "unusual and highly contrived" *expériences* as well as those that present "themselves spontaneously to our senses." Thus the term seems to refer both to observations and to experiments.

52. Descartes, *Discourse on the Method*, in *Philosophical Writings* 1: 144. Garber (*Descartes' Metaphysical Physics*) has pointed out that this and similar passages do not display the type of rationalism characterized by the nineteenth-century conception. See

Clarke, "Descartes' Philosophy of Science," 280–81, for a discussion of the tension between passages like this one and others in which Descartes insisted upon strict deductions from a certain foundation.

53. Descartes, *The Principles of Philosophy*, in *Philosophical Writings* 1: 189. Once again, in the Haldane and Ross edition (1: 215), *expériences* is translated as "experiments."

54. Descartes to Mersenne, 10 May 1632, in Descartes, *Philosophical Writings* 3: 38. See Clarke, "Descartes' Philosophy of Science," 272–73, on Descartes's distrust of experiments done by others. See also Geneviève Rodis-Lewis, "Descartes' Life and the Development of His Philosophy," in *The Cambridge Companion to Descartes*, ed. John Cottingham (Cambridge: Cambridge University Press, 1992), 34, for Descartes's references to Bacon; and Graham Rees, "The Transmission of Bacon Texts: Some Unanswered Questions," in *Francis Bacon's Legacy of Texts*, ed. William A. Sessions (New York: AMS Press, 1990), 311–23, on the dissemination of Bacon's works on the Continent.

55. See Horton, "Defence of Bacon"; and Urbach, *Bacon's Philosophy of Science*, for detailed analyses of Bacon's "interpretations."

56. Bacon, *New Organon*, bk. 2, aph. xx, in *Works of Bacon* 8: 210.

57. Ibid., aph. xix, in *Works of Bacon* 8: 210. His "first vintage" is discussed in detail on 210–18.

58. Descartes, *Discourse on the Method*, pt. 5, in *Philosophical Writings* 1: 132–34.

59. Descartes, *Rules for the Direction of the Mind*, in *Philosophical Writings* 1: 43–44. He noted that there was no problem with using suppositions that need not be believed "if it is obvious that they detract not a jot from the truth of things, but simply make everything much clearer" (40). See Clarke, "Descartes' Philosophy of Science," 268–70, for a more detailed discussion of Descartes's attitude toward false hypotheses.

60. Bacon, *De augmentis*, bk. 3, in *Works of Bacon* 8: 507.

61. Descartes, *Principles of Philosophy*, "Author's Letter," in *Philosophical Writings* 1: 186. These metaphors were epistemologically significant. See Garber, *Descartes' Metaphysical Physics*, 30–39, for a discussion of Descartes's conception of the epistemological interconnection of all truths; and Clarke, "Descartes' Philosophy of Science," 265, on the way in which Descartes judged the acceptability of hypotheses by how well they fit within his system of knowledge. Bacon's conception of a pyramid also entails the conception that the knowledge of nature will be interconnected, but it is not clear if his reasons for believing so were epistemological or ontological. He had also used the metaphor of a tree to describe his idea of the interconnection of knowledge in his *De augmentis* (bk. 3, in *Works of Bacon* 8: 471). As we will see in chapter 4, Boyle also believed that knowledge was interconnected, but his conception was based upon his ontological belief in the interconnection of natural processes.

62. Descartes, *Discourse on the Method*, in *Philosophical Writings* 1: 150. See Garber, *Descartes' Metaphysical Physics*, 53–58; and Clarke, "Descartes' Philosophy of Science," 272–75, on the ways in which Descartes's epistemological goal was similar to that of the school philosophers.

63. Descartes, *Principles of Philosophy*, in *Philosophical Writings* 1: 288. The significance of this passage is discussed by Phillip R. Sloan in "Descartes, the Skeptics, and

the Rejection of Vitalism in Seventeenth-Century Physiology," *Studies in History and Philosophy of Science* 8 (1977): 1–28.

64. Descartes, *Oeuvres de Descartes*, 12 vols., ed. C. Adam and P. Tannery (Paris: Léopold Cerf, 1897–1910), 1: 196, as cited in Blake, "Experience in Descartes' Theory of Method," 86.

65. Descartes, *Oeuvres de Descartes* 1: 100, as cited in Ernan McMullin, "Conceptions of Science in the Scientific Revolution," in *Reappraisals of the Scientific Revolution*, ed. David C. Lindberg and Robert S. Westman (Cambridge: Cambridge University Press, 1990), 43. See also Descartes's letters to both Mersenne and Beeckman, in which he made much the same claim: *Philosophical Writings* 3: 13–17, 26–28, 45. Some of Descartes's letters were published in three volumes by his disciple Claude Clerselier in 1657. Boyle read these letters, as is apparent from his references to them in his "Hydrostatical Discourse," in *Works* 3: 598; and in his *Hydrostatical Paradoxes*, in *Works* 2: 793.

66. Descartes, *Rules for the Direction of the Mind*, in *Philosophical Writings* 1: 48. He illustrated his meaning by an account of how one would proceed to discover truths about the nature of the magnet. First, one "carefully gathers together all the available observations concerning the stone in question; then he tries to deduce from this what sort of mixture of simple natures is necessary for producing all the effects the magnet is found to have" (49). Given such a passage, it could be argued that Descartes's approach was more experimental than that which is normally attributed to him. We will see that Boyle did not perceive it as such, however. Immediately following this passage, Descartes went on to note that knowledge consists "simply in the putting together of self-evident facts."

67. Bacon, *New Organon*, bk. 1, aph. xcviii, in *Works of Bacon* 8: 133.

68. Ibid., preface, in *Works of Bacon* 8: 32.

69. Ibid., bk. 2, aph. lii, in *Works of Bacon* 8: 347–48.

70. Bacon, *Parasceve; or, Preparative Towards a Natural and Experimental History*, in *Works of Bacon* 8: 361.

71. In the preface to *New Organon* Bacon wrote that all "premature human reasoning which anticipates inquiry, and is abstracted from the facts rashly and sooner than is fit, is by me rejected" (*Works of Bacon* 8: 37).

72. Bacon, *New Organon*, bk. 1, aph. lxix, in *Works of Bacon* 8: 100.

73. Ibid., aph. xcviii, in *Works of Bacon* 8: 134.

74. Bacon, *Advancement of Learning*, bk. 2, in *Works of Bacon* 6: 267–68.

75. Bacon, *New Organon*, preface, in *Works of Bacon* 8: 32.

76. Bacon, *Advancement of Learning*, bk. 2, in *Works of Bacon* 6: 276. Compare this with his discussion of the Idols of the Tribe in *New Organon*, bk. 1, aph. xli, where he noted that "the human understanding is like a false mirror" that "distorts and discolors the nature of things" (*Works of Bacon* 8: 77).

77. Bacon, *New Organon*, bk. 1, aph. xxxviii, in *Works of Bacon* 8: 76.

78. Ibid., bk. 1, aph. c, in *Works of Bacon* 8: 135–36.

79. Ibid., aph. l, in *Works of Bacon* 8: 83.

80. Ibid., aph. xcix, in *Works of Bacon* 8: 135.

81. Ibid.

82. Bacon, *Parasceve*, in *Works of Bacon* 8: 355. Pérez-Ramos (*Bacon's Idea of Science*) has

said that this facet of Bacon's philosophy displays a "constructivist" stance toward knowledge. This concept will be discussed in more detail below.

83. Bacon, *New Organon*, bk. I, aph. lxx, in *Works of Bacon* 8: 100.

84. Bacon, *Advancement of Learning*, bk. 2, in *Works of Bacon* 6: 187–88. The nature of Descartes's method, where truths are established by logical connections within a grand system, forced him to admit that "if what I have written" on any topic "turns out to be false, then the rest of my philosophy is entirely worthless" in his letter to Mersenne of 9 February 1639. Descartes, *Philosophical Writings* 3: 134. He expressed a similar sentiment in earlier letters to Mersenne and to Beeckman (41, 46).

85. Laudan, in "Clock Metaphor and Hypotheses," presented Descartes's method as more hypothetical than have traditional interpretations in the history of science. Then, by arguing that Boyle's perception of the Cartesian system would have been similar, he concluded that Boyle owed his methodology to the influence of Descartes. Rogers, in "Descartes and the Method of English Science," criticized Laudan's interpretation by arguing that Descartes's philosophy was not as hypothetical as Laudan claims. See McMullin, "Conceptions of Science in the Scientific Revolution"; McMullin also argues for a more traditional view of Cartesian methodology. I have argued for a direct Baconian influence on Boyle in "Robert Boyle's Baconian Inheritance."

86. Quoted in Laudan, "Clock Metaphor and Hypotheses," 50–51 n. 19, from an unpublished note in which Boyle refers to Descartes as *"le plus ingenieu."* Laudan concludes from this that "Boyle's admiration for Descartes is certainly undisguised."

87. Boyle, *Certain Physiological Essays*, "Proëmial Essay," in *Works* 1: 311.

88. Boyle, *Origin of Forms and Qualities*, preface, in *Works* 3: 12, 11–12.

89. Boyle, *Certain Physiological Essays*, "Proëmial Essay," in *Works* 1: 300. In chapter 8 we will see that Boyle was also opposed to systems because of the rhetorical effect that they produced. Systematic writing could give the impression that all the necessary work had been completed and thus could serve to discourage others from furthering the inquiry.

90. Bacon, *New Organon*, bk. 1, aph. cxxiv, in *Works of Bacon* 8: 156. See also Bacon, *Advancement of Learning*, bk. 1, in *Works of Bacon* 6: 131.

91. Boyle, *Certain Physiological Essays*, "Proëmial Essay," in *Works* 1: 301.

92. Ibid., 301–2.

93. Ibid., 303. In the same passage he noted that "it has been truly observed by a great philosopher, that truth does more easily emerge out of error than confusion." This was Bacon's precept as set down in *New Organon*, bk. 2, aph. xx, in *Works of Bacon* 8: 210.

94. Boyle, *Certain Physiological Essays*, "Proëmial Essay," in *Works* 1: 303. Boyle's discussion of the usefulness of hypotheses is markedly dissimilar from Descartes's.

95. Boyle, *Usefulness of Natural Philosophy*, pt. 1, essay 4, in *Works* 2: 46–47.

96. Bacon, *New Organon*, bk. 1, aph. xlix, in *Works of Bacon* 8: 82.

97. Ibid., aph. cxxiv, in *Works of Bacon* 8: 156.

98. Boyle, *Usefulness of Natural Philosophy*, pt. 1, essay 4, in *Works* 2: 46. See Alexander, *Ideas, Qualities, and Corpuscles*, chap. 1, for a discussion of the Baconian elements in Boyle's philosophy.

99. Boyle, *Things Said to Transcend Reason*, 6th advice, in *Works* 4: 467. This notion of "assured knowledge" tied Boyle's criticism of mathematics to his criticism of systems in general. The supposed completeness of the demonstrations presented by such writers could be misleading. As both Clarke ("Descartes' Philosophy of Science") and Garber (*Descartes' Metaphysical Physics*, chap. 2) have pointed out, Descartes's method was not as mathematical as it often appears at first sight. But for my purposes, it is important to see that Boyle did find it to be overly mathematical, and there is support for such a reading. See John Cottingham, introduction to *The Cambridge Companion to Descartes*, ed. John Cottingham (Cambridge: Cambridge University Press, 1992), 14, on Descartes's "program for mathematicizing physics"; and Descartes's letter to Mersenne, in Descartes, *Philosophical Writings* 3: 124, where he praised Galileo for his use of "mathematical methods."

100. Letter from Boyle to Samuel Hartlib, in *Works* 1: xxxix. The nonacceptance of Copernican astronomy was not unusual even at this late date.

101. Boyle, *New Experiments Physico-Mechanical*, conclusion, in *Works* 1: 117, 121.

102. Laudan ("Clock Metaphor and Hypotheses," 28) sees such expressions as "lipservice which most British methodologists paid to Bacon."

103. Boyle, *Certain Physiological Essays*, "Proëmial Essay," in *Works* 1: 307. The "underbuilder" or "under-laborer" image was becoming popular at this time. Bacon spoke of the need for such modesty in the preface to his *New Organon*, Galileo described himself as digging in the quarries for the materials upon which others would build in his *Two New Sciences*, and Locke would follow this precedent in his *Essay on Human Understanding*.

104. Boyle, *Experimenta et observationes physicae*, advertisement, in *Works* 5: 567.

105. Thomas Hobbes, *Physical Dialogue*, translation of *Dialogus physicus de natura aeris*, trans. Simon Schaffer, in Shapin and Schaffer, *Leviathan and the Air-Pump*, 379.

106. Letter from Spinoza to Oldenburg, April 1662, in Henry Oldenburg, *The Correspondence of Henry Oldenburg*, 11 vols., ed. A. Rupert Hall and Marie Boas Hall (Madison: University of Wisconsin Press, 1965), 1: 462. The work to which Spinoza referred was Boyle's *Certain Physiological Essays*, particularly "A Physico-Chymical Essay . . . touching the differing Parts and Redintegration of Salt-Petre."

107. Boyle, *Certain Physiological Essays*, "Proëmial Essay," in *Works* 1: 308.

108. Ibid., 309, 310. Boyle's notion of "intermediate causes" will be discussed in more detail in chapters 3, 4, and 6.

109. Bacon, *New Organon*, in *Works of Bacon* 8: 53.

110. Boyle, *Usefulness of Experimental Philosophy*, pt. 2, essay 1, sec. 7, in *Works* 3: 423.

111. Bacon, *New Organon*, bk. 1, aph. cxxiv, in *Works of Bacon* 8: 157. See also Bacon, *Parasceve*, in *Works of Bacon* 8: 363–64.

112. Pérez-Ramos, *Bacon's Idea of Science*, esp. chaps. 10, 11, and 12.

113. Boyle, *Usefulness of Experimental Philosophy*, pt. 2, essay 1, sec. 7, in *Works* 3: 423.

114. Royal Society, *Boyle Papers* 28: 293.

115. Letter from Boyle to Marcombes, in *Works* 1: xxxiv. As we will see, Boyle's conception of usefulness was both epistemological and utilitarian. Gilbert Burnet, bishop of Salisbury, wrote in notes for a planned biography of Boyle that his "first design in philosophy was to recommend the usefulness of Experimental Philosophy to the gentry who being disgusted with the dry notions of that taught in schools and

knowing no other were prejudiced against all philosophy" (BL, Add. MS 4229, fol. 60v). See Harwood in Boyle, *Early Essays*, pp xl–xli, for a discussion of Boyle's early commitment to an ethics, composed of modesty and toleration, designed to serve the common good. Boyle was writing *The Usefulness of Natural Philosophy*, part 1, at the same time that he was composing his "Aretology."

Chapter Two

1. *Empiricism* is a term that tends to be used rather loosely, but historically, from ancient times to the present, empiricism has normally been identified with the position that knowledge claims must be restricted to the actual and the observable. See, for example, Lloyd, *Science, Folklore, and Ideology*, for an account of empiricism in the ancient medical sects; Popkin, *History of Skepticism*, for an account of the empiricism of Gassendi and Mersenne; and van Fraassen, *Scientific Image*, for the latest contribution to empiricist science. See Kockelmans, "Problem of Truth," for one of the most recent criticisms of empiricist assumptions.

2. Van Fraassen (*Scientific Image*, 1–2), for example, characterizes Boyle as a failed empiricist because he advocated a corpuscular explanation of the macroworld. Shapin and Schaffer (*Leviathan and the Air-Pump*) present Boyle as a naive empiricist when they say, for example, that his method "regards experimentally produced matters of fact as self-evident and self-explanatory" (225).

3. Boyle's experimentalism has been compared with Gassendi's probabilistic empiricist approach in Osler, "Locke and the Changing Ideal." In "Providence and Divine Will" Osler claims that "Gassendi's mitigated skepticism and nominalist ontology became characteristic of English science as represented in the works of Boyle and Newton" (560). At the other extreme, Laudan, in "Clock Metaphor and Hypotheses," characterizes Boyle's method as influenced by the hypothetical rationalism of Descartes. In both cases the conclusion is the same: Boyle was a skeptical probabilist.

4. We saw in the last chapter that Boyle sought to discover the invisible corpuscles that were causally responsible for manifest effects in nature. We will explore this in more detail in chapters 4 and 5.

5. Even when the nonempirical elements are recognized, the references to experience are still taken as indicative of an empiricist tendency to which the nonempirical must be reconciled.

6. See Paul H. Kocher, "Francis Bacon on the Science of Jurisprudence," *Journal of the History of Ideas* 18 (1957): 3–26; and Bernard McCabe, "Francis Bacon and the Natural Law Tradition," *Natural Law Forum* 9 (1964): 111–21. The attempt to interpret "law" univocally creates confusion. McCabe, for example, finds it puzzling that Bacon was apparently unconcerned with natural law in the legal tradition, since he was interested in the "existence of ascertainable natural laws in the cosmological sense" (111). But for Bacon, natural laws in science were to be discovered from particulars, whereas natural laws in the legal sense were the rational first principles discoverable from and grounded upon individual reason. Therefore, it is not surprising that Bacon was opposed to natural laws in the legal sense, just as he appears to be in McCabe's analysis.

7. Boyle, in *Free Inquiry* (sec. 2, in *Works* 5: 170–71), noted that because inanimate things lack intelligence, they cannot, strictly speaking, make their motions conformable to laws. Chief Justice Hale also drew a distinction between laws of nature, which were "fixed and unalterable," and laws of men, which were not. See Matthew Hale, *Considerations Touching Amendments or Alterations of Laws*, in Edmund Heward, *Matthew Hale* (London: Robert Hale and Co., 1972). By the eighteenth century these distinctions became blurred, however. Blackstone, for example, stated that a "law signifies a rule of action and it is applied indiscriminately to all kinds of action whether animate, or inanimate, rational or irrational." See William Blackstone, *Commentaries on the Laws of England*, vol. 1 (1765; reprint, Chicago: University of Chicago Press, 1979), 38. Boyle's assessment of the nature and importance of scientific laws will be discussed more fully in the next two chapters.
8. See Jane E. Ruby, "The Origins of Scientific 'Law,'" *Journal of the History of Ideas* 47 (1986): 341–60.
9. Boyle, for example, spoke of the "testimony of nature" and said that "matters of fact ought to be brought to trial" in his *Hydrostatical Paradoxes*, preface, in *Works* 2: 742, 744. He spoke of "judicious and illustrious witnesses" in his *New Experiments*, experiment 17, in *Works* 1: 34. Bacon spoke of the "inquisition of things" in *New Organon*, in *Works of Bacon* 8: 35. He noted that "nature exhibits herself more clearly under the trials and vexations of art than when left to herself" in *De augmentis*, bk. 2, chap. 2, in *Works of Bacon* 8: 415.
10. Hunter ("Conscience of Robert Boyle") suggests that Boyle's beliefs concerning casuistry contributed to his diffident and scrupulous attitude in natural philosophy. Barbara J. Shapiro has given the fullest treatment of the relationship between English common law and science, in her "Law and Science in Seventeenth-Century England," *Stanford Law Review* 21 (1969): 727–65, and her *Probability and Certainty*. She has shown an important link between the two realms in their shared notion of "moral certainty." She does not, however, provide a detailed discussion of the notion of experience, which was the ground for this type of certainty. Neal Gilbert, in *Renaissance Concepts of Method* (New York: Columbia University Press, 1960), discusses the relevance of legal methodology for English science, but he restricts his attention to the type of law taught at the universities and does not examine that which was practiced by common lawyers such as Bacon at the Inns of Court.

    The seventeenth century was not the only time that legal practices had an important influence on science. See G. E. R. Lloyd, *Magic, Reason, and Experience* (Cambridge: Cambridge University Press, 1979). Lloyd maintains that the legal system of the polis of ancient Greece was important for the rise of science per se.
11. Julian Martin, *Francis Bacon, the State, and the Reform of Natural Philosophy* (Cambridge: Cambridge University Press, 1992); Shapin and Schaffer, *Leviathan and the Air-Pump*; Charles Whitney, *Francis Bacon and Modernity* (New Haven, Conn.: Yale University Press, 1986). Markku Peltonen, in "Politics and Science: Francis Bacon and the True Greatness of States," *Historical Journal* 35 (1992): 279–305, has argued, against both Martin and Whitney, that there are significant dissimilarities between Bacon's political and scientific programs. I have argued against Shapin and Schaffer's political interpretation of Boyle's use of the legal analogy in "Scientific Experi-

ment and Legal Expertise": The Way of Experience in Seventeenth-Century England," *Studies in History and Philosophy of Science* 20 (1989): 19–45. Peter Dear, in "Narratives, Anecdotes, and Experiments," has taken exception to my criticism of Shapin and Schaffer as well as to my analysis of the methodological aspects of the legal analogy. He finds a significant disanalogy between the types of facts dealt with by the law and by experiment. According to Dear, the fact in a legal trial "is already well known; the question of whether such things happen at all is not at stake. But an experimental event is by its very nature *novel;* its establishment is a matter of determining the reality of a kind of event that is not already generally accepted as occurring" (163). This criticism is wide of the mark for a number of reasons. All analogies, by definition, have a negative aspect as well as a positive one, and my task has merely been to uncover the positive ways in which the law could be likened to experimental reasoning. More important, however, we will see in chapter 7 that the notion of the novelty of an experimental event is misleading. Most often, Boyle's experiments were designed to reproduce a familiar phenomenon in an artificial environment in order to isolate the causal variables responsible for its production.

12. Blackstone, in his *Commentaries,* cited Locke's claim that it would be a "strange absurdity" to suppose that "gentlemen of independent estates and fortune" would be ignorant of the law (7). Royal Society, *Boyle Papers* 36: fols. 20r–21r, contains a list of legal documents, mostly concerned with the transfer of property, that were "left in a Wooden Box with Mr. Dury." There are copies of legal documents of various kinds scattered throughout the *Papers.* For a discussion of Boyle's dealings with the law, see Maddison, *Life.* Since the legal analogy is methodological and not substantive, acquaintance with legal procedures is sufficient for my purposes. See Boyle, *Early Essays,* 256–81, on the number of legal handbooks and dictionaries in Boyle's possession at the time of his death.

13. The details of this discussion are taken from J. H. Baker, *An Introduction to English Legal History* (London: Butterworths, 1971); idem, ed., *The Reports of John Spelman,* vol. 2 (London: Selden Society, 1978); Harold J. Berman, *Law and Revolution: The Formation of the Western Legal Tradition* (Cambridge: Harvard University Press, 1983); William Holdsworth, *A History of English Law,* 4th ed., 12 vols. (London: Methuen, 1936); E. W. Ives, *The Common Lawyers of Pre-Reformation England* (Cambridge: Cambridge University Press, 1983); John H. Langbein, *Prosecuting Crime in the Renaissance* (Cambridge: Harvard University Press, 1974); Brian P. Levack, *The Civil Lawyers in England, 1603–1641* (Oxford: Clarendon Press, 1973); Wilfred R. Prest, *The Inns of Court under Elizabeth I and the Early Stuarts, 1590–1640* (Totowa, N.J.: Rowman and Littlefield, 1972); and W. C. Richardson, *A History of the Inns of Court* (Baton Rouge: Louisiana State University Press, 1975). Roman law was taught at English universities for use in the courts of the Admiralty and the Church. Common law was used for all other legal matters and was taught at the Inns of Court.

14. Ives, *Common Lawyers,* 160. The foundation of the law was considered to reside in the Writs, which were arranged in groups within the Register, and the Yearbooks, which were records of past cases. Baker (*Reports of Spelman,* 27–28) argues that there was not, in fact, as much continuity in the records as the lawyers claimed, but the

issue here is how the lawyers justified their procedures, not whether this justification was completely warranted.

15. Bacon, *The Maxims of the Law*, in *Works of Bacon* 14: 182–83.

16. Ibid., 181. In his *De augmentis* (bk. 8, aph. lxxxv, in *Works of Bacon* 9: 338), Bacon noted that one should "make the rule from existing law." *De augmentis* is not a treatise on the common law per se. It was addressed to the reform of learning in all areas, including the law, and reflected Bacon's preference for the common law procedure of historical precedent when dealing with civil matters. Compare Holdsworth's discussion in *History of English Law* 5: 485–87.

17. On the flexibility of the lawyers' approach to precedent, see Ives, *Common Lawyers*, 156; Baker, *Reports of Spelman* 163; and Holdsworth, *History of English Law* 4: 285.

18. Bacon, *Maxims of the Law*, in *Works of Bacon* 14: 285.

19. See Baker, *Reports of Spelman*, 161–63.

20. See Bacon, *Works of Bacon* 10: 240–46, for his notes on the Commission for the Union and his preface to the Report of the Commission.

21. Ibid., 243, 328.

22. Bacon, *Works of Bacon* 7: 639–79. McCabe, in "Bacon and the Natural Law Tradition," 116, notes that "this proceeding was clearly a test case." The case was in part orchestrated by the Crown as a way to get the law passed. See also Peltonen, "Politics and Science," 293–94.

23. Quoted in Holdsworth, *History of English Law* 4: 226. Coke was lord chief justice of the Court of Common Pleas and for many years was Bacon's rival for royal support. He was the author of *The Institutes*, which, together with the works of Bacon, came to be used by some revolutionaries in the 1640s as their grounds for asserting the authority of Parliament. This argument could be made since the common law was seen as a constraint on the absolute power of the sovereign. See Thomas Andrew Green, *Verdict according to Conscience* (Chicago: University of Chicago Press, 1985), 163; Levack, *Civil Lawyers*, 123, 144; and David Ogg, *England in the Reign of Charles II*, vol. 1 (Oxford: Clarendon Press, 1955). Lawyers themselves were divided in their allegiance during the civil war, and a number of prominent legalists had been leaders of the opposition to Crown policy. See Richardson, *History of the Inns of Court*, 290–91.

24. Coke quoted in Holdsworth, *History of English Law* 4: 226. The Roman lawyers' legal debates bore a strong similarity to the disputations on texts held at the universities. See Levack, *Civil Lawyers*, 150.

25. See Richardson, *History of the Inns of Court*, 91–150, for the fullest account of education at the Inns. See also Baker, *Reports of Spelman*, 161–63; and Ives, *Common Lawyers*, 37–38, 158–61, on the "common erudition" learned at the Inns.

26. Quoted in Richardson, *History of the Inns of Court*, 148–49. Coke's arguments are quite similar to Bacon and Boyle's arguments against the Aristotelians. Somewhat paradoxically, however, the argument is actually derived from those that Aristotle had set forth for different modes of demonstration in his *Nicomachean Ethics*. This influence will be seen more clearly in Hale's discussion below.

27. Precedent was only a guide, because it was not strictly binding. See Baker, *Reports of Spelman*, 161–63.

28. See Barbara J. Shapiro, "Sir Francis Bacon and the Mid-Seventeenth-Century Movement for Law Reform," *American Journal of Legal History* 24 (1980): 331–62, for a detailed account of the type of reforms proposed during the revolutionary period.

29. Thomas Hobbes, *Dialogue between a Philosopher and a Student of the Common Laws of England*, in Thomas Hobbes, *The English Works of Thomas Hobbes*, 11 vols., ed. William Molesworth (London: John Bohn, 1839–45), 6: 1–160.

30. Ibid., 4. Hobbes directed his attention to Coke without noting that Bacon, while differing on a number of points from Coke, held the same type of attitude toward common law procedures in general.

31. Ibid., 5, 8, 26. Hobbes's definitions of *reason* and *law* were not those that a common lawyer would accept, and therefore the subsequent arguments that depended upon these definitions would also not be acceptable. Compare his discussion in *Leviathan*, chap. 19, where he maintained that judges have to be only men of common sense, not students of the law.

32. This was Hobbes's form of argument throughout the *Dialogue*. See *English Works* 6: 63, 119, 125, 127, 132, 137. Hobbes concluded that Coke employed the "weakest" of reasoning (144).

33. Matthew Hale, "Reflections by the Lord Chief Justice Hale on Mr. Hobbes His Dialogue of the Law," in Holdsworth, *History of English Law* 5: 503. I have modernized Hale's spelling. Hale's response came in two parts: the first (500–506) concerns Hobbes's notion of reason; the second (506–13) argues against Hobbes's notion of sovereignty. Boyle was acquainted with a number of Hale's associates, in particular John Wilkins and Gilbert Burnet. Burnet was Hale's biographer and had planned to write a biography of Boyle. See the notes that he made for this purpose in BL, Add. MS 4229, fols. 60r–63v. See More, *Life and Works*, for more details on this relation.

34. Hale, "Reflections," in Holdsworth, *History of English Law* 5: 501–2.

35. See Baker, *Reports of Spelman*, 29.

36. Hale, "Reflections," in Holdsworth, *History of English Law* 5: 506.

37. Ibid., 503. Hale reflected Coke's sentiments when he wrote that when one is "long and industriously studied" in the law, although there will be no "infallibility in his judgment . . . he will be much better fitted for right judgment" than "he that hath no other stock to trade upon than the bare exercise of his faculty of reason" (505).

38. Ibid., 503.

39. Ibid., 502.

40. Bacon, *De augmentis*, in *Works of Bacon* 9: 232.

41. Hale, "Reflections," in Holdsworth, *History of English Law* 5: 503, 505. The inability of a particular person to grasp the reason of the law is not justification for rejecting it as unreasonable. Hale stressed that this experience is a constraint on the decisions of judges, which otherwise would be arbitrary (506). This type of argument could be interpreted as an attempt by a member of a particular interest group to preserve the status quo, but Hale's arguments were not simply reactionary. See Barbara J. Shapiro, "Law Reform in Seventeenth-Century England," *American Journal of Legal History* 19 (1975): 280–312, on the reforms suggested by the Hale Commission. As Shapiro notes, Hale was opposed to the "superstitious veneration" of

tradition (298). The law had to be modified continuously as the circumstances warranted. Hale also wanted reforms in the law so that it would be more accessible. He argued that "one large and authentical abridgement of the books and tracts of the law reduced under apt and alphabetical titles" ought to be prepared; Hale, *Considerations Touching Amendments*, in Heward, *Matthew Hale*, 164.

42. Boyle considered legal proof to be of the type that he called moral demonstration. Common lawyers believed that a training in logic and dialectic was a necessary preliminary to the study of law, but when Abrahm Fraunce, a member of Gray's Inn at the time of Bacon, published his *Lawyer's Logicke* in 1588, it was viewed as little more than an English translation of the dialectic of Ramus and received little attention. See Prest, *Inns of Court*, 132–46; and Richardson, *History of the Inns of Court*, 147–49. Lisa Jardine, in *Francis Bacon: Discovery and the Art of Discourse* (Cambridge: Cambridge University Press, 1974), has argued that Bacon's "Great Instauration" should be viewed as being in line with the dialectical tradition. Her argument, however, does not do much more than establish that he was motivated by his aversion to dialectic, which would be a typical response of a common lawyer.

43. Langbein, *Prosecuting Crime in the Renaissance*, 211; Holdsworth, *History of English Law* 5: 169–75.

44. Langbein, *Prosecuting Crime in the Renaissance*, 237–39; Holdsworth, *History of English Law* 5: 173–87. Torture was used because a confession carried the value of 1. England also used torture for serious crimes, but there were no apparent criteria to determine when or why it would be used.

45. See Holdsworth, *History of English Law* 5: 195–96, on the open, public nature of trials. See Green, *Verdict according to Conscience*, 110–11; and Baker, *Reports of Spelman*, 92–100, on the necessity of oral testimony. By the seventeenth century written depositions had been introduced into English law, but they were not part of the official record and were not binding. An example of all these factors can be found in the 1649 "Trial of Lieutenant-Colonel John Lilburne," in T. B. Howell, comp., *State Trials*, vol. 4 (London: Longman, 1816), 1270–1470. Also in this trial one can see the great amount of freedom to speak that the defendant had. In Roman procedures, the defendant was extremely limited in what he could say and when he could say it. See Langbein, *Prosecuting Crime in the Renaissance*, 237; and Holdsworth, *History of English Law* 5: 169–75.

46. See Langbein, *Prosecuting Crime in the Renaissance*, 251, for the distinction between "matters of fact" and "matters of law." The jury, for example, would find the defendant guilty of homicide, and the bench would decide whether it was murder or manslaughter and would pass sentence accordingly. Shapin and Schaffer (*Leviathan and the Air-Pump*, 342) discuss this distinction as a "social boundary" that reflected the particular concerns of "the polity that emerged in the Restoration." See Sargent, "Scientific Experiment and Legal Expertise," for criticism of this interpretation. The category of "matters of fact" will be discussed in more detail in chapter 6 below.

47. Baker, *Reports of Spelman*, 107–12; Holdsworth, *History of English Law* 2: 108–17, 3: 596–634; Green, *Verdict according to Conscience*, 106–7, 142. The jurors were told to judge "according to the evidence and your conscience."

48. Holdsworth, *History of English Law* 3: 633. The right of the accused to challenge

the testimony of the accusers turned the trial into a "proper test," according to Green, *Verdict according to Conscience,* 136.

49. Matthew Hale, *The History of Common Law,* as quoted in Barbara J. Shapiro, "'To a Moral Certainty': Theories of Knowledge and Anglo-American Juries, 1600–1850," *Hastings Law Journal* 38 (1986): 161.

50. The assessment of credibility became official in the 1661 Clarendon Act, when the rule was changed from "two lawful witnesses" to "two lawful and credible witnesses" being necessary for conviction for treason. Shapin and Schaffer (*Leviathan and the Air-Pump,* 327) suggest that Boyle's use of the two-witness rule also indicates the incursion of Restoration political concerns into his scientific thought. But the two-witness rule dated back to biblical times, was official in England from the reign of Edward VI, and was also used on the Continent, where the testimony of two witnesses was sufficient for proof, since each carried a probability of 0.5. During the revolutionary period, religious radicals supported the two-witness rule on biblical grounds. See B. J. Shapiro, "Law Reform," 290–91. She notes that the Hale Commission assimilated some of the proposals of the radicals in "modified form." The Clarendon Act reinforced the new requirement that the witnesses be credible in addition to being lawful.

51. This was according to Chief Justice Vaugham in Bushnell's Case of 1670, as quoted in B. J. Shapiro, "To a Moral Certainty," 164.

52. Baker, *Reports of Spelman,* 124.

53. Bacon, *New Organon,* bk. 1, aph. l, in *Works of Bacon* 8: 83. In *De augmentis* (bk. 2, chap. 2, in *Works of Bacon* 8: 410), Bacon described "Mechanical and Experimental History" as that which puts nature "in constraint," in contrast with "History of Generations," where nature "follows her ordinary course of development."

54. Bacon, *New Organon,* "Plan of the Work," in *Works of Bacon* 8: 47. In *De augmentis* (bk. 5, in *Works of Bacon* 9: 72), Bacon referred to "methods of experimenting" as "Learned Experience."

55. Bacon, *New Organon,* "Plan of the Work," in *Works of Bacon* 8: 47.

56. Ibid., bk. 1, aph. l, in *Works of Bacon* 8: 82. Note that Bacon's use of *observation* here is not limited to sense perception. See also aph. i of bk. 1, where he spoke of things that are observed "in thought" (*Works of Bacon* 8: 67).

57. Ibid., bk. 1, aph. l, in *Works of Bacon* 8: 83.

58. Ibid., "Plan of the Work," in *Works of Bacon* 8: 44. He said much the same in bk. 1, aph. l.

59. Ibid., "Plan of the Work," in *Works of Bacon* 8: 47–48.

60. Ibid., bk. 1, aph. li, in *Works of Bacon* 8: 83.

61. Ibid., "Plan of the Work," in *Works of Bacon* 8: 43.

62. Boyle, *Christian Virtuoso,* appendix to pt. 1, in *Works* 6: 705.

63. Boyle, *Usefulness of Natural Philosophy,* pt. 1, essay 1, in *Works* 2: 9. For Bacon, the spider represented "reasoners" who "make cobwebs out of their own substance" (*New Organon,* bk. 1, aph. xcv, in *Works of Bacon* 8: 131). Boyle expanded somewhat upon the spider's activities, but the conclusion is the same: the experimentalist is to use both sense and reason.

64. Boyle, *Things Said to Transcend Reason,* 5th advice, in *Works* 4: 462.

65. Boyle, *Christian Virtuoso,* appendix to pt. 1, in *Works* 6: 707.

66. Boyle, *Excellency of Theology*, sec. 5, in *Works* 4: 59.
67. Boyle, *Christian Virtuoso*, appendix to pt. 1, in *Works* 6: 715. We will see in chapters 5 and 6 below that reason also played an important role in the collection of observation reports.
68. Boyle, *Christian Virtuoso*, pt. 1, in *Works* 5: 524.
69. Boyle, *Usefulness of Natural Philosophy*, pt. 1, essay 1, in *Works* 2: 9. Boyle discussed a number of ways in which one's own or others' observation reports must be assessed for credibility. These will be discussed in chapter 6 below.
70. Boyle, *Christian Virtuoso*, pt. 1, in *Works* 5: 539.
71. Boyle, *Hydrostatical Paradoxes*, preface, in *Works* 2: 742.
72. Boyle, *Christian Virtuoso*, pt. 1, in *Works* 5: 539. Compare this discussion with Bacon's on the idols of the mind, in *New Organon*, aph. xxxviii-lxviii, in *Works of Bacon* 8: 76–99.
73. Boyle, *Christian Virtuoso*, pt. 1, in *Works* 5: 539–40.
74. Ibid., 513, 540.
75. Boyle, *Certain Physiological Essays*, "Of Unsucceeding Experiments," in *Works* 1: 352–53.
76. Boyle, *The Reconcileableness of Reason and Religion*, sec. 7, in *Works* 4: 179–80.
77. Ibid., 181.
78. Ibid., sec. 8, in *Works* 4: 182. Boyle's description of a trial is a bit condensed, but basically right. It would be the jury that would find the accused guilty, based upon the concurrence of probabilities, and then the judge would pass the sentence of death upon the indictment. See Green, *Verdict according to Conscience*, 180–81. Boyle rehearsed his discussion of moral demonstration in a number of unpublished notes. See, for example, Royal Society, *Boyle Papers* 5: fols. 18v–19v, for two slightly different formulations. In both he noted that moral demonstration consisted of a set of "collateral Arguments" that are all "partial, and some of them perhaps indirect arguments" that when taken together "make and become one total Argument . . . that directly proves the grand and ultimate conclusion" (fol. 18v).
79. Boyle, *Reconcileableness of Reason and Religion*, sec. 8, in *Works* 4: 182. The concurrence produces a type of hypothetical necessity: if one wants to be rational, then one ought to assent to the conclusion.
80. Ibid., 183. Boyle's first formulation of the wager argument appeared in his "Aretology", written about 1647. See Boyle, *Early Essays*, 181. Lorraine Daston, in *Classical Probability in the Enlightenment* (Princeton, N.J.: Princeton University Press, 1988), 63, notes that Boyle's notion of concurrence involved the "additional conviction carried by convergent evidence." She correctly points out that the determination of probabilities was useful as a guide for action.
81. Boyle, *Reconcileableness of Reason and Religion*, sec. 8, in *Works* 4: 183.
82. See Ian Hacking, *The Emergence of Probability* (Cambridge: Cambridge University Press, 1975), esp. chap. 2, for Hacking's discussion of this dual aspect of probability. Because Hacking focuses upon the appeal to authority that was often associated with the qualitative aspect of probability, however, he concludes that our ordinary notion of probability did not exist until the seventeenth century. See Douglas Lane Patey, *Probability and Literary Form* (Cambridge: Cambridge Univer-

sity Press, 1984), appendix A, 266–72, for a detailed criticism of Hacking; also see chaps. 1 and 2 for Patey's alternative account that draws upon the history of probability in the law, rhetoric, and the "low" sciences of medicine and chemistry. The influence of these latter disciplines on Boyle's notion of probable proof will be discussed in the next chapter.

83. Boyle, "The Excellency and Grounds of the Mechanical Hypothesis," in *Works* 4: 77.

84. Boyle, *Excellency of Theology*, sec. 5, in *Works* 4: 59.

85. Royal Society, *Boyle Papers* 35: fol. 202r, quoted in Westfall, "Unpublished Boyle Papers," 117. This third criterion is reminiscent of the testing procedure suggested by Carneades. As Ralph Doty describes it, a person seeing something coiled on the floor of a barn at night would prod the coiled object with a stick to determine whether it was a rope or a snake. This would be an experiment "aptly devised," since background knowledge would inform one that snakes react differently to prodding than do coils of rope. It is significant that Boyle chose Carneades, rather than one of the more radical Pyrrhonists, as his spokesman in *The Sceptical Chymist.* See Ralph Doty, "Carneades: A Forerunner of William James's Pragmatism," *Journal of the History of Ideas* 47 (1986): 133–38. See also Patey, *Probability and Literary Form,* 15–16, for his discussion of Carneades's method of probability.

86. The notion of the usefulness of knowledge is also embedded within this criterion. If a hypothesis is true, then it should be effective in the sense that it should produce works.

87. Boyle, "Cosmical Suspicions," in *Works* 3: 318.

88. Hobbes, *Physical Dialogue*, 379. Hobbes also opposed the establishment of an experimental community because he saw it as one more instance of a group attempting to establish itself as an authority in opposition to the absolute authority of the sovereign. Shapin and Schaffer (*Leviathan and the Air-Pump*) have looked at these debates as ones about the determination of the proper social order. This was certainly an issue for both philosophers, and we will see more detail about how Boyle viewed the problem in chapters 6, 7, and 8 below. My purpose here, however, is to show the equally important, and interesting, epistemic aspects of the debates.

89. Hobbes, *Physical Dialogue*, 358. In his dedication to Sorbière he allowed that the experimentalists might advance science, but only when "they have either discovered the true science of motion for themselves or else they have accepted mine" (347).

90. Ibid., 362. This was the same type of tactic Hobbes used in *Leviathan,* whereby he sought to derive all principles of politics from his definitions of reason and society. Boyle sought definitions that were to be tested by experience and not set up by an arbitrary act of reason.

91. Boyle, *New Experiments*, 2d ed., "Examen of Mr. Hobbes's *Dialogus Physicus*," in *Works* 1: 186–242.

92. Ibid., chap. 6, in *Works* 1: 241.

93. Boyle, *Hydrostatical Paradoxes*, preface, in *Works* 2: 742. In *New Organon*, bk. 1, aph. xcvi, Bacon had maintained that mathematics "ought only to give definiteness to natural philosophy, not to generate or give it birth" (in *Works of Bacon* 8: 132). See

Peter Dear, "Jesuit Mathematical Science and the Reconstitution of Experience in the Early Seventeenth Century," *Studies in History and Philosophy of Science* 18 (1987): 133–75, for a discussion of the distinction between mathematical and physical sciences. Dear traces what he finds to be a transformation in the term *experience* from the Aristotelian notion of a universal evident statement used as a premise in a scientific demonstration to the notion in the seventeenth century of a discrete historical event. Dear argues that, with this transformation, expertise and witnessing became fundamental to the establishment of facts that could be used as evident suppositions in scientific argument. Boyle's notion of experience does not seem to fit well within this discussion. The facts established by experience are not discrete events but are regularities, for example, that animals will die ten times faster in a pump from which most of the air has been removed than in one left full. And these regularities are not to be used as suppositions in scientific explanation. Rather, they are the facts to be explained by science—the natural effects whose causes are to be discovered. If Dear is right, it would seem that there were (at least) two experimental traditions: one derived from the mathematical sciences and one from the "low" sciences of law, chemistry, and medicine.

94. Boyle, *Certain Physiological Essays*, "Of Unsucceeding Experiments," in *Works* 1: 347. In the preface to his *Hydrostatical Paradoxes* Boyle said that naturalists should be "somewhat diffident of conclusions" that "eminent writers . . . pretend to have mathematically demonstrated" (*Works* 4: 742), and in *Origin of Forms and Qualities* he stressed that he wished to "speak physically of things" (*Works* 3: 25). In the preface to the second edition of *New Experiments Physico-Mechanical* Boyle said that he meant to "devise experiments and to enrich the history of nature," not to establish principles (*Works* 1: 121). The air was one of the most important objects of investigation, particularly because of its necessity for life. In *New Organon*, bk. 1, aph. 1, Bacon noted that the air was one area of study that had been seriously neglected (*Works of Bacon* 8: 83).

95. Boyle, *Things Said to Transcend Reason*, 1st advice, in *Works* 4: 450, italics mine.

96. Boyle, *Origin of Particular Qualities*, advertisements, in *Works* 4: 234, italics mine. Compare this with the discussion of probability in John Locke, *An Essay Concerning Human Understanding*, ed. Peter H. Nidditch (Oxford: Clarendon Press, 1979), bk. 4, chap. 16, sec. 6, p. 662. Locke states that probabilities based upon the nature of things and the testimony of reliable witnesses "rise so near to *Certainty*, that they govern our Thoughts as absolutely . . . as the most evident demonstration." Locke was here talking about probabilities concerning "matters of fact," and he acknowledged that when one turns to hypotheses about unobservables, probability is more difficult to achieve. Yet he maintained that "wary Reasoning from Analogy [e.g., the principle that like effects have like causes] leads us often into the discovery of Truths, and useful Productions, which would otherwise lie concealed" (bk. 4, chap. 16, sec. 12, pp. 666–67). On Locke's debt to Boyle, see Alexander, *Ideas, Qualities, and Corpuscles*. Boyle's use of analogy as an argument form will be discussed in chapters 6 and 7 below.

97. In the preface to his "Examen of Hobbes" (*Works* 1: 186), Boyle said that he had decided to respond despite the fact that he had been "informed by learned men . . . that my publishing anything against his objections would not be necessary,

nor much expected." In a letter to Robert Moray, dated 14 July 1662, Huygens wrote that he was at first disappointed to see that Boyle had spent his time responding to such "frivolous" objections but was pleased to find that the task had led Boyle to include a number of additional experiments that were quite beneficial to learning. The letter is printed in Maddison, *Life,* 107.

98. Boyle, "Examen of Hobbes," in *Works* 1: 186. Boyle's reference to Hobbes's *Mathematicae hordiernae* is in *Works* 1: 241.

99. Boyle's translation of Hobbes's *Dialogus physicus,* in *Works* 1: 232. Schaffer's translation of the *Dialogus physicus* differs little and only in insignificant ways from Boyle's (Hobbes, *Physical Dialogue,* 347, 358).

100. Shapin and Schaffer, *Leviathan and the Air-Pump,* 173–74. Earlier, Shapin and Schaffer maintain that Hobbes had shown that "an alternative and superior explanation could be proffered" of Boyle's experiments (111–12) and state that the "resonance with the 'Duhem-Quine' thesis is intentional" because they see Hobbes's objections as having provided a "concrete exemplar of this 'modern' thesis" (112 n. 5). But there is some ambiguity in their discussion. Presumably, they take Hobbes to be illustrating the inconclusive nature of experimental evidence by means of his alternative theoretical interpretation, but clearly, in order to be a "concrete exemplar," the challenge needs to be more than simply made. It must actually account for all of the experiments. As Boyle conclusively showed, Hobbes's alternative theory could not do this. As for the "modernity" of the underdetermination thesis, we will see in chapter 7 that Boyle had his own formulation and resolution of the problem.

101. Boyle, *New Experiments,* "To the Lord of Dungarvan," in *Works* 1: 10. Following the practice of Pascal and others who had performed the Torricellian experiments, he referred to this visibly empty space as an "experimental vacuum." See Pascal's letter to Perier, 15 November 1647, in Blaise Pascal, *The Physical Treatises of Pascal,* trans. I. H. B. Spiers and A. G. H. Spiers (New York: Columbia University Press, 1937), 9.

102. There were forty-three numbered experiments, but under each title Boyle included accounts of the numerous repetitions and variations that had been made. Boyle's notion of the spring of the air as an intermediate cause that eventually received factual status will be discussed in chapter 6. My concern here is simply to illustrate the epistemological aspects of the debate, especially the evidential relations between experimental results and theory, not the validity of the factual claims made during the debate.

103. Experiments 10–12 involved candles of wax and tallow along with "matches" and burning coals. Experiment 41 consisted of a long series of trials made with mice and small birds.

104. Boyle, *New Experiments,* in *Works* 1: 107. Since antiquity it had been recognized that air was necessary for life. Boyle suggested a "vital quintessence," in opposition to those who held that respiration was only needed for the cooling of the blood. Hooke also thought that there was something in the air, which he termed a "kind of nitrous quality," that was necessary to support both respiration and combustion. Robert Hooke, *Micrographia* (London: J. Martyn and J. Allestry, 1665), 113.

105. Hobbes, *Physical Dialogue,* 378. Because Boyle maintained that the air was com-

pressible, he did not have to assume any empty space outside the receiver that could be filled by the evacuated air.

106. Ibid., 370–72.

107. Boyle's translation of Hobbes's *Dialogus physicus,* in "Examen of Hobbes," chap. 5, in *Works* 1: 218; cf. Schaffer's translation (Hobbes, *Physical Dialogue,* 366–67).

108. Boyle's translation of Hobbes's *Dialogus physicus,* in "Examen of Hobbes," chap. 5, in *Works* 1: 219; cf. Schaffer translation (Hobbes, *Physical Dialogue,* 346).

109. Boyle's translation of Hobbes's *Dialogus physicus,* in "Examen of Hobbes," chap. 5, in *Works* 1: 219, 221.

110. Ibid., chap. 6, in *Works* 1: 229–30.

111. Boyle, *New Experiments,* experiment 26, in *Works* 1: 61–62. In the first attempt the pendulum moved slightly faster, but repeated trials showed that this was not constantly the case. This was just one of many times that Boyle reported a failed experiment. We will see the significance of this practice in chapters 7 and 8 below.

112. Hobbes, *Physical Dialogue,* 374.

113. Boyle, "Examen of Hobbes," chap. 5, in *Works* 1: 221.

114. Boyle, *Continuation of New Experiments Physico-Mechanical,* experiment 40, in *Works* 3: 257.

115. This is the conclusion that Christopher Hill does believe Shapin and Schaffer proved (Hill, "New Kind of Clergy," 730). Hobbes had not even discussed all of Boyle's experiments. In the *Dialogus physicus* he had *B,* the rather feeble representative of the Royal Society in the debate, graciously acknowledge Hobbes's superiority by stating: "I shall omit the rest of the experiments of the machine, which seem to be reducible to the same hypotheses of yours without difficulty" (*Physical Dialogue,* 378–79). Boyle noted that this tactic would not be acceptable to those "that take notice of the variety of those phaenomena we have set down in our treatise," in his second edition of *New Experiments* (*Works* 1: 213).

116. The concurrence of probabilities is, admittedly, a vague epistemological notion. We will see a clear example of how it was used in an actual experimental demonstration in the next chapter, and we will see Boyle's reasons for employing such a vague notion in chapters 4 and 5.

Chapter Three

1. Letter from Boyle to Hartlib, 8 May 1647, in *Works* 1: xli. In May 1648 Hartlib wrote to Boyle: "Your worthy friend and mine, Mr. *Gassend,* is reasonable well, and hath printed a book on the life of Epicurus since your going from hence" (Boyle, *Works* 6: 77). The fact that Boyle was acquainted with Gassendi's work as early as 1647 makes Kargon's thesis that Charleton was important for the dissemination of Gassendi's work in England somewhat doubtful. See Kargon, "Charleton, Boyle, and the Acceptance of Epicurean Atomism." Emerton (*Scientific Reinterpretation of Form,* 140) argues that Charleton's work was not representative of Gassendi's philosophy.

2. Letter from Boyle to Hartlib, 19 March 1647, in *Works* 1: xxxvii. Hartlib was apparently acquainted with Mersenne, as indicated by the letter from Boyle to Hartlib dated 8 May 1647: "The rise you have now to resume your former corre-

spondencies with the great *Mersennus*, I hope you will greedily embrace, he being a man truly incomparable in his own way, and the mechanics he transcends in as greatly beneficial as little understood" (Boyle, *Works* 1: xli). The account of Boyle's early years is based largely upon Boyle's "Account of Philaretus," in *Works* 1: xii–xxvi; his correspondence published in *Works*, vols. 1 and 6; and Maddison, *Life*.

3. In a letter to Lady Ranelagh (*Works* 1: xxxvi), Boyle mentioned that his experiments were being hindered by his stay at Stalbridge because the furnaces he had shipped from London repeatedly arrived broken.

4. Royal Society, *Boyle Papers*, vol. 36, fol. 86r, dated 25 January 1650, printed in Maddison, *Life*, 64. Most examples of Boyle's writings prior to 1649 are theological or ethical in nature. The first "Memorialls Philosophicall" in which he recorded chemical recipes and processes dates from 25 January 1649/50 (Boyle's birthday). Royal Society, *Boyle Papers* 28: 309–11.

5. See Maddison, *Life*; and More, *Life and Works*, on the difficulties that Boyle had at first in taking possession of his inherited estates and his sister Katherine's influence on his behalf with the Parliamentarian Party.

6. Boyle had known Petty since 1648. He had met him through Hartlib while on a brief visit to London. See Petty's letter to Boyle at Stalbridge, 21 June 1648, asking him for reports on his experimental results (Boyle, *Works* 6: 137). Petty was also a friend of Boyle's brother Lord Broghill.

7. Printed in John Fulton, *A Bibliography of Robert Boyle* (Oxford: Oxford University Press, 1961), 61. Highmore, who was a cousin of Petty's, praised Boyle for his willingness to do experiments: "You have not thought your blood and descent debased, because married to the arts." Digby and Starkey were members of a "general chemical council" at Charing Cross. In 1654 Digby was pressing Frederick Clod, Hartlib's son-in-law, to accept two years' worth of funding for work in a "universal laboratory," but Clod would not decide until "Mr. Boyle be arrived" from Ireland (Hartlib to Boyle, in Boyle, *Works* 6: 82–89). See William Newman, "Newton's *Clavis* as Starkey's *Key*," *Isis* 78 (1987): 564–74, for a brief discussion of Starkey's association with Boyle and other members of Hartlib's circle and a transcription of an early letter from Starkey to Boyle (c. 1651). See also William Newman, *Gehennical Fire: The Lives of George Starkey, an Alchemist of Harvard in the Scientific Revolution* (Cambridge: Harvard University Press, forthcoming).

8. Letter from Wilkins to Boyle, 6 September 1653, in Boyle, *Works* 6: 633–34.

9. Royal Society, *Boyle Papers*, vol. 37, fol. 194r, draft of letter from Boyle to Hartlib, 14 September 1655, printed in Maddison, *Life*, 85.

10. Boyle, *Usefulness of Experimental Philosophy*, pt. 2, "Physical Knowledge," in *Works* 3: 467. He also mentioned the discoveries made by Mersenne and Descartes (476 and throughout).

11. Boyle frequently cited Mersenne's *Les nouvelles pensées de Galilée* when discussing Galileo's mechanics. Mersenne was critical of Galileo's experimental practices in this work. For example, he wrote: "Si Galilée eust expérimente les unions de ces retours des chordes, comme i'ay fait, it eut apperceu que la chose n'est pas guère agréable"; Marin Mersenne, *Les nouvelles pensées de Galilée*, ed. Pierre Costabel and Michel-Pierre Lerner (Paris: J. Vrin, 1973), bk. 1, p. 109. Nevertheless, Boyle wrote as though Galileo had established his principles on the basis of actual ex-

perimentation. See, for example, *Usefulness of Experimental Philosophy*, pt. 2, "Physical Knowledge," in *Works* 3: 459–60. Boyle had read Galileo's original works while in Italy.

12. Galileo Galilei, *Dialogues Concerning Two New Sciences*, trans. Henry Crew and Alfonso de Salvio (New York: Dover Publications, 1954), 194. The quarry analogy is reminiscent of Bacon's miners. In his division of labor he described those "that try new experiments, such as themselves think good. These we call Pioneers or Miners"; *New Atlantis*, in *Works of Bacon* 5: 410.

13. Galilei, *Two New Sciences*, 160. This argument is similar to Bacon's stricture against constructing "sciences as one would" and is reflected in Boyle's criticisms of mathematical reasoning, especially that of Hobbes and Spinoza. See Ernan McMullin, "The Conception of Science in Galileo's Work," in *New Perspectives on Galileo*, ed. Robert E. Butts and J. C. Pitt (Dordrecht: D. Reidel, 1978), esp. 222–28.

14. Galilei, *Two New Sciences*, 153. See A. Rupert Hall, "Galileo's Thought and the History of Science," in *Galileo, Man of Science*, ed. Ernan McMullin (New York: Basic Books, 1967), 79, where Hall asserts that Galileo did not have a "decisive formative influence" on the development of experimental science. Boyle's frequent references to Galileo make this conclusion doubtful. Marie Boas Hall, "Galileo's Influence on Seventeenth-Century English Scientists," in *Galileo, Man of Science*, 405–14, discusses Galileo's theoretical contributions but not his methodological importance.

15. Galilei, *Two New Sciences*, 84, 172. Galileo presented this proposition as a "postulate." See the discussion of Galileo's adherence to an Aristotelian mode of demonstration in Ernan McMullin, introduction to *Galileo, Man of Science*, ed. Ernan McMullin (New York: Basic Books, 1967), 31–35; and idem, "Conception of Science in Galileo's Work," 229–40.

16. Galilei, *Two New Sciences*, 160.

17. Boyle, *Christian Virtuoso*, pt. 1, prop. 1, in *Works* 5: 527. In *Excellency of Theology* (sec. 5, in *Works* 4: 61), Boyle discussed how the limits of pumps had to be found out by experience, much as Galileo had described (*Two New Sciences*, 16). Other references to Galileo occur throughout Boyle's published and unpublished works. See, for example, *New Experiments*, in *Works* 1: 81; *Certain Physiological Essays*, in *Works* 1: 304; and "Aretology," in Boyle, *Early Essays*, 36. The way in which Boyle used Galileo's work as an example of how to write a scientific treatise of limited scope will be discussed in chapter 8.

18. Boyle, *Christian Virtuoso*, pt. 1, prop. 1, in *Works* 5: 527.

19. Ibid.

20. In the introduction to his *Hydrostatical Paradoxes* (*Works* 2: 749), Boyle wrote that the practice of "abstracting some little niceties" of actual weights was appropriate, especially at this early stage of investigation, if such practice would bring one closer to the truth. See Ernan McMullin, "Galilean Idealization," *Studies in History and Philosophy of Science* 16 (1985): 247–73, for a discussion of the varieties of idealization employed by Galileo.

21. Pascal, *Physical Treatises*, 31, 75. See Garber, *Descartes' Metaphysical Physics*, 137–43, on Descartes's role in suggesting Pascal's Puy-de-Dôme experiment and his alternative explanation of this phenomenon.

22. Pascal, *Physical Treatises*, 75.

23. Boyle, *Usefulness of Experimental Philosophy*, pt. 2, "Of Men's Great Ignorance," sec. 1, in *Works* 3: 479, 477.

24. Boyle, *Hydrostatical Paradoxes*, preface, in *Works* 2: 739, 740. Galileo (*Two New Sciences*, 166–67) wrote that he would "investigate and demonstrate some of the properties of accelerated motion" but that he would not investigate its cause. See Stillman Drake, *Cause, Experiment, and Science* (Chicago: University of Chicago Press, 1981), i, on Galileo's rejection of the Aristotelian search for causes in favor of the "modern" search for laws. We will see in chapter 4 that Boyle's continued search for causes makes him more of an Aristotelian than is often realized.

25. Boyle, *Hydrostatical Paradoxes*, paradox 3, scholium, in *Works* 2: 763. He repeated this goal in his *Medicina hydrostatica*, in *Works* 5: 460. See McMullin, "Conceptions of Science in the Scientific Revolution," 57–61, for a discussion of the difference between mathematical and physical accounts of nature.

26. Boyle, *Hydrostatical Paradoxes*, introduction, in *Works* 2: 746. Conant ("Boyle's Experiments," 59) states that Boyle was not a mathematician. Wiener ("Experimental Philosophy of Boyle") states that Boyle had little mathematical background. And A. Rupert Hall, in *The Scientific Revolution*, 2d ed. (Boston: Beacon Press, 1983), 178, states that "Boyle had no natural aptitude towards the mathematization of nature." Steven Shapin, in "Boyle and Mathematics," has given a detailed account of the ways in which these three studies failed to capture Boyle's ambivalence on the topic of mathematics. In works such as *Hydrostatical Paradoxes*, Boyle does display his knowledge of mathematics, and in *Usefulness of Experimental Philosophy* (*Works*, vol. 3), an entire essay is devoted to the usefulness of mathematics. The absence of mathematical demonstrations in his work should be attributed to his philosophical arguments against the mathematical model of reasoning and to his interest in the qualitative aspects of nature.

27. Boyle, *Hydrostatical Paradoxes*, preface, in *Works* 2: 741.

28. Ibid., introduction, in *Works* 2: 745–46. Boyle had no problems with Pascal's second treatise on the air. Earlier, in his "Examen of Hobbes," (*Works* 1: 203), he had called Pascal a "happy promoter of experimental learning."

29. Boyle, *Hydrostatical Paradoxes*, introduction, in *Works* 2: 746, 745–46. Gassendi also derided scientists who conducted experiments in their imaginations. See Pierre Gassendi, *The Selected Works of Pierre Gassendi*, ed. Craig Brush (New York: Johnson Reprint Corp., 1972), 116. Gassendi's motivation for this criticism differed from Boyle's, however. As described by Brush, Gassendi's criticism was generated by his empirical notion of demonstration, which consisted of pointing toward a purported appearance (76–77).

30. Boyle, *Hydrostatical Paradoxes*, introduction, in *Works* 2: 746. Here again Boyle maintains that successful practice can be a sign of the truth of a hypothesis.

31. Ibid., 745.

32. Ibid., paradox 2, scholium, in *Works* 2: 759.

33. Ibid., preface, in *Works* 2: 743–44.

34. Letter from Spinoza to Oldenburg, April 1662, in Oldenburg, *Correspondence* 1: 462. A. Rupert Hall and Marie Boas Hall discuss this correspondence in light of the corpuscular philosophy in "Philosophy and Natural Philosophy: Boyle

and Spinoza," in *Mélanges Alexandre Koyre*, vol. 2 (Paris: Harman, 1964), 240–56. Spinoza maintained that all inquiry had to begin with the *cogito* because it was the only truth that could be known with certainty. Baruch Spinoza, *Principles of Cartesian Philosophy*, trans. Harry E. Wedeck (New York: Philosophical Library, 1961).

35. Spinoza to Oldenburg, April 1662, in Oldenburg, *Correspondence* 1: 463.

36. Oldenburg to Spinoza, 3 April 1663, in Oldenburg, *Correspondence* 2: 41. "Ordinary experiments" were those that happened accidentally or those done by empirics who lacked a knowledge of causes.

37. Spinoza to Oldenburg, 17 July 1663, in Oldenburg, *Correspondence* 2: 92. It is unlikely that either Bacon or Boyle would have thought that this proposition had been adequately proved, since it was presented by Bacon as an axiom of the first vintage.

38. Oldenburg to Spinoza, 4 August 1663, in Oldenburg, *Correspondence* 2: 103–4.

39. Boyle, "Excellency and Grounds of the Mechanical Hypothesis," in *Works* 4: 72. These are the intermediate causes and emergent axioms discussed earlier.

40. Boyle, *Certain Physiological Essays*, "Some Specimens," in *Works* 1: 355–56.

41. A corpuscular conception of matter within the alchemical tradition clearly influenced the content of Boyle's corpuscular philosophy. See Antonio Clericuzio, "Carneades and the Chemists: A Study of the *Sceptical Chymist* and Its Impact on Seventeenth-Century Chemistry," in *Robert Boyle Reconsidered*, ed. Michael Hunter (Cambridge: Cambridge University Press, 1994), 79–90; idem, "Redefinition of Boyle's Chemistry"; Newman, "Boyle's Debt to Corpuscular Alchemy"; and Principe, "Boyle's Alchemical Pursuits." My concern here is with the methodological and epistemological aspects of the chemical tradition. The content of Boyle's corpuscular philosophy will be discussed in chapter 4.

42. Contrary to popular belief, this epithet does not appear on Boyle's tombstone. Boyle's original resting place, in St. Martin-in-the-Fields, London, contained no memorial inscription. In the 1720s the original church was demolished, and no record has survived concerning the whereabouts of Boyle's remains. Maddison (*Life*, 196–97) has traced the origin of the epithet to an 1849 issue of the *British Quarterly Review*.

Herbert Butterfield and Thomas S. Kuhn have expressed doubts about whether Boyle deserves the title, because chemistry progressed very little in the following century. Butterfield, *The Origins of Modern Science* (New York: Free Press, 1957), 141–50; Kuhn, "Boyle and Structural Chemistry."

43. Bacon, *New Organon*, bk. 2, aph. viii, in *Works of Bacon* 8: 177; Boyle, *Certain Physiological Essays*, "Some Specimens," in *Works* 1: 356.

44. Boyle, *Christian Virtuoso*, pt. 1, in *Works* 5: 523–24.

45. Boyle, *Certain Physiological Essays*, "Some Specimens," in *Works* 1: 358. Boyle's interest in chemistry was in part motivated by his belief that it would contribute to advances in the corpuscular philosophy. As he wrote near the end of his life, "I cultivated chemistry not so much for itself, as for the sake of natural philosophy" (*Works* 1: cxxx). He often noted that his great wealth freed him from the need to pursue chemistry for economic gain.

46. Boyle, *Certain Physiological Essays*, "Some Specimens," in *Works* 1: 358. In *A Chymical*

Paradox of 1680, Boyle wrote that he doubted that fire always yielded a true analysis of bodies, noting that it might sometimes act as a "material cause" (*Works* 4: 500).

47. Boyle, *Certain Physiological Essays*, "Some Specimens," in *Works* 1: 358.

48. Boyle, *Origin of Forms and Qualities*, "Production and Reproduction of Forms," in *Works* 3: 52. This is clearly an instance of the "constructive" notion of knowledge examined in chapter 1. Boyle's defense of the epistemic virtues of artificiality will be discussed in chapter 7.

49. See Patey, *Probability and Literary Form*, 35–40, on how the reasoning from signs used by the ancient medical theorists was taken up by chemists in the sixteenth century. This type of reasoning is based upon an inference from effects to the hidden causes via an indicator, which is the sign. Such signs were also predictive. By enabling a naturalist to identify a substance, they also allowed an inference to be made about what a substance would do in specific circumstances. For a good account of the ancient approaches to medicine of the Dogmatists, Empiricists, and Methodists, see Lloyd, *Science, Folklore, and Ideology*.

50. Boyle, *Sceptical Chymist*, preface, in *Works* 1: 460. He made a similar criticism of the chemists' way of writing in *Usefulness of Natural Philosophy* (pt. 2, appendix, in *Works* 2: 232). Boyle's discussions concerning the rational grounding of chemistry have led to some confusion in the secondary literature. As Principe has effectively argued in "Boyle's Alchemical Pursuits," Boyle never lost interest in the more mystical or spiritual aspects of traditional alchemical methods and goals. In some of his works he actually wrote in the obscure way for which he had criticized the chemists. When he spoke of the applications of chemistry to natural philosophical inquiry, however, he argued for the need for a more straightforward reporting style.

51. Boyle, *Sceptical Chymist*, preface, in *Works* 1: 461. According to Arthur Edward Waite, the alchemists believed that "truth lies in obscurity, and the wise never write more 'deceitfully' than when their words are plain, nor more veridically than when their words are dark"; Waite, *The Secret Tradition in Alchemy* (New York: Samuel Weiser, 1969), 339. More will be said about Boyle's views on writing in chapter 8.

52. Boyle, *Experimental History of Colours*, preface, in *Works* 1: 664. This work is often discussed as if it were a work on optics, but it is really more a work on the physicochemical properties of colored objects. In the theoretical discussion in part 1, Boyle touched upon optical theory, but the experiments in parts 2 and 3 are devoted almost entirely to the investigation of how mixtures of chemical substances could produce color changes in various objects.

53. Boyle, *Sceptical Chymist*, part 3, in *Works* 1: 510.

54. Ibid., conclusion, in *Works* 1: 584. Some artisans had an interest in keeping their art obscure in order to protect the monetary gain that they obtained from such practice. Boyle had no such economic interest in chemistry. He did respect the fact that some had to make a living from their expertise, however, and he was careful in his reports not to reveal tradesmen's entire processes that he had learned, unless the process was so complex that its replication would not be worth the time. See, for example, his letter printed in Boyle, *Works* 1: cxxx.

55. Boyle, *Sceptical Chymist*, conclusion, in *Works* 1: 584.

56. Boyle, *Experiments on the Mechanical Origin of Qualities,* "Of the Imperfections of the Chemists' Doctrine of Qualities," chap. 1, in *Works* 4: 274, 275.
57. Ibid., 273.
58. Ibid., chap. 6, in *Works* 4: 278–79.
59. Boyle, *Experiments on the Mechanical Origin of Qualities,* "Reflections upon the Hypothesis of Alcali and Acidum," chap. 1, in *Works* 4: 284. He also noted his preference for Helmont in his preface to *Sceptical Chymist* (*Works* 1: 461).
60. Boyle, *Experiments on the Mechanical Origin of Qualities,* "Reflections upon the Hypothesis of Alcali and Acidum," chap. 2, in *Works* 4: 285.
61. Boyle, *Sceptical Chymist,* part 3, in *Works* 1: 510.
62. Boyle, *Experiments on the Mechanical Origin of Qualities,* "Reflections upon the Hypothesis of Alcali and Acidum," chap. 2, in *Works* 4: 284.
63. Ibid., chap. 4, in *Works* 4: 287. We will see how he used signs to determine the nature of phosphorus in his essay "On the Icy Noctiluca" in chapter 8.
64. Ibid. He also argued against their other tests that concerned the heat produced or the taste of the substances. He maintained, for example, that it would be the mechanical affections of matter, not the material ingredients themselves, that caused the substance to be hot to the touch.
65. Boyle, *Sceptical Chymist,* pt. 5, in *Works* 1: 544. Boyle would develop a number of successful indicator tests. See M. B. Hall, *Boyle on Natural Philosophy,* 88–89, on how Boyle's experimental work was devoted to clearing up confusions over different types of substances and how he developed tests that used the work that he had done on the physico-chemical properties of bodies revealed by their colors.
66. Boyle, *Experiments on the Mechanical Origin of Qualities,* "Of the Imperfections of the Chemists' Doctrine of Qualities," chap. 3, in *Works* 4: 275.
67. Boyle, *Sceptical Chymist,* preface, in *Works* 1: 463. See Douglas McKie's discussion of the destructive nature of Boyle's chemical work in his introduction to Boyle, *Works* 1: xiv*.
68. Boyle, *Experiments on the Mechanical Origin of Qualities,* "Reflections upon the Hypothesis of Alcali and Acidum," chap. 5, in *Works* 4: 288. In the conclusion to *Sceptical Chymist,* Boyle noted that his whole work was a "discourse against the received doctrines of elements" in general (*Works* 1: 584). Kuhn ("Boyle and Structural Chemistry") has argued that Boyle set back the progress of chemistry because he refused to accept the existence of elements. But as M. B. Hall has pointed out in *Boyle and Seventeenth-Century Chemistry,* the elements that Boyle denied were the universal ones that the chemists posited as existing in any body whatsoever.
69. Boyle, *Experiments on the Mechanical Origin of Qualities,* "Of the Imperfections of the Chemists' Doctrine of Qualities," chap. 6, in *Works* 1: 279. Emerton, in *Scientific Reinterpretation of Form,* notes that Boyle often used the crystalline forms of salts as hints for the salts' identification and for understanding their underlying corpuscular structure.
70. Boyle, "Excellency and Grounds of the Mechanical Hypothesis," in *Works* 4: 74.
71. Boyle, *Experiments on the Mechanical Origin of Qualities,* "Of the Imperfections of the Chemists' Doctrine of Qualities," chap. 5, in *Works* 4: 277.
72. Ibid., "Reflections upon the Hypothesis of Alcali and Acidum," chap. 6, in *Works* 4: 290.

73. Ibid., "Of the Imperfections of the Chemists' Doctrine of Qualities," chap. 6, in *Works* 4: 278.
74. Ibid., "Reflections upon the Hypothesis of Alcali and Acidum," chap. 5, in *Works* 4: 288–89. He said of this objection that it was "of that importance, that, though there were no other, this were enough to shew, that the hypothesis, that is liable to it, is insufficient for the explication of qualities."
75. Ibid., "Of the Imperfections of the Chemists' Doctrine of Qualities," chap. 3, in *Works* 4: 275.
76. Ibid., chap. 7, in *Works* 4: 281. M. B. Hall, in *Boyle and Seventeenth-Century Chemistry*, has likened these primary concretions to today's concept of a molecule.
77. Boyle, "Excellency and Grounds of the Mechanical Hypothesis," in *Works* 4: 73.
78. Boyle, *Sceptical Chymist*, appendix, in *Works* 1: 589.
79. Boyle, *Experiments on the Mechanical Origin of Qualities*, "Reflections upon the Hypothesis of Alcali and Acidum," chap. 8, in *Works* 4: 291.
80. In 1659 Oldenburg wrote about these attempts to reconcile the corpuscular and chemical doctrines prior to Boyle's publications in letters to Borel and to Southwell. Oldenburg, *Correspondence* 1: 322, 348–49.
81. Royal Society, *Boyle Papers*, 16: fol. 215r.
82. Boyle, "Account of Philaretus," in *Works* 1: xx. See also Boyle, *Medicinal Experiments*, in *Works* 5: 315–16, for Boyle's account of his ill health; and the "Aretology," in Boyle, *Early Essays*, for his early interests in medicine. Boyle compiled a vast collection of remedies, some of which were published in 1688 as *Medicinal Experiments*, subtitled *Receipts Sent to a Friend in America*. Two additional volumes, taken from his notes, were published after his death, in 1692 and 1695. Many times recipes were given in more general works, such as his *Usefulness of Experimental Philosophy* (pt. 2, in *Works*, vol. 3). His study of medicine caused him to be something of a valetudinarian. See William Petty's letter to him from Dublin, dated 15 April 1653, in which Petty warned Boyle about "practicing upon yourself with medicaments (though specifics) not sufficiently tried by those that administer or advise them." Petty also warned that there was a "distemper . . . incident to all that begin the study of diseases" to have a "continual fear, that you are always inclining or falling into one or other" (*Works* 6: 138–39).
83. See Sir George Clark, *A History of the Royal College of Physicians of London*, vol. 1 (Oxford: Clarendon Press, 1964); Allen G. Debus, *The English Paracelsians* (London: Oldbourne Book Co., 1965); Kenneth Dewhurst, *John Locke (1632–1704): Physician and Philosopher* (London: Wellcome Historical Medical Library, 1963); and Robert G. Frank Jr., *Harvey and the Oxford Physiologists* (Berkeley and Los Angeles: University of California Press, 1980), on physicians' reactions to the uses of chemistry. For recent work on the relationship of medicine to the new philosophy, see Harold J. Cook, "The New Philosophy and Medicine in Seventeenth-Century England," in *Reappraisals of the Scientific Revolution*, ed. David C. Lindberg and Robert S. Westman (Cambridge: Cambridge University Press, 1990), 397–436; John Henry, "Medicine and Pneumatology: Henry More, Richard Baxter, and Francis Glisson's *Treatise on the Energetic Nature of Substance*," *Medical History* 31 (1987): 15–40; and Miguel A. Sanchez-Gonzalez, "Medicine in John Locke's Philosophy," *Journal of Medicine and Philosophy* 15 (1990): 675–95.

84. Boyle, *Usefulness of Natural Philosophy*, pt. 2, essay 3, in *Works* 2: 91–92.
85. Ibid., essay 5, chap. 2, in *Works* 2: 117.
86. Ibid., chap. 9, in *Works* 2: 154. Boyle had firsthand knowledge of such criticisms. Many of his friends and acquaintances, such as Lower, Locke, Petty, Highmore, Sydenham, and Stubbe, had been trained as physicians. Boyle himself received an honorary M.D. from Oxford in 1665.
87. Ibid., 153. Boyle was particularly critical of the "vulgar *Methodus Medendi*" because he questioned the "safeness" of "the nature of those helps they usually imploy," such as "*Bleeding, Vomiting, Purging, Sweating,* and *spitting,*" which tend to "weaken or discompose where they are employed but do not certainly cure afterwards" (Royal Society, MS 199, fol. 120r). Henry Stubbe, physician and experimentalist, wrote to Boyle in May 1670 that while he admired Boyle's work, he feared that "others make much use of it to our [physicians'] prejudice" (*Works* 1: xciv).
88. Boyle, *Usefulness of Natural Philosophy*, pt. 2, essay 5, chap. 9, in *Works* 2: 152.
89. Ibid., chap. 5, in *Works* 2: 127; chap. 7, in *Works* 2: 138.
90. Boyle was in favor of changes in medical treatment, but he warned that it would be dangerous to alter recipes in a haphazard manner. One should discover through chemical trials which ingredients were efficacious. He cited Bacon in his *Usefulness of Natural Philosophy*, pt. 2, saying that altering recipes without full knowledge left too much to judgment and not enough to experience (Boyle, *Works* 2: 244).
91. Boyle, *Usefulness of Natural Philosophy*, pt. 2, essay 3, "Semeiotical Part of Physick," in *Works* 2: 89, 91. See Walter Pagel's account of uroscopy in *Paracelsus* (Basel: S. Karger, 1958), 54.
92. Boyle was aware of the physician's daily work. He would often accompany Dr. Sydenham to visit patients. See Thomas Birch, "The Life of the Honourable Robert Boyle," in Boyle, *Works* 1: lxxxvi; and More, *Life and Works*, 128.
93. Quoted in Patey, *Probability and Literary Form*, 44. I have modernized the spelling and punctuation. See Sydenham's dedication of this work to Boyle, printed in Boyle, *Works* 1: lxxxvi. See also Kenneth Dewhurst, *Dr. Thomas Sydenham (1624–1689): His Life and Original Writings* (Berkeley and Los Angeles: University of California Press, 1966); and G. G. Meynell, *A Bibliography of Dr. Thomas Sydenham (1624–1689)* (Folkestone, Eng.: Winterdown Books, 1990).
94. Quoted in Patey, *Probability and Literary Form*, 44.
95. Ibid., 50–51.
96. Hale, "Reflections on the Common Law," in Heward, *Matthew Hale*, 503.
97. Boyle, *Certain Physiological Essays*, "Of Unsucceeding Experiments," in *Works* 1: 340. In his manuscripts, Boyle was careful to note the desired dosage and the circumstances that could alter the dosage given. See, for example, Royal Society, *Boyle Papers*, vols. 25 and 27; and Royal Society, MS 199, fol. 116r. In the latter manuscript he noted that "purges that are not Drastic, may yet be unsafe" if the "patient is unfit for the purge by his nature."
98. Boyle, *Usefulness of Natural Philosophy*, pt. 2, essay 2, in *Works* 2: 76–77, 85, 77.
99. Ibid., essay 5, in *Works* 2: 167.
100. See Patey, *Probability and Literary Form*, 43; and Peter Dear, *Mersenne and the Learning of the Schools* (Ithaca, N.Y.: Cornell University Press, 1988), 110. *Conjecture* is a tech-

nical term that stands for more than a mere guess. It is a careful inference from a sign (Patey, 48–49).

101. See Frank, *Harvey and the Oxford Physiologists;* and Richard Hunter and Ida Macalpine, "William Harvey and Robert Boyle," *Notes and Records of the Royal Society of London* 13 (1958): 115–27, on Boyle's acquaintance with Harvey. It was while Boyle was in Ireland "making anatomical dissections of living animals" under the direction of William Petty that he first became convinced of the truth of Harvey's theory. See Boyle's letter to Frederick Clod, in *Works* 6: 55. The letter is not dated, but it would have been written sometime between 1652 and 1654, the period Boyle spent in Ireland.

102. See Andrew Wear, "William Harvey and the 'Way of the Anatomists,'" *History of Science* 21 (1983): 223–49, for a detailed discussion of Harvey's use of "ostensive demonstration." Jardine, in *Francis Bacon,* has argued that Galen's method of ostensive demonstration was a teaching method, not a method of discovery. Yet it seems, as Wear maintains (223), that it functioned as both. See Galen, *On Anatomical Procedures,* trans. Charles Singer (London: Oxford University Press, 1956). Although Galen maintained that ocular demonstration, particularly that of dissection, was an important teaching method (3), when he explained the actual method of dissection it seemed to be a vehicle of discovery also. He said, for example: "Facts have been discovered throughout the body, which the anatomists disregarded, shrinking detailed dissection and content with plausible ideas. . . . Thus I discovered that Nature had wrought these aforesaid muscles for important functions" (7–8). I would not go so far as Wear, however, who claims that ocular demonstration was devoid of reason (230–31); rather, it seems to be contrasted with the use of book learning and the purely rationalist form of demonstration. Galen advised his students to "first gain an exact and practical knowledge of human bones. It is not enough to study them casually or read of them only in a book: No, not even in mine" (3). I do not wish to argue the details of Harvey's own methodological thought, but it seems clear that Boyle, at least, thought that Harvey was following this method, so that for him ocular demonstration would refer to more than a mere "pointing."

The dialectic tradition was a frequent topic for methodological discussion at Padua, but I can find no indication that Boyle was aware of the intricacies of these discussions. The only reference he made to medicine on his visit to Padua was to an exhibit he saw of a man's veins and arteries laid out in their natural order on a board. For the role of dialectic in the medical tradition, see Gilbert, *Renaissance Concepts of Method,* 98–196; and J. H. Randall Jr., *The School of Padua and the Emergence of Modern Science* (Padua: Editrice Antenore, 1961).

103. On the work of the critical Galenists, see Thomas S. Hall, *History of General Physiology,* vol. 1 (Chicago: University of Chicago Press, 1975); Walter Pagel, *New Light on William Harvey* (New York: S. Karger, 1976); and Owsei Temkin, *Galenism: Rise and Decline of a Medical Philosophy* (Ithaca, N.Y.: Cornell University Press, 1973). Harvey's work on the circulation had been a retreat from orthodox Galenism because it focused on the circulation of the blood without providing a system within which the theory would fit. From a theorist's point of view, this is similar to the destructive nature of Boyle's chemical work.

104. William Harvey, *De generatione*, in William Harvey, *The Works of William Harvey*, trans. Robert Willis (1847; reprint, Annapolis, Md.: St. John's College Press, 1949), 157, 270.

105. Letter from Harvey to R. Morison, M.D., of Paris, 28 April 1652, in *Works of Harvey*, 604, 605, 604.

106. Boyle, *Usefulness of Natural Philosophy*, pt. 1, essay 2, in *Works* 2: 22.

107. While Bacon had said that final causality belonged to the study of metaphysics, not physics, he also spoke of the "true signatures and marks set upon the works of creation." See Patey, *Probability and Literary Form*, 46. For a more detailed analysis of the way in which Boyle used such reasoning, see Lennox, "Boyle's Defense," 38–52. Boyle's views on teleological reasoning will be treated in more depth in the next chapter. As Timothy Shanahan has argued, Boyle's discussions about final causality were primarily theological in focus; Shanahan, "Teleological Reasoning in Boyle's *Disquisition about Final Causes*," in *Robert Boyle Reconsidered*, edited by Michael Hunter (Cambridge: Cambridge University Press, 1994), 177–92.

108. See Boyle, *Usefulness of Natural Philosophy*, pt. 1, essay 4, in *Works* 2: 36–49, for his criticisms of the Aristotelians' speculations about final causes.

109. Ibid., pt. 2, essay 5, in *Works* 2: 175.

110. See Lennox, "Boyle's Defense," 48.

111. Boyle, *Usefulness of Natural Philosophy*, pt. 1, essay 5, in *Works* 2: 50.

112. See Lennox, "Boyle's Defense," 41, 45.

113. This is what Boyle meant when he said that the "modus" had not been "fully explicated." Gassendi had been one of the last to accept the circulation of the blood. He finally acquiesced in the circulation but preferred Descartes's theory that the function of the circulation was for the distribution of heat from the heart. See G. S. Brett, *The Philosophy of Gassendi* (London: Macmillan, 1908), 111; and Robert Willis, introduction to *The Works of William Harvey* (Annapolis, Md.: St. John's College Press, 1949), li.

114. Boyle, *Usefulness of Natural Philosophy*, pt. 2, essay 1, in *Works* 2: 69. This work was written in the 1650s but was not published until 1663.

115. Harvey, *De motu cordis*, in *Works of Harvey*, 9–105; Descartes, *Discourse on the Method*, pt. 5, in *Philosophical Writings* 1: 134–39.

116. Descartes, *Discourse on the Method*, pt. 5, in *Philosophical Writings* 1: 134, 137.

117. In *Excellency of Theology* (sec. 3, in *Works* 4: 42–45) and *Disquisition about Final Causes* (sec. 1, in *Works* 5: 399), Boyle noted that all of Descartes's principles depended upon teleological conjectures about God's original design. Chemistry was a crucial element in the full proof of the circulation theory. Frank (*Harvey and the Oxford Physiologists*, 19) maintains that Harvey was "aloof from chemistry and doubted the applicability of chemistry to biology." This is a rather paradoxical statement given the subsequent career of his theory. It is not clear, however, how aloof Harvey actually was. Boyle, in *Usefulness of Natural Philosophy* (pt. 2, essay 5, chap. 11, in *Works* 2: 167), indicated that Harvey did at least try some of the alchemists' suggestions. He mentioned the particularly gruesome one of laying a dead man's hand on the afflicted part of a patient's body. See also Walter Pagel, *William Harvey's Biological Ideas* (New York: S. Karger, 1967), 99–101, for Harvey's letter to

John Beverwijck, wherein he displays his familiarity with the doctrines of Paracelsus and other iatrochemists.

118. See Boyle, *Usefulness of Natural Philosophy*, pt. 2, essay 1, in *Works* 2: 69ff., for an account of the many experiments Boyle performed for the proof of Harvey's theory.

119. Ibid., 69. Boyle used similar experiments against Hobbes's notion that the heart was the "seat of sense." He showed, for example, that a warm-blooded animal might survive the loss of its heart for a short while, although it could not survive the "exemption or spoiling of the brain" (70).

120. Descartes himself had performed similar experiments. See his letter to Plempius, dated 15 February 1638, in Descartes, *Philosophical Writings* 3: 79–85. Descartes discussed an experiment on a severed fish heart yet rejected the notion that it refuted his theory. He claimed that "I could always judge, or, as often happened, see with my own eyes, that some remaining drops of blood had fallen from parts higher up into the part where the pulse was occurring" (80). See 84, for the importance that Descartes placed on his ability to explain the color of arterial blood. See also his letter to Plempius, 23 March 1638, 92–96, for his response to a second set of objections.

121. Richard Lower, *Tractatus de corde* (1669), trans. K. J. Franklin, *Early Science in Oxford*, ed. R. T. Gunther, vol. 9 (London: Dawson's, 1968), 164. The observation was communicated to Boyle by Lower in a letter dated 24 June 1664, printed in Boyle, *Works* 6: 472–73. Discoveries such as this led Boyle to compile a vast number of observations and experiments on the human blood, published as *The Natural History of Human Blood* in 1684.

122. Royal Society, *Boyle Papers* 17: fol. 40r. In *Usefulness of Natural Philosophy* (pt. 1, essay 2, in *Works* 2: 22), Boyle noted that although some supposed that "*Columbus, Cesalpius, Padre Paulo,* and Mr. *Warner*" had "some notion of the circulation," it was Harvey who in fact discovered this "so advantageous . . . use of the valves of the heart [and] nimble circular motion of the blood." It must be stressed that the circulation was, and still is, unobservable by simple sense perception. In the seventeenth century it had to be inferred; today we are only able to "see" the process by the use of sophisticated technical apparatuses.

123. Royal Society, *Boyle Papers* 17: fol. 40r.

124. Boyle, *Usefulness of Natural Philosophy*, pt. 1, essay 1, in *Works* 2: 5.

125. Boyle, *New Experiments*, 2d ed., preface to "Defence of the Doctrine," in *Works* 1: 122.

126. Boyle was also involved in the attempt to explain the action of specific medicines as a result of their corpuscular makeup. See his "Essay on the Reconcileableness of Specific Medicines to the Corpuscular Philosophy," in *Works* 5: 1–31. Because of his dual interest in the chemical and mechanical traditions, it is difficult to determine whether he should be classified as an iatrochemist or an iatromechanist. For a discussion of iatromechanism, see Theodore Brown, "The College of Physicians and the Acceptance of Iatro-Mechanism in England, 1665–1695," *Bulletin of the History of Medicine* 44 (1970): 12–30; and Sloan, "Descartes, the Sceptics, and the Rejection of Vitalism."

Chapter Four

1. Spinoza was clearly a "necessarian." Leibniz and Descartes are sometimes categorized as such, but we will see below that there are problems with such attributions. *Voluntarism*, a term not used by Boyle, encompasses a more complex set of positions. See, for example, Klaaren, *Religious Origins;* McGuire, "Boyle's Conception of Nature"; Popkin, *History of Skepticism;* Timothy Shanahan, "God and Nature in the Thought of Robert Boyle," *Journal of the History of Philosophy* 26 (1988): 547–69; B. J. Shapiro, *Probability and Certainty;* and van Leeuwen, *Problem of Certainty.* Both Osler, in "Providence and Divine Will," and Lisa T. Sarasohn, in "Motion and Morality: Pierre Gassendi, Thomas Hobbes, and the Mechanical World View," *Journal of the History of Ideas* 46 (1985): 363–79, describe Gassendi's "empiricism" as a product, in part, of his voluntarist conception of God's creation. But see Joy, *Gassendi the Atomist,* for a different view of Gassendi's epistemology. Dear, in *Mersenne and the Learning,* notes that Mersenne's empiricism was a product of his voluntarism.

2. Particularly relevant to this issue is the vast literature on the "Merton thesis." Two helpful anthologies that survey this literature are I. Bernard Cohen, ed., *Puritanism and the Rise of Modern Science* (New Brunswick, N.J.: Rutgers University Press, 1990); and Webster, *Intellectual Revolution.*

3. Works in this area overlap with the studies cited above. Jacob and Jacob, in "Anglican Origins of Modern Science," for example, used McGuire's interpretation of Boyle's ontological commitment to the passivity of matter (McGuire, "Boyle's Conception of Nature") and argued, against the Merton thesis, that Boyle's polemics against atheism can be seen as an indication of his Anglican allegiances. J. R. Jacob made a similar point in his 1978 article "Boyle's Atomism." Richard Olson, "On the Nature of God's Existence, Wisdom, and Power: The Interplay between Organic and Mechanistic Imagery in Anglican Natural Theology, 1640–1740," in *Approaches to Organic Form,* ed. Frederick Burwick (Dordrecht: D. Reidel, 1987), 1–48, follows the Jacobs' thesis. Olson argues that Boyle's voluntarism also contributed to his political outlook: Boyle, asserts Olson, adopted a mechanical philosophy "precisely because the passivity of matter presumed by mechanists ensured the continuing need for a divine energizing and conserving presence" (1–2). Shapin and Schaffer (*Leviathan and the Air-Pump*) offer a similar analysis. Michael Hunter, "Science and Heterodoxy: An Early Modern Problem Reconsidered," in *Reappraisals of the Scientific Revolution,* ed. David C. Lindberg and Robert S. Westman (Cambridge: Cambridge University Press, 1990), 437–68, provides a more detailed and balanced account of the actual threat of atheism as perceived by Boyle and his contemporaries. Henry, in "Occult Qualities and the Experimental Philosophy," argues that Boyle actually advocated the activity of matter in part because he believed that Cartesian passivity was theologically dangerous. Jan Wojcik, in "The Theological Context of Boyle's *Things above Reason,*" in *Robert Boyle Reconsidered,* ed. Michael Hunter (Cambridge: Cambridge University Press, 1994), 139–55, gives a good account of Boyle's somewhat ambivalent attitude toward the controversy surrounding the Calvinist doctrine of predestination and shows

how Boyle's position was closest to that of nonconformists such as Baxter, Owen, and Ferguson and opposed to that put forward by Joseph Glanvil.

4. See McGuire, "Boyle's Conception of Nature"; and Shanahan, "God and Nature," for more detail on how these metaphysical and epistemological positions are thought to be associated.

5. See Henry, "Occult Qualities and the Experimental Philosophy"; and Garber, *Descartes' Metaphysical Physics*, 299–305, for discussions concerning the limits of Descartes's voluntarism. In his notes on Spinoza's philosophy, Leibniz wrote that "God necessarily exists, but he produces things freely. And while the power of things is produced by God, it is distinct from divine power, and things themselves operate, even if they may have received their forces for acting [from elsewhere]." Gottfried Wilhelm Leibniz, *Philosophical Essays*, ed. and trans. Roger Ariew and Daniel Garber (Indianapolis: Hackett Publishing, 1989), 275. We will see below that there are a number of similarities between Leibniz and Boyle on this point.

6. See Dear, *Mersenne and the Learning*, on Mersenne's belief in the universal harmony present in the world despite its radical contingency.

7. See Shanahan, "God and Nature," for a discussion of the many permutations at the theological level and particularly for Shanahan's account of "concurrentism," which was a notion of God's causal activity that proposed a middle ground between necessarianism and voluntarism. Boyle's eclecticism can also be seen in the various works that he read on religion. In a partial listing of books in the possession of his servant John Warre at the time of Boyle's death (published in Boyle, *Early Essays*), there are works by Augustine, Baxter, Calvin, Comenius, Diodati, Erasmus, Ferguson, Luther, Owen, and Polhill, among others. Also, in a commonplace book (Royal Society, MS 187, fol. 30v), there is a list of books "for the collection making by Mr. Boyle of books tending to the proofs and defence of the truth of the Christian religion" that includes works by Augustine, Tertullian, Aquinas, Charleton, Confucius, and Stillingfleet.

8. Latitudinarians had an eclectic approach to theology and advocated tolerance through a reasonable mediation of the claims of various religious sects. They agreed with High Church Anglicans that dissent was dangerous, but they were opposed to the use of force and sought instead a rational and universal basis for assent. See Hill, *World Turned Upside Down*; James R. Jacob, "Boyle's Circle in the Protectorate: Revelation, Politics, and the Millennium," *Journal of the History of Ideas* 38 (1977): 131–40; and Westfall, *Science and Religion*.

9. It is risky at best to discuss Boyle's religious sympathies in any straightforward manner. He regularly attended an Anglican church, and his advocacy of toleration would seem to make him a latitudinarian, but there was also a strong Calvinist slant to many of his writings, which could be attributed to the influence of his tutor, Isaac Marcombes, and his later association with Hartlib and Dury, that would seem to make him a Puritan. His aversion to oath-taking, as indicated by his refusal to accept the presidency of the Royal Society, however, would seem to make him a dissenter. See Burnet's account of Boyle's life (BL, Add. MS 4229, fol. 60v), where he noted that while Boyle was a member of the Church of England, "yet he was not keen on any of our debates and loved virtue and goodness

in all people." Because of his eclecticism, any attempt to assimilate his views to a
particular religious sect will likely fail to capture the complexity of his thought.

10. This is the safest way to proceed, because in addition to the difficulties presented
by Boyle's eclecticism, he also never composed a complete metaphysical system.
That would have been clearly opposed to all that he had said about the prematu-
rity of system building. Boyle's excessive "scrupulosity," as discussed in detail in
Hunter, "Conscience of Robert Boyle" and "Casuistry in Action," provides yet
another reason for resisting the temptation to put Boyle's thought into one particular category.

11. Boyle, "Account of Philaretus," in *Works* 1: xxii.

12. Boyle, *Usefulness of Natural Philosophy*, pt. 1, in *Works* 2: 57, 15, 30.

13. Ibid., essay 5, in *Works* 2: 62.

14. Ibid., essay 2, in *Works* 2: 15. See Hunter, "Science and Heterodoxy"; and Wojcik,
"Theological Context of Boyle's *Things above Reason*," for the perceived threat of
atheism and how it was linked with issues in natural philosophy. Boyle was not
so much supporting the church by his writings as trying to show why the church
should support him. His concern was certainly not unique to England. See E. J.
Dijksterhuis, *The Mechanization of the World Picture* (Oxford: Clarendon Press, 1961),
435–36, on how Boyle sought to Christianize atomism in the manner that Gassendi had followed before him.

15. Boyle, *Usefulness of Natural Philosophy*, pt. 1, essay 2, in *Works* 2: 15.

16. Ibid., essay 3, in *Works* 2: 34.

17. Boyle, *Reconcileableness of Reason and Religion*, sec. 7, in *Works* 4: 180.

18. Boyle, *Usefulness of Natural Philosophy*, pt. 1, essay 3, in *Works* 2: 31. He quoted
Bacon's saying that no one should "think or maintain, that a man can search too
far, or be too well studied, in the book of God's word, or in the book of God's
works" (58).

19. Ibid., essay 5, in *Works* 2: 53. This is really an abstract consideration about the
wonder of the body itself, not that it always works properly. Boyle, as we have
seen, was sickly most of his life.

20. Ibid., 55.

21. Ibid., essay 2, in *Works* 2: 18, 29, 18. Boyle met Menasseh ben Israel, a rabbi and
Old Testament scholar, in Amsterdam in 1648 and learned the Hebrew language
from him. See R. E. W. Maddison, "Studies in the Life of Robert Boyle, F. R. S.
Part VI: The Stalbridge Period, 1645–55, and the Invisible College," *Notes and
Records of the Royal Society* 18 (1963): 104–24.

22. Boyle, *Usefulness of Natural Philosophy*, pt. 1, essay 5, in *Works* 2: 50.

23. Ibid., essay 3, in *Works* 2: 31.

24. Ibid., essay 5, in *Works* 2: 56–57.

25. Ibid., essay 3, in *Works* 2: 32. Bacon, in the preface to his "Great Instauration," had
described man as the "servant and interpreter" of nature, and in his *New Organon*
he noted that his own work "comes from God's goodness and returns to his glory"
(*Works of Bacon* 8: 53).

26. Boyle, *Usefulness of Natural Philosophy*, pt. 1, essay 3, in *Works* 2: 32–33. Man is not
a "priest" in the sense of the established church, but merely as the representer

of the rest of creation to God. In *Final Causes* (sec. 1, in *Works* 5: 401), Boyle discussed this issue in the context of his argument against Descartes's rejection of the use of final causes in natural philosophy. Boyle believed that most natural philosophers of the past, not only experimenters, could be called priests of nature. Shapin and Schaffer (*Leviathan and the Air-Pump*, 310–19) give a detailed account of Hobbes's adverse reaction to this type of discourse. As their discussion concludes, however, it is not clear whether they are summarizing Hobbes or presenting their own view when they say, for example, that the notion of experimenters as "priests of nature" implied that their work could have "direct effects on the establishment of religion" (319). While Shapin and Schaffer are right that others would use Boyle's science for such purposes, this was not Boyle's concern in either place where he discussed the notion.

27. Boyle, *Usefulness of Natural Philosophy*, pt. 1, essay 3, in *Works* 2: 33.

28. Ibid., essay 5, in *Works* 2: 57. In essay 3 Boyle noted that according to the hermetic corpus "there can be no religion more true of just, than to know the things that are and to acknowledge thanks for all things to him, that made them" (*Works* 2: 31). He also noted that Galen had maintained that any "good anatomist" would have a strong "invitation to believe, and admire an omniscient author of nature" (*Works* 2: 52–54).

29. Ibid., essay 5, in *Works* 2: 63.

30. Ibid., essay 1, in *Works* 2: 10. Boyle added the speculation that Aristotle may have borrowed from the work of Solomon.

31. Ibid., essay 2, in *Works* 2: 29. Mersenne's pious attitude toward the study of nature was also a product of Augustine's influence. See Dear, *Mersenne and the Learning*, 115, 223.

32. Boyle, *Usefulness of Natural Philosophy*, pt. 1, essay 5, in *Works* 2: 62.

33. Ibid., essay 3, in *Works* 2: 30.

34. Boyle, *Final Causes*, sec. 1, in *Works* 5: 400.

35. Boyle, *Usefulness of Natural Philosophy*, pt. 1, essay 2, in *Works* 2: 15. Man honors God by contemplating nature, because in this way "the great works are not without witness" (16). Boyle often cited Cicero and Seneca in such contexts, yet he opposed the general Stoic conception of nature on the same grounds as he disapproved of other "naturists" who supposed that the world was a "living animal." On the influence of Stoic philosophy in the sixteenth and seventeenth centuries, see Peter Barker, "Stoic Contributions to Early Modern Science," in *Atoms, "Pneuma," and Tranquility*, ed. Margaret J. Osler (Cambridge: Cambridge University Press, 1991), 135–54; idem, "Jean Pena (1528–58) and Stoic Physics in the Sixteenth Century," *Southern Journal of Philosophy* 23 (1985): 93–107; and Peter Barker and Bernard R. Goldstein, "Is Seventeenth-Century Physics Indebted to the Stoics?" *Centaurus* 27 (1984): 148–64. Boyle's type of determinism is Stoic in general outline but appears to owe its specific formulation more to the Christian influence of Augustine.

36. Boyle, *Usefulness of Natural Philosophy*, pt. 1, essay 2, in *Works* 2: 20–21. Boyle cited Gassendi here as his authority on astronomy.

37. Thomas Tymme, *A Dialogue Philosophical* (London, 1612), as quoted in Debus,

English Paracelsians, 89. Interestingly, this passage appeared in Tymme's dedication to Chief Justice Coke.

38. Boyle, *Usefulness of Natural Philosophy*, pt. 1, essay 2, in *Works* 2: 22–23. Boyle's argument here is similar to that which John Ray would develop in more detail in *The Wisdom of God Manifested in the Works of the Creation* (London, 1727).

39. Boyle, *Usefulness of Natural Philosophy*, pt. 1, essay 3, in *Works* 2: 30–31. The structure of living bodies, discovered through anatomical dissection, could also lead one to acknowledge God's wise design. Galen, a "heathen," when he had considered "the exquisite structure of the human body" had broken "forth into very elevated and even pathetical celebrations of God" and said that his books were composed as "hymns to the creator's praise" (*Works* 2: 52).

40. Boyle, *Christian Virtuoso*, pt. 1, in *Works* 5: 514.

41. Boyle, *Usefulness of Natural Philosophy*, pt. 1, essay 3, in *Works* 2: 30.

42. Ibid., 34.

43. Ibid., essay 5, in *Works* 2: 58. Boyle cited Bacon's *Advancement of Learning*, bk. 1. The quotation in *Works of Bacon* 6: 96, reads: "A little or superficial knowledge of philosophy may incline the mind of man to atheism, but a farther proceeding therein doth bring the mind back again to religion." Cf. Bacon, *The Essayes or Counsels, Civil and Morall*, essay 16, "Of Atheism," where he wrote that "a little philosophy inclineth man's mind to atheism, but depth in philosophy bringeth men's minds about to religion" (*Works of Bacon* 12: 132).

44. Boyle, *Usefulness of Natural Philosophy*, pt. 1, essay 5, in *Works* 2: 57–58.

45. The first part of *The Christian Virtuoso* was published in 1690. The appendix to it and a second part were published for the first time in Birch's 1744 collection of Boyle's works. Boyle did not often discuss religion in his scientific works, but he thought it was appropriate to discuss natural philosophy in his theological works. The works cited in the text were listed by him as theological works in Royal Society, *Boyle Papers* 35: fol. 187r; and Royal Society, MS 185, fols. 1v, 3r.

46. Boyle, *Usefulness of Natural Philosophy*, pt. 1, essay 5, in *Works* 2: 62.

47. Boyle, *Free Inquiry*, sec. 1, in *Works* 5: 161. By calling this work a "free inquiry," Boyle meant that it was to be considered as speculative. Also, of course, by using the term *vulgar* in his title, Boyle merely meant to refer to that conception of nature that was "commonly" received. J. R. Jacob ("Boyle's Atomism") has described this work as an "exercise in Anglican polemics," but it is not clear how it supports Anglicanism in particular. Boyle's opposition to an animate nature is taken by Jacob as an indication of Boyle's desire to allow for the action of spirit in the world and thus to support the Christian worldview. Boyle's work was certainly used by others for such purposes, but in the context of his own writings, Boyle's actual argument was that natural phenomena are physically, not spiritually, caused in the world.

48. Boyle, *Free Inquiry*, sec. 2, in *Works* 5: 169. This is an Aristotelian type of definition of nature. Boyle's goal was quite similar to Aristotle's: he merely advocated a different method by which it could be achieved.

49. Ibid. Boyle listed eight different senses in all. He made fine distinctions within the two physical senses, but these were the four general categories. For the first

usage, *natura naturans*, Boyle advised philosophers to use the word *God* if that is what they meant.

50. Ibid., sec. 7, in *Works* 5: 221.

51. Ibid., sec. 4, in *Works* 5: 175.

52. Ibid., sec. 8, in *Works* 5: 247.

53. Ibid., 246. This passage is reminiscent of his criticisms of Scholastic philosophy discussed in chapter 1.

54. Ibid., 249.

55. Boyle, "Excellency and Grounds of the Mechanical Hypothesis," in *Works* 4: 73.

56. Boyle, *Free Inquiry*, sec. 4, in *Works* 5: 177.

57. Ibid., italics mine. In support of his nominalist interpretation of Boyle's ontological view of a world composed of "non-related particulars," McGuire ("Boyle's Conception of Nature," 525) quotes from Boyle's first definition of nature in general and joins to it a portion of his definition of the nature of particular bodies. The nature of a particular body could be described as a "convention of the mechanical affections" of its parts, but Boyle did not use this notion to describe nature in general. In addition, McGuire makes no mention of the second definition of nature in general given here. As it was meant by Boyle as an amplification of his discussion, it should not be ignored.

58. Boyle, *Free Inquiry*, sec. 4, in *Works* 5: 178.

59. Ibid., sec. 7, in *Works* 5: 221.

60. Ibid., sec. 4, in *Works* 5: 177, 178. Given these two definitions, it is easy to see why Boyle chose to describe his philosophy as mechanical. In a sense, Boyle's discussion of individual bodies is similar to that developed later by Leibniz. If (as Leibniz would say, and as we will see Boyle does say) every property of a thing is essential to its identity and some of its properties are external relations, then individual bodies are not nonrelated particulars but are defined by their *inherent* relations to other bodies. See Gottfried Martin, *Leibniz: Logic and Metaphysics*, trans. K. J. Northcott and P. G. Lucas (Manchester: Manchester University Press, 1964), 129–54; and Benson Mates, *The Philosophy of Leibniz: Metaphysics and Language* (Oxford: Oxford University Press, 1986), 170–97.

61. Boyle, *Free Inquiry*, sec. 2, in *Works* 5: 169, 170.

62. Ibid., 170. It is not often realized that it was Helmont's notion of a law about which Boyle expressed dissatisfaction. McGuire ("Boyle's Conception of Nature," 538) notes that Boyle did not approve of Helmont's notion that the "laws of nature emanated from God." But in the passage from which the quotation in the text is taken, Boyle's concern was not with emanation but with the fact that for Helmont the laws were somehow "obeyed" by matter, which would entail the same type of notion of an intelligent nature that Boyle was trying to eliminate.

63. McGuire ("Boyle's Conception of Nature," 536) makes this association, and many historians who advocate an empiricist interpretation of Boyle's epistemology, such as Osler, J. R. Jacob, and Shapin and Schaffer, have followed McGuire's analysis.

64. Boyle, *Free Inquiry*, sec. 4, in *Works* 5: 175. Boyle called the law of the land a "notional rule."

65. Ibid., sec. 7, in *Works* 5: 222; cf. sec. 2, in *Works* 5: 170–71.

66. Ibid., sec. 7, in *Works* 5: 222. It must be admitted that Boyle was quite vague in

his discussions of metaphysical topics, but this is as it should be. He was not attempting to construct an entire metaphysical theory but merely to make speculative suggestions about how one could view nature in such a way as to preserve the intelligibility of such notions as laws and causes while avoiding the pitfalls of deism or animism.

67. Ibid., sec. 2, in *Works* 5: 170.

68. Boyle, "A Paradox of the Natural and Preternatural State of Bodies," in *Works* 3: 782. Boyle discussed how "preter-natural" things only appear to be "contrary to nature" in *Free Inquiry*, sec. 7, in *Works* 5: 220.

69. Boyle, *Free Inquiry*, sec. 7, in *Works* 5: 219. Galileo's law of acceleration and Newton's law of universal gravitation would be examples of general laws; Boyle's ideal gas law would be a "custom of nature."

70. Boyle, "Cosmical Suspicions," in *Works* 3: 318, 323.

71. Ibid., 323. See Henry, "Boyle and Cosmical Qualities," for a discussion of how these relational properties introduce a nonmechanical or "occult" aspect into Boyle's corpuscular philosophy. This aspect has been attributed to the influence of the corpuscular tradition within alchemy. See Clericuzio, "Redefinition of Boyle's Chemistry"; Newman, "Boyle's Debt to Corpuscular Alchemy"; and Principe, "Boyle's Alchemical Pursuits."

72. If Boyle believed in radical contingency, then to be consistent he should have been more like Hume, who denied the efficacy of design arguments because the concept of design seemed to imply a strict causal determinism not verifiable by our senses.

73. Boyle, *Free Inquiry*, sec. 2, in *Works* 5: 170.

74. Boyle, "Cosmical Suspicions," in *Works* 3: 319.

75. Boyle, "Excellency and Grounds of the Mechanical Hypothesis," in *Works* 4: 68–69. Boyle's argument, here and elsewhere, for the superiority of corpuscular hypotheses is that the two principles, matter and motion, will be sufficient to explain all processes.

76. Royal Society, *Boyle Papers* 1: fol. 36r. He argued that the Aristotelians, Epicureans, and Cartesians were all guilty of errors in their philosophy because they failed to appreciate this distinction. See, for example, Boyle, *Usefulness of Natural Philosophy*, pt. 1, essay 4, in *Works*, vol. 2.

77. Boyle, *Origin of Forms and Qualities*, "The Relative Nature of Physical Qualities," in *Works* 3: 23.

78. This phrase is used by Popkin in his *History of Skepticism* as a way to describe Gassendi and Mersenne's "mitigated skepticism," a forerunner of today's strict empiricism.

79. Boyle, *Usefulness of Natural Philosophy*, pt. 1, essay 4, in *Works* 2: 39. See also Royal Society, *Boyle Papers* 14: fol. 1v, where Boyle discussed how tenses are mixed in the Bible because it was written by God, who "knows no tenses but the present." In *Final Causes* (sec. 3, in *Works* 5: 413), Boyle noted that God's "piercing sight is able to penetrate the whole universe, and survey all the parts of it at once." He is "able not only to see the present state of things . . . , but to foresee all the effects" that particular bodies will produce. This position is similar to the Calvinist notion of predestination, and Boyle acknowledged that it seemed to create a conflict with

the idea that humans have free will. See his discussions of this problem in *Things above Reason*, in *Works* 4: 408–9, 466; and *Greatness of Mind Promoted by Christianity*, in *Works* 5: 548. See Wojcik, "Theological Context of Boyle's *Things above Reason*," on Boyle's "conditional" notion of predestination, where God has foreknowledge of all that will happen but is not the direct cause of human sin and error. Ultimately Boyle believed that God's prescience was compatible with the existence of free will but that humans were not capable of fully understanding the compatibility.

80. Boyle, *Free Inquiry*, sec. 1, in *Works* 5: 163.

81. Ibid.

82. Ibid. In this passage Boyle is talking not about causality but about necessity. We will see that his notion of necessity fits well with the causal activity of matter.

83. Ibid., sec. 4, in *Works* 5: 179. This discussion seems to indicate, once again, an Augustinian influence.

84. Boyle, *Origin of Forms and Qualities*, "The Relative Nature of Physical Qualities," in *Works* 3: 34.

85. Boyle, *History of Fluidity and Firmness*, in *Works* 1: 443–57. See Clericuzio, "Redefinition of Boyle's Chemistry"; and Henry, "Occult Qualities and the Experimental Philosophy." Clericuzio and Henry do not agree on which parts of matter are active, but both agree that for Boyle matter is active. Shapin and Schaffer (*Leviathan and the Air-Pump*, 202) seem to fall into the same misunderstanding as Boyle's contemporaries. They quote Boyle's *Free Inquiry* (*Works* 5: 210) that "motion does not belong essentially to matter" and conclude that Boyle believed that "parcels of matter" were also not "capable of self-movement." But the context of Boyle's discussion about the motion of matter (*Works* 5: 209ff.) was an argument against the Aristotelian distinction between "natural" and "violent" motions, where violent motions are those caused by external agents. To show the inherent difficulties with such a dichotomy, Boyle noted that if one believes that "the motions of *most* bodies . . . were impressed on them" by "extrinsical impellers," then the motions of most bodies would be violent and not natural. If, on the other hand, one believes in the "Epicurean hypothesis" that "every indivisible corpuscle has actual motion, or an incessant endeavour to change places," then there would need to be "no other principle of motion than that unloseable endeavour of the atoms" that compose bodies, and all motion would thus be natural. Whatever the case, Boyle concluded that "the only occasion I had to mention it [motion] here was to shew, that the vulgar distinction of it into natural and violent is not so clear and well grounded as to oblige us to admit, (what it supposes) that there is such a being [nature] as the naturists assert" (*Works* 5: 211).

86. Boyle, *History of Fluidity and Firmness*, "Intestine Motions," sec. 4, in *Works* 1: 447.

87. Ibid., sec, 1, in *Works* 1: 444; Leibniz, *Philosophical Essays*, 295. Leibniz noted that Boyle's experiments could be used as "experience" against Locke's contention but added that he believed that "reason alone supports this, and it is one of the proofs I use for refuting atoms." There are a number of other places in his essays where Leibniz showed his approval of Boyle. See, for example, his essay "Against Barbaric Physics" and his fifth letter to Clarke, both in *Philosophical Essays*, 317–18, 344. Leibniz first met Boyle when he visited England in 1673. There is a marked dissimilarity between Boyle and Leibniz at the epistemological level, of course,

but their ontological views about physical causality are similar. See Hidé Ishiguro, "Pre-established Harmony versus Constant Conjunction: A Reconsideration of the Distinction between Rationalism and Empiricism," in *Rationalism, Empiricism, and Idealism*, ed. Anthony Kenny (Oxford: Clarendon Press, 1986), 61–85, for a detailed discussion of Leibniz's nonlinear conception of causality, how he believed that "the understanding of the global structure of things adds to our understanding of the processes or movements of things in it," and how such a conception is different from Hume's notion of constant conjunction. Ishiguro supports this complex view of causality in Leibniz by noting his intense interest in a number of different areas of study. This, of course, is a trait that Leibniz also shared with Boyle. See also A. Rupert Hall, *Henry More: Magic, Religion, and Experiment* (Oxford: Blackwell, 1990), 191–92, on the ways in which Boyle shared more similarities with Leibniz than with Newton.

88. Ariew and Garber, in Leibniz, *Philosophical Essays*, 155, 156–57. Leibniz said that the notion had to be enlarged, but his subsequent discussion differs little from Boyle's. McGuire ("Boyle's Conception of Nature," 539) notes that Leibniz's "On Nature Itself" was written as a response to Christopher Sturm's defense of Boyle's *Free Inquiry*. The only reference Leibniz made to Boyle in this work was favorable, however, and the description that Leibniz gave of Sturm's arguments was not like Boyle's at all. In the same passage, McGuire also claimed that in this essay Leibniz "indicated similarities between his view and those of van Helmont." It is not exactly clear what McGuire means by this. Earlier, however, he maintained that for Leibniz and Helmont, nature was "an emanation from Divine being" (533). Leibniz gave no indication in this essay that he believed in any type of emanation theory, and in his notes on Spinoza's philosophy, he explicitly rejected such a theory (Leibniz, *Philosophical Essays*, 275). Leibniz made a similar point in his essay "Two Sects of Naturalists," arguing against the Stoic notion that "blind necessity determines him [God] to act" (282). There is a striking similarity between Boyle and Leibniz on this point also.

89. The metaphor is normally associated with those who maintained that the book was written in the language of mathematics. But it had also been used by the chemists, as can be seen in the quote by Tymme in Debus, *English Paracelsians*, 89. See above at note 37.

90. Boyle, *Free Inquiry*, sec. 5, in *Works* 5: 195. He noted that this was a question that was "not wont to be discussed."

91. Ibid., 196.

92. Ibid., sec. 8, in *Works* 5: 251–52.

93. Ibid., 252. Cf. Leibniz, "Against Barbaric Physics," in *Philosophical Essays*, 319: "The laws of motion and nature have been established, not with absolute necessity, but from the will of a wise cause, not from a pure exercise of will, but from the fitness [convenientia] of things." Leibniz said much the same in his essay on Locke, in *Philosophical Essays*, 297.

94. Boyle, *Free Inquiry*, sec. 8, in *Works* 5: 252. Boyle allowed that miraculous events could occur, but while supernatural, they need not be "contranatural." Further, these types of supernatural events were outside the realm of natural philosophy, so a discussion of them properly belonged to theology and metaphysics. See

Boyle, *Final Causes*, sec. 4, in *Works* 5: 420, for the distinction between physico-theology (metaphysics) and physics. The occurrence of a miracle did not require that the laws of nature be contravened. God could use secondary causes to produce the miracle. As Boyle said in "Excellency and Grounds of the Mechanical Hypothesis," even "if an angel himself should work a real change in the nature of a body, it is scarce conceivable to us men, how he could do it without the assistance of local motion; since, if nothing were displaced, or otherwise moved than before . . . it is hardly conceivable, how it should be in itself other, than just what it was before" (in *Works* 4: 73). See also Royal Society, MS 199, fols. 126v–125r; and discussions of this issue in Alexander, *Ideas, Qualities, and Corpuscles*; Marie Boas Hall, "Matter in Seventeenth-Century Science," in *The Concept of Matter in Modern Philosophy*, ed. Ernan McMullin (Notre Dame, Ind.: University of Notre Dame Press, 1978), 76–99; Hunter, "Science and Heterodoxy"; and Keith Hutchison, "Supernaturalism."

95. Boyle, *Free Inquiry*, sec. 5, in *Works* 5: 199.

96. Ibid., sec. 8, in *Works* 5: 251.

97. Boyle sought to revise traditional atomism, and Leibniz rejected atomistic explanations, but both were motivated by the same types of considerations. In his *Discourse on Metaphysics* Leibniz also expressed his approval of teleological reasoning and his opposition to pure geometry. In his essay "On Nature Itself" Leibniz called for a "philosophy midway between the formal and material, a system that correctly joins and preserves both" (*Philosophical Essays*, 167). This is certainly reminiscent of the type of project that Boyle hoped to achieve by joining together the mechanical and chemical disciplines.

98. Boyle, *Origin of Forms and Qualities*, "Author's Discourse," in *Works* 3: 4, 5. The *Origin* was written at the same time that Boyle was writing his draft of *A Free Inquiry*.

99. Ibid., "Of Substantial Forms," in *Works* 3: 48. The seminal principles in bodies are those which allow the body to act. This is much like the notion that Leibniz discussed in "On Nature Itself," when he maintained that "the law God laid down left some trace of itself impressed on things," so that "we must admit that a certain efficacy has been placed in things, a form or a force, something like what we usually call by the name 'nature,' something from which the series of phenomena follow in accordance with the prescript of the first command" (*Philosophical Essays*, 158–59).

100. Boyle, *Origin of Forms and Qualities*, advertisements to sec. 2, in *Works* 3: 75. Again, Boyle's argument is much like the type of argument that would be developed by Leibniz. In his essay "On Body and Force" Leibniz noted that "motions are in matter." Although "in origin they [motions] ought to be attributed to God, the general cause of things, . . . in particular cases, they ought to be attributed to the force God placed in things" (*Philosophical Essays*, 253). In his essay "On Nature Itself" Leibniz noted, much as Boyle had done, that while "primary matter is merely passive," "secondary matter," that which is contrived into bodies, is not (162). See Clericuzio, "Redefinition of Boyle's Chemistry"; and Henry, "Boyle and Cosmical Qualities," for discussions of Boyle's "active corpuscles."

101. Boyle, "Of the Atmospheres of Consistent Bodies," appended to *Continuation of New Experiments*, in *Works* 3: 278.

102. Boyle, *Origin of Forms and Qualities*, "The Relative Nature of Physical Qualities," in *Works* 3: 27, 28, 36.
103. Ibid., "Of Substantial Forms," in *Works* 3: 47, italics mine.
104. Ibid., "Publisher to the Reader," in *Works* 3: 2, and Boyle's preface, in *Works* 3: 12.
105. Ibid., "The Relative Nature of Physical Qualities," in *Works* 3: 27, 28.
106. Ibid., 36. Compare Mates, *Philosophy of Leibniz*, 197. Mates notes that for Leibniz accidents are "in the domain of reality." They are "ways of being" of substances. Of course Boyle did not go on to such elaborate speculations as monads and entelechies, but I am not trying to show that Boyle had the same metaphysical views as Leibniz, merely that one could have a nominalist criticism of other philosophers and still retain a realistic account of the qualities and causal powers of bodies.
107. Boyle, *Origin of Forms and Qualities*, "Of Substantial Forms," in *Works* 3: 41.
108. Ibid., preface, in *Works* 3: 11.
109. Boyle, *Free Inquiry*, sec. 2, in *Works* 5: 170.
110. Boyle, *Origin of Forms and Qualities*, preface, in *Works* 3: 11.
111. Ibid., "The Relative Nature of Physical Qualities," in *Works* 3: 23. Primary qualities are those that matter has by virtue of being matter, such as size and shape, while secondary qualities are those that bodies have by virtue of the configuration of the matter within them. See Alexander, *Ideas, Qualities, and Corpuscles*, for a more detailed account of Boyle's distinction.
112. Boyle, *Origin of Forms and Qualities*, "The Relative Nature of Physical Qualities," in *Works* 3: 23.
113. Boyle, *History of Particular Qualities*, "Cosmical Qualities of Things," chap. 1, in *Works* 3: 306. Boyle went on to note that because these qualities result from "the determinate fabrick of the grand system or world they are parts of," he would call them "cosmical" or "systematical qualities."
114. Boyle, *Origin of Forms and Qualities*, "The Relative Nature of Physical Qualities," in *Works* 3: 22, 32. Corruption is the opposite process. The matter is not destroyed; only its manner of existence has been altered.
115. Ibid., 23. Boyle went on to note that because the human body has "several external parts, as the eye, the ear, etc., each of a distinct and peculiar texture, whereby it is capable to receive impressions from the bodies about it, and upon that account is called an organ of sense; we must consider, I say, that these sensories may be wrought upon by the figure, shape, motion and texture of bodies without them after several ways, some of those external bodies being fitted to affect the eye, others the ear, others the nostrils, etc."
116. Ibid., preface, in *Works* 3: 11. He noted that Descartes "has something concerning some qualities; but though, for reasons elsewhere expressed, I have purposely forborne to peruse his system of philosophy, yet I find by turning over the leaves that he has left most of the other qualities untreated of; and of those that are more properly called sensible, he speaks but very briefly and generally, rather considering what they do upon the organs of sense, than what changes happen in the objects themselves, to make them cause in us a perception sometimes of one quality and sometimes of another."
117. Ibid.

118. Ibid., 18; the "*Excursion*" is on 18–37. See Henry, "Boyle and Cosmical Qualities."
119. Boyle, *Origin of Forms and Qualities,* "The Relative Nature of Physical Qualities," in *Works* 3: 18. Boyle identified Tubalcain as the first smith. Also, he used an analogy with an artificial engine for the same purpose (29). He noted that a clock, for example, has certain powers because of the order in which its parts have been arranged.
120. Ibid., 18, 19.
121. Boyle, *History of Particular Qualities,* "Cosmical Qualities of Things," chap. 1, in *Works* 3: 306.
122. Boyle, *Origin of Forms and Qualities,* "The Relative Nature of Physical Qualities," in *Works* 3: 20.
123. Boyle, *History of Particular Qualities,* "Cosmical Qualities of Things," chap. 2, in *Works* 3: 307.
124. Studies that present Boyle as a voluntarist or empiricist are therefore questionable, especially when they attempt to support such interpretations by claiming that Boyle believed in the passivity of matter or had a conception of the world as consisting of a set of unrelated particulars, as does the study produced by McGuire. Other studies, such as those by J. R. Jacob and Shapin and Schaffer, that follow McGuire's analysis and attempt to provide sociological explanations for Boyle's having had such beliefs are, of course, equally problematic.
125. Boyle, *Origin of Forms and Qualities,* preface, in *Works* 3: 12. He credited Bacon and Mersenne with having begun the work required for such histories.

Chapter Five

1. According to More (*Life and Works,* 110), the motto was taken from Horace's "Nullius addictus jurare in verba magistri," which translates "I am not bound over to swear to any master's dictates." I agree with More that rendering the motto as "Nothing in words" "plays fast and loose with the Latin language" and only vaguely expresses the Royal Society's wish not to accept any claim on the basis of authority.
2. Boyle would have been aware of Mersenne's criticisms of Galileo's thought experiments, because he had read Mersenne's *Les nouvelles pensées Galilée.*
3. Boyle admitted, in his *Hydrostatical Paradoxes,* that he had no problems with Pascal's conclusions, only with the way in which he had arrived at them.
4. Hobbes (*Physical Dialogue,* 349–50) evidently thought that Harvey's theory was acceptable on the grounds that it was compatible with his theory of simple circular motion. He could not have consistently accepted it as experimentally proved, since his dialogue was expressly designed to deny the efficacy of such proof. In a similar manner, Spinoza accepted Bacon's theory that heat is a form of motion, but he believed that it had been sufficiently proved by reason alone and that Boyle's experiments to confirm it were superfluous.
5. Letter from Spinoza to Henry Oldenburg, September 1661, in Oldenburg, *Correspondence* 1: 426–28.
6. Boyle, *Occasional Reflections,* sec. 4, discourse 11, in *Works* 2: 412.
7. Boyle, *History of Cold,* "An Examen of Antiperistasis," in *Works* 2: 662.

8. Boyle, *Experiments on the Mechanical Origin of Qualities*, "Reflections upon the Hypothesis of Alcali and Acidum," chap. 6, in *Works* 4: 289.

9. Ibid., chap. 7, in *Works* 4: 291.

10. Boyle, *Usefulness of Natural Philosophy*, pt. 1, essay 2, in *Works* 2: 29. In his "Excellency and Grounds of the Mechanical Hypothesis" Boyle also used the text metaphor, and in his conclusion he likened the world to a "letter written in cyphers" (*Works* 4: 78).

11. Boyle, *Experiments on the Mechanical Origin of Qualities*, "Imperfection of the Chemist's Doctrine of Qualities," chap. 4, in *Works* 4: 277.

12. Boyle, in *Usefulness of Natural Philosophy* (especially in pt. 1, essay 5), gave many examples of how the book was written for our instruction.

13. Boyle, *Origin of Forms and Qualities*, advertisements to sec. 2, in *Works* 3: 75.

14. Boyle was personally acquainted with many of the leading theologians of his day and had read many of the works of the church fathers. References to all these thinkers are scattered throughout his published works and letters. Menasseh ben Israel was also a correspondent of Henry Oldenburg's. See Oldenburg, *Correspondence* 1: 123–25, where Oldenburg discussed the "great rabbi's attempts to reconcile passages in the Old Testament." Oldenburg mentioned in this letter that he had met the rabbi at the home of Boyle's sister Lady Ranelagh. In his "Account of Philaretus" Boyle mentioned that he had lodged with Diodati while in Italy (*Works* 1: xxiv), and in *Usefulness of Natural Philosophy* he referenced Diodati's notion that nature's words "speak in our tongues" (pt. 1, in *Works* 2: 29). See Maddison, "Stalbridge Period" and "Grand Tour," for more information on Boyle's associations with Jean Diodati and Menasseh ben Israel. See More, *Life and Works*; and Thomas Birch's introduction to Boyle, *Works* 1: lvi, for a more detailed account of Boyle's contacts with theologians and linguists at Oxford and in London.

15. Boyle, "Essay of the Holy Scripture," in Royal Society, *Boyle Papers* 7: fols. 1r–94r. Part of this draft is printed in Boyle, 1: xlviii–l. Michael Hunter and Jan Wojcik plan to produce an edition of this manuscript text. The hermeneutic approach to Scripture was not new to the seventeenth century; it had begun at least as early as the sixteenth. See Robert S. Westman, "The Copernicans and the Churches," in *God and Nature*, ed. David C. Lindberg and Ronald L. Numbers (Berkeley and Los Angeles: University of California Press, 1986), 89, on the Protestant reformers' belief that the "Bible's individual stories needed to be woven together into one cumulative 'narrative web.'" Bacon had written about the interpretation of Scripture in the *Advancement of Learning* (bk. 2, in *Works of Bacon* 6: 400–408).

16. Boyle, *Usefulness of Natural Philosophy*, pt. 1, essay 5, in *Works* 2: 58.

17. Ibid., essay 2, in *Works* 2: 19, 20. See also Royal Society, *Boyle Papers* 7: fol. 9r, on the need to gain knowledge in all of the sciences in order to understand the Bible; and 8: fol. 124v, on how Christ used the characteristics of animals to teach moral truths.

18. Boyle, *Usefulness of Natural Philosophy*, pt. 1, essay 4, in *Works* 2: 41, 48. He "freely admitted" that they were subtle philosophers (41).

19. Ibid., 42–43, 42.

20. Ibid., essay 5, in *Works* 2: 59. He noted that no one had as yet "satisfactorily made out how matter can move itself" (42). Boyle's argument here is not that matter in

the actual world has no motive force but that its motive force must have been given to it in the beginning.

21. Ibid., essay 2, in *Works* 2: 19. Boyle's appeal to such an "accommodation theory" of interpretation and his beliefs about the relation between the two books are very similar to the beliefs that Galileo expressed in his "Letter to the Grand Duchess Christina." Both thinkers made extensive use of Augustine in the justification of their positions.

22. Boyle, *Usefulness of Natural Philosophy*, pt. 1, essay 2, in *Works* 2: 19.

23. Boyle, *Excellency of Theology*, sec. 1, in *Works* 4: 18. Cf. Boyle, *Usefulness of Natural Philosophy*, pt. 1, essay 5, in *Works* 2: 58. In *Christian Virtuoso* he distinguished between theology and "theologers": theology refers to the system of truths communicated by divine revelation and collected in the Bible (*Works* 6: 702).

24. In *Christian Virtuoso* Boyle maintained that the Bible could reveal "truths, that human reason left to itself would never have attained to the knowledge of" (appendix to pt. 1, conference 4, in *Works* 6: 708). See also Boyle, *Reconcileableness of Reason and Religion*, in *Works* 4: 155. Boyle noted that some things are beyond reason, "and in these it is, that I think it as well her [reason's] duty to admit revelation, as her happiness to have it proposed to her."

25. In *Excellency of Theology* Boyle maintained that "the book of Scripture discloses to us much more of the attributes of God, than the book of nature" (sec. 1, in *Works* 4: 7).

26. Ibid., 12. Boyle agreed with Descartes that through reason we could conclude that we have a soul that differs from the body, but he did not agree that we could then conclude that the soul was immortal. This sentiment clearly conflicts with the claim made by J. R. Jacob (*Robert Boyle*, 179) that Boyle's mechanical philosophy was used to confirm the doctrine of the immortality of the soul.

27. Boyle, "Excellency and Grounds of the Mechanical Hypothesis," in *Works* 4: 78, 68.

28. Boyle, *Free Inquiry*, sec. 1, in *Works* 5: 165. Nicholas H. Steneck, in "Greatrakes the Stroker: The Interpretation of Historians," *Isis* 73 (1982): 161–77, is critical of the analyses offered by the Jacobs, but he agrees that somehow Boyle's version of the mechanical philosophy was meant as a way to support the Anglican Church. He maintains, for example, that Boyle's interest in the miracles attributed to Greatrakes reveals his belief that supernaturalism was an important element in natural philosophy. This analysis seems too simplistic. Hutchison, in "Supernaturalism," presents a more accurate picture of the relation between science and religion when he notes that Boyle was opposed to the invocation of God in discussions of natural philosophy.

29. Boyle, *Free Inquiry*, sec. 1, in *Works* 5: 165.

30. Boyle, *Origin of Forms and Qualities*, "Author's Discourse," in *Works* 3: 7. As Boyle said later, it was necessary from a metaphysical standpoint to postulate that God was the original motive force, "but the world being once framed and the course of nature established, the naturalist . . . has recourse to the first cause, but for its general and ordinary support and influence, whereby it preserves matter and motion from annihilation or desition" (*Works* 3: 48).

31. Boyle, *Usefulness of Natural Philosophy*, pt. 1, essay 5, in *Works* 2: 61. Boyle went on

to describe how God rewards us when we use our talents: "The imploying the little knowledge I have in the service of him I owe it to, may invite him to increase that little, and make it less despicable." He remained critical of the alchemists for using arguments from Scripture that he found to be "not very cogent, and somewhat irreverent" (*Works* 2: 101), and he also expressed his opposition to the more mystical of the group for claiming to have received some type of divine revelation that guided them in their work: "I dare not affirm, with some of the Helmontians and Paracelsians, that God discloses to men the great mystery of chymistry by good angels, or by nocturnal visions" (*Works* 2: 61).

32. In *Christian Virtuoso* he noted that libertines were those who thought that "a virtuoso ought not to be a Christian" (pt. 1, preface, in *Works* 5: 508).

33. Boyle, *Usefulness of Natural Philosophy*, pt. 1, essay 3, in *Works* 2: 35.

34. Ibid.

35. Dijksterhuis, *Mechanization of the World Picture*, 443. See More, *Life and Works*, 232; and J. R. Jacob, *Robert Boyle*, on Boyle's belief that all knowledge is interrelated.

36. Boyle, *Excellency of Theology*, sec. 1, in *Works* 4: 18.

37. Boyle, "Account of Philaretus," in *Works* 1: xxiii. Boyle went to live at the home of his tutor, Isaac Marcombes, in Geneva after he had been stranded while on a tour of the Continent because of the depletion of his father's fortune during the Irish conflicts.

38. Ibid., xxii. See also Boyle's letters to Francis Tallents, February 1647, in *Works* 1: xxxv; and to John Mallet, 2 March 1652, in BL, Add. MS 32093, fol. 293r, in which he wrote that he was afraid that the sectarian controversies would lead to no religion at all.

39. Letter from Boyle to John Dury, May 1647, in *Works* 1: xxxix. Boyle's sister Katherine was Dury's niece by marriage. She might also have been the author of a letter in Royal Society, *Boyle Papers* 14: fols. 28r–42r, written in opposition to the cruelty and injustice of a law that placed religious nonconformists in prisons that were infected with the plague.

40. Letter from Boyle to John Dury, May 1647, in *Works* 1: xl. See Hunter, "Science and Heterodoxy" and "Conscience of Robert Boyle"; and Wojcik, "Robert Boyle and the Limits of Reason," for discussions of Boyle's concerns about atheism and for details concerning the figures at whom Boyle's criticisms were directed.

41. Boyle, *Considerations Touching the Style of the Holy Scriptures*, introduction, in *Works* 2: 257. This work is based in large part upon Boyle's unpublished "Essay of the Holy Scripture" and notes in his commonplace book, Royal Society, MS 186, fol. 127v–143r. The *Style* is composed of an epistle dedicatory, a preface to the reader, and a lengthy introduction followed by individual responses to eight numbered objections against the Scripture. The criticism of the Bible hit close to home for Boyle. The *Style* was dedicated to his brother Lord Orrery, and according to Burnet's account of Boyle's life, it was "part of a larger design for raising the scriptures in his brother's esteem" (BL, Add. MS 4229, fol. 60r).

A discussion of the influences on Boyle's hermeneutic views would require an entire work by itself. It can be seen in his writings that he was aware of, and partially approved of, the works of Menasseh ben Israel, Spinoza and his critics, Urban VIII, Augustine, Galileo, Diodati, Calvin, and Luther. In the *Style*, however, he stated that he was discussing an aspect of biblical scholarship that he had not found elsewhere. My purpose here is not to conjecture about the origin of his ideas but to set out clearly what they were in order to show how they influenced his experimental method. Wojcik, in "Theological Context of Boyle's *Things above Reason*," provides a good account of the background surrounding debates over the interpretation of Scripture in Boyle's day.

42. Boyle, *Style of the Holy Scriptures*, 6th objection, in *Works* 2: 282.

43. Ibid., introduction, in *Works* 2: 263.

44. Ibid., 2d objection, in *Works* 2: 270.

45. Ibid., 3d objection, in *Works* 2: 274.

46. Ibid., 2d objection, in *Works* 2: 271.

47. Ibid., 1st objection, in *Works* 2: 265.

48. Royal Society, *Boyle Papers* 7: fol. 10r. Cf. Boyle, *Style of the Holy Scriptures*, 2d objection, in *Works* 2: 271.

49. Boyle, *Style of the Holy Scriptures*, 7th objection, in *Works* 2: 289.

50. Boyle, *Things Said to Transcend Reason*, 2d advice, in *Works* 4: 452.

51. Boyle, *Style of the Holy Scriptures*, 6th objection, in *Works* 2: 281.

52. Royal Society, *Boyle Papers* 3: fol. 92v. Boyle expressed the same attitude in *Excellency of Theology*, in *Works* 4: 26; and "Account of Philaretus," in *Works* 1: xxiii.

53. Boyle, *Excellency of Theology*, sec. 1, in *Works* 4: 16.

54. Boyle, *Style of the Holy Scriptures*, 3d objection, in *Works* 2: 275. See also 266–67 for his opposition to the imposition of "metaphysical subtleties" on the Bible.

55. Ibid., 277, 276. See also Boyle, *Excellency of Theology*, sec. 1, in *Works* 4: 17. He retained this attitude throughout his life, as can be seen from his last will and testament (*Works* 1: clxvii), where he established a lecture series for the proof of the truth of Christianity in general.

56. Royal Society, MS 186, fols. 128v, 127v–128r. In *Christian Virtuoso* he listed four conditions that two propositions must meet in order to produce a real contradiction (appendix to pt. 1, in *Works* 6: 703).

57. Boyle, *Style of the Holy Scriptures*, introduction, in *Works* 2: 259. Boyle was an accomplished linguist. At an early age he was able to converse in English, French, Italian, and Latin. In the 1650s he studied Hebrew and Greek in order to read the Bible in its original languages, and he had begun the study of Arabic when his eyesight failed. See his letter to John Mallet, dated January 1653, in *Works* 1: l–lii, for his discussion and assessment of the usefulness of various books on biblical grammar.

58. Boyle, *Style of the Holy Scriptures*, introduction, in *Works* 2: 259.

59. Boyle, "Essay of the Holy Scripture," in Royal Society, *Boyle Papers* 7: fols. 12r–13r.

60. Boyle, *Style of the Holy Scriptures*, introduction, in *Works* 2: 259.

61. Ibid., 1st objection, in *Works* 2: 265.

62. Ibid.

63. Ibid., 7th objection, in *Works* 2: 283.

64. Boyle, *Christian Virtuoso*, appendix to pt. 1, conference 4, in *Works* 6: 709. Boyle did not write at length about innate ideas as Locke would do later, but in this passage and others, he clearly indicated that the mind has certain "inbred notions" that are known prior to experience. See Boyle, *Reconcileableness of Reason and Religion*, sec. 7, in *Works* 4: 180.

65. Boyle, *Christian Virtuoso*, appendix to pt. 1, conference 4, in *Works* 6: 709.
66. Ibid., conference 3, in *Works* 6: 691.
67. Ibid., conference 4, in *Works* 6: 709.
68. Ibid., 712. In *Style of the Holy Scriptures* (introduction, in *Works* 2: 256–57), he noted that he was concerned not with the objections of atheists but with those of theists who granted that the Bible was true, from which he could conclude that the passages could not be "contradictory to themselves." He did not claim that every passage in the text was literally true, only that the message contained within the text would be coherent and true. See More, *Life and Works*, chap. 9. While his emphasis upon activity and determinism seems to indicate a Calvinist, or at least a Puritan, influence, his rejection of a literal interpretation of the Bible betrays a more Anglican attitude toward theology.
69. Boyle, *Style of the Holy Scriptures*, 3d objection, in *Works* 2: 275. See Patey, *Probability and Literary Form*, 24–25, for a discussion of the importance of consistency as an internal property in literary criticism. Patey finds Locke to be original in his attempt to make probability epistemically continuous with knowledge (27), but it would seem that Locke may have been indebted to Boyle for the idea.
70. See, for example, Boyle, *Christian Virtuoso*, appendix to pt. 1, conference 4, in *Works* 6: 712. The text cannot contradict itself or any absolute truth known by reason. See Boyle's discussion of "local reason" in Royal Society, *Boyle Papers* 1: fol. 7r.
71. Boyle, *Style of the Holy Scriptures*, 3d objection, in *Works* 2: 275, 273. He also noted that the "omission or misplacing of parentheses" or of "points of interrogation" can make the text "appear less discursive." Most punctuation had been added in translation, and "perhaps none of them" had "been so happy, as to leave no room for alterations, that may deserve the title of corrections and amendments" (272).
72. Ibid., 276.
73. Ibid.
74. An example of such a reconciliation can be found in *Christian Virtuoso*, pt. 2, sec. 2, prop. 2, in *Works* 6: 778–81. Boyle discussed a passage in 1 Corinthians that seemed to contradict a truth of experience and showed how a reconciliation could be achieved through a different translation of one preposition. See also his critical reaction to a metaphorical interpretation of Daniel in *Works* 1: cxi.
75. Boyle, *Style of the Holy Scriptures*, 7th objection, in *Works* 2: 293.
76. Ibid., 1st objection, in *Works* 2: 268.
77. Boyle, *Christian Virtuoso*, pt. 2, sec. 2, prop. 5, in *Works* 6: 785–86.
78. Boyle, *Style of the Holy Scriptures*, 1st objection, in *Works* 2: 267. He noted that the "longest and industriousest life will still leave undiscovered mysteries in it [the Bible]" (293). In his *Seraphic Love* (*Works* 1: 289), Boyle maintained that we would only fully understand these mysteries in heaven.
79. Boyle, *Things Said to Transcend Reason*, 5th advice, in *Works* 4: 465.
80. Ibid.
81. Boyle often used the term *explication* rather than *explanation*. He was extremely careful in his use of words, and this choice should be seen as one more indication that his experimental practice was designed as a method of interpretation.
82. Boyle, *Christian Virtuoso*, pt. 2, sec. 2, aph. 21, in *Works* 6: 796.
83. Royal Society, *Boyle Papers* 9: fol. 17r.

84. Ibid., 4: fol. 6r.
85. Boyle, *Christian Virtuoso,* appendix to pt. 1, conference 4, in *Works* 6: 700.
86. Ishiguro, in "Pre-established Harmony," discusses how Leibniz's conception of universal harmony functioned as a system of explanation (71). Boyle's conception differed. Harmony functioned as an ontological assumption that would guide his research, but because of God's unlimited power and wisdom, experiential evidence would have to be collected before one could speculate about the relations that actually held in the universe as a whole.
87. Boyle, *Usefulness of Natural Philosophy,* pt. 1, essay 5, in *Works* 2: 54.
88. Boyle, *Usefulness of Experimental Philosophy,* pt. 2, preamble, in *Works* 3: 401.
89. Bacon made a similar point in his *New Organon,* bk. 1, aph. lxxvii, when he described "true consent" as "that which consists in the coincidence of free judgments, after due examination," and not in a "prejudgment" based upon the authority of a philosophical system (*Works of Bacon* 8: 108). In aph. lxxvi he had also noted that because these systems of the individual disciplines were in conflict with each other, it was a sure "sign" that they were all faulty (107).
90. Royal Society, *Boyle Papers* 4: fol. 6r. See also his essay "Of Unsucceeding Experiments," in *Certain Physiological Essays,* in *Works* 1: 351, where he noted that contradictions in our observations of nature are often "but seeming."
91. In the *Style of the Holy Scriptures* (6th objection, in *Works* 2: 281–82), Boyle noted that prejudice created a problem for the interpretation of all texts, such as those of Aristotle, Homer, and Virgil, because the author's words could often be "tortured into a confession of what was never in his thoughts." He was afraid that his own texts would be misinterpreted, and he admonished his readers to take account of "the intire series of my words," in his "Proëmial Essay" in *Certain Physiological Essays,* in *Works* 1: 314.
92. Boyle, *Christian Virtuoso,* pt. 1, in *Works* 5: 539.
93. This conception of Boyle's epistemology can be found in McGuire, "Boyle's Conception of Nature"; Osler, "Locke and the Changing Ideal"; and Shapin and Schaffer, *Leviathan and the Air-Pump.*
94. Boyle, *Certain Physiological Essays,* "Proëmial Essay," in *Works* 1: 303.
95. Boyle, *Origin of Forms and Qualities,* preface, in *Works* 3: 13.
96. Boyle, *Experiments and Notes, etc., about the Production of Qualities,* advertisements, in *Works* 4: 234.
97. Dijksterhuis (*Mechanization of the World Picture,* 437) makes this point about Boyle's work on the mechanical explication of qualities.
98. Boyle, *Certain Physiological Essays,* "Proëmial Essay," in *Works* 1: 303, 307.
99. Boyle, *Origin of Forms and Qualities,* advertisements to sec. 2, in *Works* 3: 75.
100. Boyle, *Christian Virtuoso,* appendix to pt. 1, conference 3, in *Works* 6: 694. *Imagination* here refers to the supposed actual image in the mind received from a sense perception. This notion will be discussed in more detail in chapter 7.
101. Boyle, *Things Said to Transcend Reason,* 6th advice, in *Works* 4: 469.
102. Boyle, *Christian Virtuoso,* pt. 1, in *Works* 5: 523.
103. Boyle, *Essays of Effluviums,* "Of the Great Efficacy of Effluviums," chap. 1, in *Works* 3: 677.
104. Boyle, *Christian Virtuoso,* pt. 1, in *Works* 5: 523.

105. Boyle, *History of Particular Qualities*, chap. 4, in *Works* 3: 303.
106. Boyle, *Christian Virtuoso*, pt. 1, in *Works* 5: 539.
107. Boyle, *Usefulness of Natural Philosophy*, pt. 1, essay 5, in *Works* 2: 50.
108. Royal Society, *Boyle Papers* 22: 203. This theme is often repeated in Boyle's published works: see "Cosmical Suspicions," in *Works* 3: 306; and *Excellency of Theology*, in *Works* 4: 36.
109. Royal Society, *Boyle Papers* 22: 203. Boyle's belief in the interconnections of things can be traced as far back as an early essay composed at Stalbridge: Royal Society, MS 197, fols. 4r–43r, printed in Boyle, *Early Essays*, 185–202. The title of this essay is illegible. Editor John Harwood has published it under the title "The Doctrine on Thinking," but it was less a doctrine than a set of practical "directions" (fol. 5r) for meditation. See fol. 40r, where Boyle discussed "the secret alliances of things" that are "so strange, and, to us, so extravagant"; and fol. 42r, for his discussion of the "connections of truths that link them betwixt themselves."
110. Boyle, "Excellency and Grounds of the Mechanical Hypothesis," in *Works* 4: 77.
111. Boyle, *Excellency of Theology*, sec. 5, in *Works* 4: 59.
112. Royal Society, *Boyle Papers* 1: fol. 3r.
113. Boyle, *Certain Physiological Essays*, "Proëmial Essay," in *Works* 1: 314. This is similar to his injunction against accepting the literal meaning of passages in the Bible. See Bacon, *New Organon*, bk. 1, aph. lxxxii, in *Works of Bacon* 8: 115, for Bacon's claim that experience must be "duly ordered and digested." Boyle's conception of a "factual proposition" will be discussed in detail in chapter 6.
114. Boyle, *Christian Virtuoso*, pt. 1, in *Works* 5: 539. He also maintained that "philosophy" would often have to be used to correct the information of the senses.
115. Royal Society, *Boyle Papers* 36: fol. 13v.
116. Boyle, *Christian Virtuoso*, pt. 1, in *Works* 5: 539.
117. Boyle, *Certain Physiological Essays*, preface to "Some Specimens," in *Works* 1: 356.
118. Royal Society, *Boyle Papers* 9: fol. 111r. This is similar to what he had to say about the acquisition of knowledge in scriptural study leading to higher levels of understanding; see Boyle, *Style of the Holy Scriptures*, 7th objection, in *Works* 2: 293. In *Of the High Veneration Man's Intellect Owes to God* Boyle spoke of knowledge as a "progressive or discursive thing" (*Works* 5: 150). Boyle was less of a rationalist than Leibniz, but he was not therefore a strict empiricist who would maintain that all we can attain is a probabilistic science of appearances. Boyle's fallibilism, that should not be confused with such a probabilism, will be discussed in more detail in the following chapters.
119. Royal Society, *Boyle Papers* 25: 297.
120. Ibid., 1: fol. 37r. Cf. Boyle, *Certain Physiological Essays*, "Proëmial Essay," in *Works* 1: 311. We will see that he also believed that the details of the method itself would change over time.

Chapter Six

1. Boyle, *Usefulness of Natural Philosophy*, pt. 2, essay 2, in *Works* 2: 76.
2. Boyle, *Certain Physiological Essays*, "Proëmial Essay," in *Works* 1: 302.
3. See Boyle, *Usefulness of Natural Philosophy*, pt. 1, essay 4, in *Works* 2: 36, on the

imperfections of natural histories; and *Excellency of Theology*, sec. 5, in *Works* 4: 55–60, on how the inconstancy of scientific doctrine was to be corrected by observations. See also *Things Said to Transcend Reason*, 4th advice, in *Works* 4: 460. Boyle said that he wished to avoid the practice of those who make hasty extrapolations from "a small number of things," whereby the mind, not being used to its full advantage, merely "gratifies both its vanity, and its laziness."

4. Dear, "Miracles, Experiments, and the Ordinary Course of Nature," 676. See also Daston, "Factual Sensibility," 466, where Daston refers to a fact as "a datum of experience, as distinguished from the conclusions that may be based upon it"; and 467, for her discussion of how "their [Boyle's and his contemporaries'] passionate 'inspection of particulars' and their pontillistic vision of reality created the factual sensibility." Jan V. Golinski, in "A Noble Spectacle: Phosphorus and the Public Culture of Science in the Early Royal Society," *Isis* 80 (1989): 11–39, makes frequent reference to Boyle and "matters of fact" but does not state what he means by this phrase. His numerous references to Shapin, "Pump and Circumstance," would indicate, perhaps, that he follows the formulation of the factual discussed below at note 8.

5. Boyle, *The History of Air*, title 6, in *Works* 5: 617. Another "fact" of this type was given later, when Boyle discussed how "it is by long observation and often repeated experience, found certain" that sleeping on the "shores all night at *Johanna*" causes illness (727). Although different from the positivist conception of a fact, these facts concerning regularities are often found in our ordinary usage today.

6. Royal Society, *Boyle Papers* 35: fol. 45r. The legal connotation is not surprising given the importance of legal adjudication procedures for Boyle's methodology. It indicates that a fact was a statement, the truth of which had been ascertained to a high degree of probability and thus could serve an evidential function. Daston, in "Factual Sensibility," 466–67, noted the legal influence on Bacon's discussion of facts.

7. Boyle, *Experimental History of Colours*, pt. 1, chap. 1, in *Works* 1: 668. Boyle also used the term *phaenomena* to refer to natural effects; see, for example, the quotation in chapter 2 above at note 96. In chapter 1 above (at note 97) we saw Bacon's discussion of how he wished to make discoveries about "the world, such as it is in fact." In Boyle's paraphrase of Bacon's discussion (chapter 1, at note 98), he said that he desired to discover "how things . . . are really produced." See Bacon, *Sylva sylvarum*, in *Works of Bacon* 4: 151–483, and 5: 7–169, for Bacon's numerous discussions of facts that follow this usage. He wrote that "as they are not to mistake the causes of these operations; so much less are they to mistake the fact, or effect"; and "those effects which are wrought by the percussion of the sense, and by things in fact, are produced likewise in some degree by the imagination" (5: 57). See Pérez-Ramos, *Bacon's Idea of Science*, for the most extended discussion of Bacon's conception of a fact. Pérez-Ramos notes, for example, that "the *factum*" is "the actualization or instantiation of Nature's effects as a result of manipulatory skills" (161).

8. Shapin and Schaffer, *Leviathan and the Air-Pump*, 139. On 23–24 they note the epistemic aspect of the factual category. Their identification of the factual as a linguistic category of descriptive statements can be found in a number of places. See

49, 67, and 71. They maintain that for Boyle "causal inquiry was to be tactically segregated from the main tasks of the natural philosopher; hypotheses about causes were conjectural and should be regarded as distal to fact production" (147). They are right that, performatively speaking, facts and hypotheses are to be kept separate, but it is certainly not the case, as they conclude, that "Boyle said that he wished to ban the search for causal notions from mechanical and experimental philosophy" (203).

9. Because they focus their analysis almost exclusively upon Boyle's *New Experiments*, Shapin and Schaffer (*Leviathan and the Air-Pump*, 50) maintain that the "epistemological status" of the spring of the air remained ambiguous, since sometimes Boyle referred to this concept as a causal hypothesis, while at other times he "operationally treated the spring of the air as a matter of fact." Clearly the ambiguity arises from their attempt to make the factual coextensive with descriptive statements. Some of the other confusions identified by them are uninteresting. They found it troubling, for example, that sometimes Boyle referred to the spring as a notion, while other times he called it a doctrine or a hypothesis. But these terms are epistemically synonymous—all refer to speculative claims that lack factual status. Shapin and Schaffer are correct that Boyle treated the spring and the spring's cause as "fundamentally different explanatory items: the former was 'evinced' by the experiments; the latter was not" (51). As we will see, though, Boyle did not leave the distinction between the two notions "unexplained," nor did he offer "no criteria for identifying in what way they were entitled to such radically different treatments" (51).

10. Boyle, *New Experiments*, "To the Reader," in *Works* 1: 2.

11. Boyle, *Certain Physiological Essays*, "Proëmial Essay," in *Works* 1: 308.

12. Ibid., 309.

13. Boyle, *New Experiments*, experiment 1, in *Works* 1: 12. Throughout his works he retained the belief that the discovery of subordinate causes should come first and insisted that explanations constructed out of such causes represented an important stage in our attempt to understand nature, even though he clearly recognized that other philosophers, such as Hobbes, rejected any causal claims that did not make reference to general laws governing the motion of the least parts of matter.

14. Boyle, *New Experiments*, preface to "Defence of the Doctrine," in *Works* 1: 123.

15. Ibid., experiment 2, in *Works* 1: 20. The first experiment (10 ff.), for example, gave an account of the "phaenomena exhibited to us by the engine" that "doth constantly and regularly offer itself to our observation" such as depends "upon the fabric of the engine itself, and not upon the nature of this or that particular experiment." In a similar manner, experiments 4, 5, and 6 concerned a series of tests on bladders enclosed within the receiver of the engine.

16. Ibid., conclusion, in *Works* 1: 117. See also Boyle's prefatory letter, "To the Lord of Dungarvan," in *Works* 1: 5–10.

17. Royal Society, *Boyle Papers* 5: fol. 18v. This passage is repeated with slight changes at fol. 19v.

18. Boyle, *Things Said to Transcend Reason*, 6th advice, in *Works* 4: 468–69.

19. Ibid., 3d advice, in *Works* 4: 455.

20. Boyle, *New Experiments*, experiment 1, in *Works* 1: 11. From this passage one can

see that Boyle's vagueness about the term *pressure* is a result of the fact that in his account both the weight of the ambient air and the spring itself could exert pressure. As he said in his first *Continuation of New Experiments*, *the spring* was a term used by him to illustrate that the air has an "active and a resistant force or strength" that allows it to "produce certain effects" (*Works* 3: 183). See A. R. Hall, *Henry More*, 182–85, for a detailed discussion of Boyle's concept of the spring of the air.

21. Boyle, *History of Air*, title 2, in *Works* 5: 614–15. While this history was published after his death, the quotation in the text is taken from the second title, "A Short Answer to a Question about the Nature of the Air Given by Mr. Boyle to Mr. H. Oldenburg," written many years earlier.

22. Royal Society, MS 191, fol. 26v. Boyle wrote this note in a more discursive fashion for *Christian Virtuoso* (pt. 1, in *Works* 5: 511) and added a discussion there of Bacon's use of analogies. In his *Free Inquiry* (sec. 5, in *Works* 5: 194–95), Boyle noted that we can learn through analogy because natural processes are interrelated. The use of analogical reasoning in the works of Newton and Locke has been widely discussed, but this practice was quite common. Boyle also noted Descartes's practice of analogical reasoning when he wrote that "Monsr Des Cartes somewhere says that he scarce thought that he understood anything in Physics, but what he could declare by some apt similitude, of which indeed he has many in his writings, as where he compares the particles of fresh water to little eels and the corpuscles of salt in the seawater to little rigid strands" (Royal Society, MS 191, fol. 26v). Boyle added that a "well applied comparison much helps the imagination by illustrating things scarce discernable, so as to represent it by things much more familiar and easy to be apprehended" (fol. 27r).

23. Letter from Boyle to Hartlib, published as title 13 in *History of Air*, in *Works* 5: 642.

24. Boyle, *Certain Physiological Essays*, "Proëmial Essay," in *Works* 1: 306. He ended by noting that the "industry" of "new experiments" would "enrich us with the other" half. See also his *History of Blood*, preface, in *Works* 4: 595, where he noted that current knowledge about the blood consisted "more of observations than experiments."

25. Boyle, *Experimenta et observationes physicae*, advertisements, in *Works* 5: 568–69. In chap. 5, scholium to experiment 3, in *Works* 5: 594, he noted that while things were there delivered as matters of fact, "it should not be thence inferred" that they "were lighted on by chance." See also *Sceptical Chymist*, appendix, in *Works* 1: 589; *Experimental History of Colours*, preface, in *Works* 1: 664; Royal Society, MS 187, fol. 152v; and Royal Society, MS 189, fol. 43r.

26. We will see below that Boyle actively sought observations about strange or odd occurrences as tests of generally accepted theories. In the next chapter I will discuss Boyle's distinction between probatory and exploratory experiments in more detail.

27. Boyle, *Origin of Forms and Qualities*, sec. 1 of "The Historical Part," in *Works* 3: 66. He noted that observations are suitable for inclusion because they are "common and familiar" (67).

28. Boyle, *Certain Physiological Essays*, "Proëmial Essay," in *Works* 1: 314.

29. Boyle, *Usefulness of Experimental Philosophy*, "Of Men's Great Ignorance," in *Works* 3: 472–73. See Royal Society, *Boyle Papers* 4: fol. 83r, for his discussion of Archimedes

and Galileo. In *Christian Virtuoso* he noted that men had been "negligent" in "observing things" and therefore there was much still to be discovered and "also usefully employed in the works of nature" (*Works* 6: 796).

30. Boyle, *History of Blood*, pt. 1, in *Works* 4: 597. See Royal Society, *Boyle Papers* 9: fol. 71r, and 18: fol. 42r, for his "Titles of the first order for the Natural History of Humane blood of Healthy Men." He discussed how one should first observe the color, taste, and odor of particular samples of blood before performing more exact experimental investigations.

31. Royal Society, *Boyle Papers* 44: fol. 29v. Laudan ("Clock Metaphor and Hypotheses," 50 n. 19) attributes the authorship of this passage to Boyle, but it is unlikely that Boyle wrote it, given the flattering way that Boyle is described here. It also has the appearance of having been copied from a longer work.

32. Observations on these subjects are recorded in a number of places in his unpublished papers. See Royal Society, *Boyle Papers* 4: fol. 45r, for questions about glaciation and echoes; 22: fols. 197r–200r, for electrical bodies; 26: fols. 75r–76v, for lime; 25: 265r, for water; 26: fol. 29r, for air; 26: fol. 46r, for blood; 26: fol. 62r, for light; 28: 1–85, for tin.

33. Ibid., 14: fols. 16v–17r.

34. Boyle, *Usefulness of Experimental Philosophy*, preamble, in *Works* 3: 400–401. See also essay 1, in *Works* 3: 404; and *Certain Physiological Essays*, "Proëmial Essay," in *Works* 1: 315. He noted on more than one occasion that he was thankful that his social station was not so high as to preclude his free association with tradesmen. He kept the addresses of a number of tradesmen and apothecaries in his notebooks; see Royal Society, MSS 187, 189, 190.

35. Boyle to Marcombes, in *Works* 1: xxx. See chapter 3 above.

36. Boyle, "The Mechanical Origin of Electricity," experiment 6, in *Works* 4: 351–52.

37. At the same time he was also busy recording observations and reflections on ethical and theological topics. See Royal Society, *Boyle Papers* 3: fol. 146r, for a collection of reflections that he began on his twenty-second birthday about things like "the loss of innocence"; and fols. 21r–42r, for a collection of extracts from theological works that he had read. He composed two "decades" of "Scripture reflections" (7: fols. 128r–33r), and he had a similar collection of "Scripture observations" from 1647 (14: fols. 1ff.). "Diurnal observations, thoughts, and collections begun at Stalbridge April 25th 1647" contains thoughts on such topics as pride, the good that can be done with riches, and the need for eloquence in moral arguments (44: fols. 94–116). He had noted that one should make the most of spare time by meditating in the "1000 little intervals" of leisure that occur throughout the day (7: fol. 291r). His *Occasional Reflections* (in *Works* 2: 323–460) was based largely upon this manuscript material. See his "Discourse Touching Occasional Meditations," in *Works* 2: 335–59, for why he meditated and his thoughts on how meditation could best be done.

38. Royal Society, *Boyle Papers* 28: 309–12. The title of this diary ended with the lines "And so, by God's permission, to be annually continued during my life." Most of the title pages to his other diaries ended with the same type of inscription. *Works* 1: xxx ff., contains a number of letters printed from earlier years with reported observations. The earliest diaries are composed almost entirely of medicinal reme-

dies, and Boyle had a two-volume medical reference work compiled (preserved in Royal Society, MSS 180 and 181) that contains an alphabetical listing of diseases neatly written in Latin. Volume 1 begins with an account of "Abdominis musculorum tumor" (MS 180, p. 1), and volume 2 ends with "Vulnera varia" (MS 181, p. 169).

39. Royal Society, *Boyle Papers* 36: fol. 86r. Moral and theological treatises are also included on this list: e.g., "Daily Reflection," "Public Spiritedness," "Theodora," "The Swearer Silenced," "Seraphic Love," and "Occasional Reflections." The last four can be found in Boyle's published works.

40. Ibid., 25: 343–62.

41. Ibid., 8: fols. 140r–148v. One "century" is recorded here. On the last page Boyle began a new heading—"Centuria II, Sept. ye 20th 1655"—but no entries are recorded. There is a numbered collection of observations entitled "Promiscuous Observations Begun the 24th of 7ber [September] 1655" in Royal Society, *Boyle Papers* 25: 153–56, that may be the missing second century. The other diary follows this collection in Boyle's papers (25: 157–60). In September 1656 he began another set of "Philosophical Collections" (25: 173–84). He returned to England during the summer of 1653, and he moved to Oxford during the fall of 1654.

42. Royal Society, *Boyle Papers* 25: 161–72. In volume 26 (fols. 96r–98r) there are loose sheets with experiments on silver that date from May and June 1657. It was about this time that Boyle began to work with Hooke and Goddard on the design of his air-pump.

43. Boyle to Oldenburg, in *Works* 6: 37. A copy of this letter can be found in Royal Society, *Boyle Papers* 35: 195. He went on to say that he had taken the advice of some friends to set down his observations in one book and give them to someone trustworthy for safekeeping. Apparently this tactic was not completely successful. In the publisher's advertisement to Boyle's *Certain Physiological Essays* (1661), it is noted that after having thus "disposed of his papers to secure them," Boyle found "that he could not himself seasonably recover them" (*Works* 1: 298).

44. Royal Society, *Boyle Papers* 22: 1. This collection was apparently put together sometime after 1664, because it includes transcribed entries from June through September 1664 concerning hydrostatic investigations. Some of the loose sheets following this collection have titles that may indicate they were part of the Memorials. "A Continuation of Philosophical Entries etc. from the XXVth of July" begins on 45. "A Physiological Notebook or Pandicta Physica Begun the 1 of January 166–" (year left blank) begins on 57. Most of these entries concern chemical and hydrostatic experiments.

45. Most of the observations also have the notation "tbd" in the margin. According to Boyle's notes on his system (Royal Society, *Boyle Papers* 25: 223), this meant that the observation had been transcribed to another place in the papers. Indeed, there are numerous places in his papers where the same observations are repeated.

46. Royal Society, *Boyle Papers* 8: fol. 64v. The way the items are listed here seems to indicate that the index was produced after *History of Cold* was published in 1665 but before the second volume of *Usefulness of Experimental Philosophy* was published in 1670. Number 15, the essay "Improbable Truths," was never published, but Boyle mentioned it as having circulated in manuscript in *Usefulness of Natural Philos-*

ophy (pt. 2, essay 5, chap. 18, in *Works* 2: 188). In *Boyle Papers* 22: 60, the index is repeated, and number 23, which is not legible in the first list, appears as "Varia Lectiones Physicae." Number 24 is left blank in the second index.

47. Royal Society, *Boyle Papers* 25: 217–443. The intended size of the collection is discussed on 224. He said that the collection comprised two tomes because he was writing it at two different places. Dating of this collection is difficult. He may have been working on it in the late 1670s, as the second entry of the second tome (225) is dated 6 September 1678. This could be either the date that the entry was made or the date that the original observation was made, since the work was composed of things "partly transcribed" from his old notes (219). In the margin of the first page he wrote, "This Preface and all the Chapters of the Paralipomena that it belongs to, are to be added, as a kind of *Appendix* to the collection entitled *Experimenta et Observationes Physicae*" (217). This work was not published until 1691, but as can be seen from the preamble and advertisements, in *Works* 5: 564–69, Boyle had sent a preliminary version of the work to Oldenburg prior to the latter's death in 1677.

48. Royal Society, *Boyle Papers* 25: 217.

49. Ibid., 222.

50. Ibid., 223.

51. Ibid., 35: fol. 187r. In addition to these collections, there are a series of observations and experiments collected for a planned second edition of *The Natural History of Human Blood* (c. 1685) in 18: fols. 11r–59r; and three centuries transcribed from observations and experiments recorded on loose sheets. "The XVI Century" is dated 19 December 1689 and contains entries on distillation and analysis of various substances, experiments on cold, "an uncommon experiment about respiration," and an elaborate series of hydrostatic experiments (21: 191–218). "The XVII Century" is not dated but contains complex chemical notes that are dated from 30 June 1690 and 21 September 1691 (21: 219–24). "The XVIII Century" is dated 16 August 1690 and includes a number of reports from foreign countries, including one dated 25 March 1691 (21: 245–300).

52. Boyle frequently cited Bacon's advice. See *Certain Physiological Essays*, "Proëmial Essay," in *Works* 1: 313; *History of Cold*, preface, in *Works* 2: 472; and *Experimenta et observationes physicae*, advertisements, in *Works* 5: 568.

53. See Royal Society, *Boyle Papers* 28: 309–12; and 25: 343–62.

54. See Oldenburg's letters to Boyle, in Boyle, *Works* 6: 156–57 and throughout, for his attempts to locate books for Boyle.

55. References to these and many other works are scattered throughout Boyle's papers and his published works. See Royal Society, *Boyle Papers* 39: fols. 1r–74r; *Usefulness of Natural Philosophy*, pt. 2, essay 4, in *Works* 2: 104–5; and *History of Air* (paper, originally printed in *Philosophical Transactions* [Jan. 1692–93]: 583), in *Works* 5: 750. See Mary B. Campbell, *The Witness and the Other World, 400–1600* (Ithaca, N.Y.: Cornell University Press, 1989), for a detailed account of travel literature.

56. The charter for the New England Corporation is printed in Boyle, *Works* 1: cli–clviii. Letters from John Eliot and others concerning the progress of the corporation are at ccv–ccxviii, and a letter from John Winthrop is at lxxi. See *Works* 6: 184–85, for a letter from Henry Oldenburg to Boyle, dated 10 December 1664,

written at the request of Lord Brouncker "to acquaint" Boyle "with your being elected into the company for the royal mines."

57. Boyle, *Experimenta et observationes physicae,* chap. 2, in *Works* 5: 575. He mentions that he was present when the ships returned from India with their delivery of diamonds (577). See Michael Hunter, "The Cabinet Institutionalized: The Royal Society's 'Repository' and Its Background," in *The Origins of Museums,* ed. Oliver Impey and Arthur MacGregor (Oxford: Clarendon Press, 1985), 161, on the collection of rarities housed at the East India House. "The Publisher to the Reader," in *History of Cold,* in *Works* 2: 463, includes a discussion about how Boyle became an "adventurer" in the Hudson's Bay Company "for the better gaining of such information" about cold.

58. Royal Society, *Boyle Papers* 27: 1–274. The numbering of the entries becomes erratic after page 159 (entry number 668), but the entries appear to be part of the same collection. Most of the entries are marked as "odd" in the margin. The collection was apparently compiled sometime in the 1660s. The first entry by date is number 239, from June 1659, but this was apparently transcribed from elsewhere, as entry number 242 is dated 27 May 1659. The latest dated entry, number 402, concerns a report about the odd death of all of the fish in a pond in Watertown, Massachusetts, received from John Eliot in 1670. This report is in the letter from Eliot printed in Boyle, *Works* 1: ccv–ccvi. See also *Boyle Papers* 26: fols. 78r–84v, for entries "transcribed out of my old Outlandish Notes," mostly concerning tin. See Royal Society, *Boyle Letters,* vol. 7, for a number of letters written to Boyle and others reporting curiosities and oddities of all kinds.

59. Royal Society, *Boyle Papers* 27: 1–3. Some of the other titles look like variations of the ones seen above, but the numbers do not match those given in the other lists. Number 6, for example, is recorded here as "Paradoxes about Flames"; number 9 is "Good Hypotheses"; and number 17 is "Cosmical Suspicions." Some of the other titles are quite curious: number 21, "Experience keeps us close"; number 22, "Learned Men"; and number 23, "Lame Method do's correct." The thirty-seventh title, "Doubts touching cold," could perhaps refer to Boyle's "Sceptical Dialogue about Cold," published in *Works* 3: 733–53; the last entry, entitled "The Plenist Hobbes the Thirty Eighth disprove," could refer to either his "Defense of the Doctrine of the Spring of the Air" or "Examen of Hobbes."

60. Royal Society, *Boyle Papers* 27: 33, 74, 86, 104, 129.

61. Ibid., 51–55, 77, 87, 90, 92, 151–53.

62. Boyle, "General Heads," printed as an appendix to *History of Air,* in *Works* 5: 733–43, originally printed in the *Philosophical Transactions,* no. 11. A draft of this work, entitled "General Heads for a natural history of a country, Great or Small, imparted by Mr. Boyle," can be found in Royal Society, *Boyle Papers* 26: fols. 47r–48v.

63. Boyle, *History of Air,* appendix, "General Heads," in Works 5: 733–34.

64. Ibid. Among the "subterraneous" topics listed by Boyle were those concerning stones, earths, and clays that could be of use to potters and those concerning minerals, particularly the ones that might have medicinal virtues. Elias Ashmole expressed similar concerns when he called for the "inspection of Particulars, especially those as are extraordinary in their Fabrick, or useful in Medicine, or applyed to Manufacture or Trade." See Arthur MacGregor, "The Cabinet of Curiosities

in Seventeenth-Century Britain," in *The Origins of Museums*, ed. Oliver Impey and Arthur MacGregor (Oxford: Clarendon Press, 1985), 152. Boyle gave an example of how questions were to be proposed in his "Articles of Enquiries Touching Mines" (Works 5: 735–43). He listed six general articles concerning the types of questions that could be asked about mines, under which he listed eighty-eight specific queries concerning these articles and eleven "Promiscuous Enquiries."

65. Boyle, *History of Air*, appendix, "General Heads," in *Works* 5: 734–35.

66. As we saw in chapter 1, the observations that Boyle made while on his tour of the Continent in part contributed to his rejection of Aristotelian philosophy. Somewhat paradoxically, however, in his *Experimenta et observationes physicae* ("Strange Reports," advertisement, in *Works* 5: 604), he said that he had been encouraged to keep records of strange reports "by the example of *Aristotle*." Bacon, in *Advancement of Learning* (bk. 2, in *Works of Bacon* 6: 184–85), also discussed the important role of oddities, or "Marvels." They could, he wrote, lead "one to correct the partiality of axioms and opinions, which are commonly framed only upon common and familiar examples."

67. See Katherine Park and Lorraine J. Daston, "Unnatural Conceptions: The Study of Monsters in Sixteenth- and Seventeenth-Century France and England," *Past and Present* 92 (1981): 20–54, on the great amount of activity such observations generated. Daston, in "Factual Sensibility," points out that by Boyle's time oddities were no longer seen as "sports of nature" because nature was no longer viewed as an entity able to "make sport" (464). She discusses how "Bacon was confident that these so-called miracles of nature would eventually 'be reduced and comprehended under some Form or fixed Law'" (465). We will see that Boyle was once again closer to the views of his famous predecessor than to those of his contemporaries described by Daston who no longer believed that oddities were "the stuff of which natural philosophy was made." See his essay "A Paradox of the Natural and Preternatural State of Bodies," in *Works* 3: 782–86, where he once again expressed his disagreement with those who believed in a distinction between "natural" and "violent" states of bodies.

68. Boyle, *Free Inquiry*, sec. 8, in *Works* 5: 252.

69. Ibid.

70. Daston ("Factual Sensibility," 465–67) discusses how the factual category was transformed from a catalog of items in support of theoretical points in the work of Galileo into "nuggets of experience detached from theory" in the work of Bacon, Boyle, and Locke. Boyle did transform the role of the factual, by giving it more than a simple confirmatory function, but as we saw above, he did not reduce the items in it to statements concerning singular occurrences, and he never claimed that such facts were theory-neutral. Facts should not be collected solely for the purpose of confirming a theory, but facts, particularly about oddities, could not be recognized as such without some theoretical knowledge. In the preface to *Experimenta et observationes physicae* (*Works* 5: 564–69), Boyle wrote at length about the necessity of cataloging *and* categorizing the information in the factual basis. This topic will be discussed in more detail in chapters 7 and 8.

71. Boyle, *Usefulness of Natural Philosophy*, pt. 1, essay 1, in *Works* 2: 9. Boyle described how "local reason" led people born in a particular place to "espouse the opinions,

true or false, that obtain there," in Royal Society, *Boyle Papers* 1: fol. 7r. Hunter ("Cabinet Institutionalized," 163–64) discusses how the Royal Society wanted its repository to be a systematic collection that could be used for philosophical purposes, but the "enterprise fell short of the grandiose ambitions" and became more like the curiosity cabinets of private individuals that were valued for their rarity alone. On the economic and social value attached to curiosity cabinets, see Daston, "Factual Sensibility," 456–59. See Robert Hooke's letters to Boyle, dated 3 February and 21 March 1666, concerning the beginning of the Royal Society collection in Boyle, *Works* 6: 504–6. Even if the collections of oddities failed to serve philosophical purposes directly, Boyle would still have found an intellectual value in them because they could "excite men's curiosity" and thus increase the number of observers contributing to the factual foundation. See Boyle, *Usefulness of Experimental Philosophy*, preamble, in *Works* 3: 395.

72. Boyle, *New Experiments*, "Examen of Hobbes," chap. 2, in *Works* 1: 193. See also his discussion in "Observations on a Diamond That Shines in the Dark," appended to *Experimental History of Colours*, in *Works* 1: 789–95, about how negative instances can overthrow generally accepted rules of nature.

73. Boyle, *Usefulness of Experimental Philosophy*, "Of Men's Great Ignorance," in *Works* 3: 473. See also his discussion in "A Paradox of the Natural and Preternatural State of Bodies," in *Works* 3: 783–84.

74. Boyle, in *Experimenta et observationes physicae* (advertisements, in *Works* 5: 569), discussed how he left room for others to make the additions or alterations as they saw fit in light of new information.

75. Boyle, *Usefulness of Natural Philosophy*, pt. 2, essay 5, chap. 10, in *Works* 2: 161–62. He had learned about these practices from Alvaro Semmedo's *History of China* (1642).

76. Royal Society, *Boyle Papers* 2: fol. 58r. He used this point as part of a line of reasoning intended to show that Descartes's a priori proof for the existence of God was actually a posteriori. According to Boyle, "Whereas indeed the rational soul with her faculties, being admitted by the Cartesians to be a creature of God, she must by them be confessed to belong to the list of those effects, whereof he is the Primary and Universal cause." Compare this passage with his *Excellency of Theology*, sec. 1, in *Works* 4: 12–15.

77. Royal Society, *Boyle Papers* 38: fol. 154r. He also noted, in his *Usefulness of Experimental Philosophy* (pt. 2, sec. 7, in *Works* 3: 422), that the lack of progress in philosophy was due to the "prejudices by which men have been hitherto imposed on about substantial forms, the unpassable bounds of nature, the essential difference betwixt natural and artificial things, etc."

78. Royal Society, *Boyle Papers* 9: fol. 13r. The work was listed as number 1.A. in his "Chapters of the Paralipomena" and appeared third under his "Topica Particularia" in the Outlandish Book.

79. Ibid., fol. 14r. A number of these points were also made in his essay "On Unsucceeding Experiments," in *Certain Physiological Essays*, in *Works* 1: esp. 243–44.

80. Royal Society, *Boyle Papers* 9: fol. 15r. The rest of the suppositions listed here concerned experiments and will be discussed in the next chapter.

81. Ibid., 18: fols. 2r–2v. "Concerning Sensation in General" appears as number 17 in

his first list and as *R* in his Outlandish Book. Boyle did not trust his own or others' senses. He spoke of the "imbecility of the visive faculty" in *Things Said to Transcend Reason*, 5th advice, in *Works* 4: 463. Notes on sensation appear in many places throughout his unpublished papers. See, for example, the Outlandish Book (Royal Society, *Boyle Papers*, vol. 27), about the phenomenon of phantom pain experienced by those who had lost a limb. At the end of volume 27, there are a number of loose sheets containing optical diagrams: see pp. 387–88, for various diagrams illustrating the way light is reflected and refracted off of flat, concave, and convex surfaces; 389, for diagrams on perspective and the reversed image produced on the retina; and 399–401, for diagrams that deal with the path of light produced in telescopes. In *Boyle Papers* 18: fols. 129r–129v, a number of experiments are listed that were performed on the nerves of animals in order to see what happens physically in sensation. See also Hooke's letter of 24 November 1664, where Hooke told Boyle that he was sending him "an instrument for refraction, whereby . . . you designed to try the refraction of the humours of the eye" (Boyle, *Works* 6: 49). In Boyle's *Excellency of Theology* (sec. 3, in *Works* 4: 42–46), he discussed Descartes's theory of sensation. While he noted that "*Des Cartes* and his followers have given the fairest account of sensation," he went on to criticize specific elements in the theory.

82. Boyle, *Things Said to Transcend Reason*, 4th advice, in *Works* 4: 460. He discussed how the systems of philosophers such as Gassendi and Aristotle could not explain all that they claimed for them. He noted that the mind "is very prone to think, that any small number of things, that it has not distinctly considered, must be of the same nature and condition with the rest, that he judged to be of the same kind."

83. Royal Society, *Boyle Papers* 1: fol. 70r.

84. Ibid., 38: fol. 154r. Here he repeated the same sentiment, writing that "'tis plain that some sensible observations require skill, to discern whether all the due conditions of sensation, relating to the organ, the object and the *medium*, concurred." These drafts are similar to his discussion in "Of Men's Great Ignorance," in *Works* 3: 472–73.

85. He discussed the various ways in which one could check for the presence of optical effects in *Certain Physiological Essays*, "Of Unsucceeding Experiments," in *Works* 1: 343, and *History of Air*, title 20, in *Works* 5: 708. He discussed the need for "long observations and repeated experiments" in many places: see *History of Air*, title 40, in *Works* 5: 727, and *Experimenta et observationes physicae*, "Strange Reports," relation 10, in *Works* 5: 609.

86. As quoted by Thomas Birch in Boyle, *Works* 1: cxlvii. Shaw also noted that Boyle showed "sagacity and industry in discovering their weak sides; always frankly proposing them, as he received them, to be confirmed or overthrown by further search and future diligence."

87. Boyle, *Certain Physiological Essays*, "Proëmial Essay," in *Works* 1: 315. Steven Shapin, in "The House of Experiment in Seventeenth-Century England," *Isis* 79 (1988): 376, refers in a footnote to the fact that Locke "advised practitioners to factor in the credit worthiness of the source by the credibility of the matter claimed by that source." He makes no reference to Boyle in this regard, and his subsequent discussion differs from mine in that his is designed to show how the "transforma-

tion of mere belief into proper knowledge was considered to consist of the transit from the perceptions and cognitions of the individual to the culture of the collective" (375). Also here, and in Shapin and Schaffer, *Leviathan and the Air-Pump* (e.g., 58, 327), discussions about credibility are confined mostly to the social status of witnesses.

88. Boyle, "Of the Temperature of the Subterraneal Regions," advertisement, in *Works* 3: 326.

89. See Boyle, *History of Air*, title 19, in *Works* 5: 698–99, for his discussion of a number of competent judges.

90. Boyle, *Experimenta et observations physicae*, "Strange Reports," relation 2, in *Works* 5: 605. Bacon (*Parasceve*, aph. 8, in *Works of Bacon* 8: 367) also wrote about the importance of giving the credentials of witnesses used in histories.

91. Boyle, "Observations about Vitiated Sight," observation 12, in *Works* 5: 451.

92. Boyle, *History of Air*, title 12, in *Works* 5: 635. Often when Boyle used the term *curious* to describe an observer, it carried the additional, earlier connotation of "careful."

93. Boyle, *Experimenta et observationes physicae*, "Strange Reports," relation 6, in *Works* 5: 607. He often included testimony from medical men. See, relation 10, in *Works* 5: 609, for the report of "an ingenious practitioner of physic"; and "Cosmical Suspicions," in *Works* 3: 317, for the report of "a very learned and experienced physician" about the plague.

94. Boyle, *Experimenta et observationes physicae*, "Strange Reports," relation 8, in *Works* 5: 608.

95. Boyle, "Cosmical Suspicions," in *Works* 3: 320. We will see in more detail in chapter 8 how Boyle used the observations made by travelers and navigators in his *History of Cold*. Foreign visitors to England also made excellent witnesses. See, for example, Boyle's account of the reports from a native of Sweden, in *History of Air*, title 11, in *Works* 5: 635.

96. Boyle, *Experimenta et observationes physicae*, "Strange Reports," relation 8, in *Works* 5: 608.

97. Boyle, *Certain Physiological Essays*, "Proëmial Essay," in *Works* 1: 313–14.

98. Boyle, *Usefulness of Experimental Philosophy*, pt. 2, sec. 2, preamble, in *Works* 3: 395. In Boyle's usage, *learned* and *illiterate* are clearly descriptive, not evaluative, terms.

99. Ibid., "The Goods of Mankind," sec. 1, in *Works* 3: 442.

100. Ibid., 442, 443.

101. Ibid., 443, 444. Boyle did note that the "learned" must observe and register these things, because the tradesmen are not concerned with the improvement of philosophy (443). In his preamble he also noted that "few tradesmen will, and can give a man a clear and full account of their own practices . . . chiefly because they omit generally to express . . . some important circumstances, which because long use hath made very familiar to them, they presume also to be known to others." To compensate, Boyle either "tried them [the practices] at home, or caused the artificers to make them in my presence" (396).

102. Ibid., 444, 445.

103. Boyle, *Certain Physiological Essays*, "Essay of the Intestine Motions," secs. 13–16, in *Works* 1: 454–55.

104. Boyle, *Usefulness of Natural Philosophy*, pt. 2, essay 5, chap. 10, in *Works* 2: 162. Boyle noted that *barbarous* was a relative term at best. See his *Occasional Reflections*, sec. 6, in *Works* 2: 450–52. In reflection 3, "Upon the Eating of Oysters," he explained that it was the "strange power of education and custom" that led the English to describe the "many nations of the Indians" as engaging in a "barbarous custom" because they "eat raw flesh," and yet "we scruple not to devour oysters alive."

105. Boyle, *Usefulness of Natural Philosophy*, pt. 2, essay 4, in *Works* 2: 106. See 103–13, for the numerous references to the unusual foods and drinks consumed in foreign lands.

106. Ibid., essay 5, chap. 10, in *Works* 2: 162.

107. Ibid., 163, 164, 162.

108. Ibid., chap. 1, in *Works* 2: 117. Boyle took this account from Gulielmus Piso, *De medicina brasiliensis libri 4* (1648).

109. Boyle, *Usefulness of Natural Philosophy*, pt. 2, essay 5, chap. 10, in *Works* 2: 163.

110. Given what we have seen on Boyle's concerns for widening the witness basis of the new philosophy, this should not be surprising, but it may be so given the extensive literature in feminist criticism that finds the roots of gender bias in seventeenth-century science. See, for example, Sandra Harding, *The Science Question in Feminism* (Ithaca, N.Y.: Cornell University Press, 1986); Evelyn Fox Keller, *Reflections on Gender and Science* (New Haven, Conn.: Yale University Press, 1985); Carolyn Merchant, *The Death of Nature* (New York: Harper and Row, 1980); and Potter, "Modeling the Gender Politics in Science." Boyle's closest intellectual confidante was his sister Katherine. If she had outlived him, she would have received Boyle's "final diary and my manuscripts" (Royal Society, MS 194, fol. 8r). See Boyle's statements concerning the intellectual equality of the genders in his preface to *The Martyrdom of Theodora*, in *Works* 5: 258–59.

111. See, for example, Boyle, *Usefulness of Natural Philosophy*, pt. 2, essay 5, chap. 20, in *Works* 2: 197, for "Lady Kent's powder"; and the appendix to sec. 1, in *Works* 2: 245, for an account of an "excellent" lady who had a cure for rickets. See also *Experimenta et observationes physicae*, chap. 4, first pentade, experiment 1, in *Works* 5: 585, for a woman's account of an attack of palsy after a fright. Elizabeth Grey, countess of Kent, was the author of a popular collection of medicinal remedies entitled *A Choice Manual of Rare and Select Secrets*. See Harwood's introduction to Boyle, *Early Essays*, lxii, for details of this work and Lady Ranelagh's private collection of medical recipes. Numerous references to the observations of women are scattered throughout Boyle's unpublished notes. See Royal Society, *Boyle Papers* 22: 62 and 27: 45; and Royal Society, MS 41 (identified as papers belonging to Boyle by Keith Moore, Royal Society archivist), fols. 17r–18r.

112. Boyle, *Experimental History of Colours*, pt. 1, chap. 2, in *Works* 1: 673.

113. Boyle, *Certain Physiological Essays*, "Essay of the Intestine Motions," sec. 9, in *Works* 1: 452–53.

114. Boyle, "Of the Mechanical Origin of Electricity," experiment 6, in *Works* 4: 351. He went on to discuss how "this repeated observation put me upon enquiring among some other young ladies, whether they had observed any such like thing; but I found little satisfaction to my question, except from one of them eminent for being ingenious" (352).

115. Boyle, "Observations about Vitiated Sight," in *Works* 5: 445–52.
116. Boyle, *Certain Physiological Essays*, "Of Unsucceeding Experiments," in *Works* 1: 345.
117. From Peter Shaw's preface to the 1738 abridgment of Boyle's works, quoted by Thomas Birch in Boyle, *Works* 1: cxlvii.
118. Boyle, *Certain Physiological Essays*, "Proëmial Essay," in *Works* 1: 313.
119. Boyle, "Cosmical Suspicions," in *Works* 3: 320.
120. Boyle, *Usefulness of Natural Philosophy*, pt. 2, essay 4, in *Works* 2: 111–12. He expressed the same type of concern about the efficacy of some "incredible medicines" (193–94).
121. Boyle, *Certain Physiological Essays*, "Of Unsucceeding Experiments," in *Works* 1: 351.
122. See Boyle's "Observations on a Diamond," annexed to the *Experimental History of Colours*, in *Works* 1: 792, for an example of how reconciliation involves both other pieces of experience and theoretical knowledge.
123. Peter Dear, "*Totius in verba*: Rhetoric and Authority in the Early Royal Society," *Isis* 76 (1985): 145–61; Shapin, "Pump and Circumstance" and "Boyle and Mathematics"; and Shapin and Schaffer, *Leviathan and the Air-Pump*—the authors of these works describe the strategy of giving a full account of the circumstances surrounding observations as "virtual witnessing." Boyle's advice on how to write reliable observational reports, and his reasons for the advice, will be discussed in detail in chapter 8.
124. Boyle, "Hydrostatical Discourse," in *Works* 3: 596–628.
125. Ibid., sec. 2, chap. 1, in *Works* 3: 608.
126. Ibid., sec. 1, chap. 2, in *Works* 3: 601.
127. Ibid., sec. 2, chap. 5, in *Works* 3: 616. The testimony was "traditional" in the sense that it was "taken for granted" in the schools and repeated, without questioning, in a number of printed works.
128. Ibid., 626.
129. Shapin and Schaffer, *Leviathan and the Air-Pump*, 218; Shapin, "Boyle and Mathematics," 64.
130. Boyle, "Hydrostatical Discourse," sec. 2, chap. 5, in *Works* 3: 621, 626.
131. Boyle, "Differing Pressures of Heavy Solids and Fluids," in *Works* 3: 647. He suggested that some of these unpleasant effects, such as nosebleeds, could be caused by the pressure of the water, but he was not fully convinced.
132. Ibid., 648.
133. Boyle, "Hydrostatical Discourse," sec. 2, chap. 5, in *Works* 3: 619. The reports that he used can be found in his unpublished notes. See, for example, his Outlandish Book, in Royal Society, *Boyle Papers* 27: 151–53, for "observations about divers obtained by questions proposed to an inquisitive traveller who was present at the famous pearl fishing at Manar between the Island of Ceylon and the neighbouring continent"; and 18: fol. 60r, for an account of divers from a surgeon, Mr. Handyside.
134. Boyle, "Differing Pressures of Heavy Solids and Fluids," in *Works* 3: 647. He also reported that a physician of his acquaintance told him that when he was actively diving he did not feel any pressure from the water, but when he allowed himself to sink slowly and was observant, he could feel pressure on his thorax.
135. Ibid.

136. Boyle, "Hydrostatical Discourse," sec. 2, chap. 5, in *Works* 3: 620, 619.

137. Ibid., 621. Boyle's discussion of the problems associated with using the human senses as reliable indicators of heat and cold will be examined in detail in chapter 8.

138. Ibid., 621, 624.

139. Ibid., 618. Boyle included two observations from "Purchase's Pilgrimage," one from a Spanish prelate in America and one from an English general in the East Indies, on the types of physical alterations that divers experienced.

140. Ibid., 619.

141. Ibid., 618.

Chapter Seven

1. Boyle, *Origin of Forms and Qualities*, preface, in *Works* 3: 14.

2. Boyle, *Usefulness of Experimental Philosophy*, pt. 2, sec. 7, in *Works* 3: 423. This argument is again reminiscent of Bacon's "constructive" notion of knowledge discussed in Pérez-Ramos, *Bacon's Idea of Science*.

3. Shapin and Schaffer (*Leviathan and the Air-Pump*, 22–23) maintain that a display of the "immense amount of labour" involved in experimental fact production can be used to question the "epistemological status of the matter of fact," because "to identify the role of human agency in the making of an item of knowledge is to identify the possibility of its being otherwise." Gooding ("Putting Agency Back into Experiment") makes a number of significant points concerning the way in which activity is involved in both observation and experiment (much as Boyle maintained, as we saw in chapter 6). I agree with Gooding that "phenomena *always* appear as outcomes of human activity" (109). It is not clear, however, that this point "shows something that received philosophies of science could not even contemplate," nor is it clear how Gooding's general sociological conclusion follows—that "all natural phenomena are bounded by human activity whose products express the culture in which it occurs" (109). See Nickles, "Good Science as Bad History," 99–105, for a detailed analysis of current debates concerning the philosophical implications of artificiality. A number of philosophers have begun to discuss the ways in which human activity plays an important epistemic role in knowledge production.

4. Boyle, *Origin of Forms and Qualities*, "Author's Discourse," in *Works* 3: 8. The full quotation of this passage is given above in chapter 1 at note 24. The other distinctions are discussed in chapters 4 and 6. According to Boyle, by distinguishing between the artificial and the natural, the Aristotelians were attempting to retain their theory of substantial forms by rejecting experimental evidence to the contrary that one could alter a natural body, and thus change its form, merely by altering the texture of its internal parts.

5. Ibid., "Reproduction of Forms," in *Works* 3: 51. In "The Curious Figures of Salts" Boyle reversed the distinction and said that the natural was also a product of artifice—of "the divine architect's geometry" (*Works* 3: 54–55). See Henry Power, *Experimental Philosophy* (1664; reprint, New York: Johnson Reprint Corp., 1966), 192, where Power discusses the "art of God"; and Robert Hooke's comparison of

the productions of art with those of nature in *A General Scheme, or Idea of the Present State of Natural Philosophy*, in Hooke, *The Posthumous Works*, 3 vols., ed. Richard Waller (1705; reprint, Hildesheim: Georg Olms, 1970), 3: 57–59.

6. Boyle, *Origin of Forms and Qualities*, "Reproduction of Forms," in *Works* 3: 51–52. He insisted that "the chymist is but a servant of nature," and in "The Curious Figures of Salts" he wrote that "there is no certain diagnostick agreed on whereby to discriminate natural and factitious bodies" (*Works* 3: 59).

7. Royal Society, *Boyle Papers* 18: fol. 5r.

8. Boyle, *Usefulness of Experimental Philosophy*, pt. 2, "The Goods of Mankind," sec. 1, in *Works* 3: 442–43. Within this extended discussion Boyle again used an example of a production made by fire: "In making of green or coarse glass, the artificer puts together sand and ashes, and the colliquation and union is performed by the action of the fire upon each body, and by as natural a way, as the same fire, when it resolves wood into ashes." Cf. Bacon, *Advancement of Learning*, bk. 2, in *Works of Bacon* 8: 410: "The artificial does not differ from the natural in form or essence, but only in the efficient."

9. See, for example, Boyle's *History of Cold*, preface, in *Works* 2: 474, and title 1, in *Works* 2: 512; and *Continuation of New Experiments*, preface, in *Works* 3: 177.

10. Boyle, *Usefulness of Experimental Philosophy*, pt. 2, "The Goods of Mankind," sec. 1, in *Works* 3: 443.

11. Boyle, *Usefulness of Natural Philosophy*, pt. 1, essay 1, in *Works* 2: 9. See also pt. 2, essay 2, in *Works* 2: 79, on the use of chemistry to discover unobserved differences; and Royal Society, *Boyle Papers* 9: fol. 25r: "The informations of Sense assisted and heightened by instruments are usually preferable to those of sense alone."

12. Boyle, *New Experiments*, "Examen of Hobbes," chap. 2, in *Works* 1: 193.

13. In *Leviathan and the Air-Pump* Shapin and Schaffer consistently present Boyle as an empiricist in contrast to Hobbes. But Hobbes's distrust of experiments and his complete reliance on sense perceptions would indicate that he was the real empiricist. Also, Hobbes's criticisms of artificiality, discussed by Shapin and Schaffer (114–15), indicate a remnant of Aristotelianism in his thought. It should not be assumed that simply because a philosopher advocates a mathematical mode of reasoning he is therefore not an empiricist at the epistemological level. As Ishiguro ("Pre-established Harmony," 81) has noted, Leibniz recognized, and criticized, the empirical element in the work of most Cartesians.

For a discussion of Hobbes's empiricist epistemology, see Paolo Mancosu and Ezio Vailati, "Torricelli's Infinitely Long Solid and Its Philosophical Reception in the Seventeenth Century," *Isis* 82 (1991): 50–70; Perez Zagorin, "Hobbes on Our Mind," *Journal of the History of Ideas* 51 (1990): 317–35; and the essays in *Perspectives on Thomas Hobbes*, ed. G. A. J. Rogers and Alan Ryan (Oxford: Clarendon Press, 1988), particularly Noel Malcolm, "Hobbes and the Royal Society" (43–66), Arrigo Pacchi, "Hobbes and the Problem of God" (171–87), G. A. J. Rogers, "Hobbes's Hidden Influence" (189–205), and Richard Tuck, "Hobbes and Descartes" (11–41).

14. Descartes, *Discourse on the Method*, pt. 4, in *Philosophical Writings* 1: 129. Boyle main-

tained that visibility to the eye is not necessary for establishing the existence of a thing.

15. Boyle, *Things Said to Transcend Reason*, 5th advice, in *Works* 4: 463.

16. Boyle, *Christian Virtuoso*, pt. 1, in *Works* 5: 538. He used the analogy in a similar manner in *Certain Physiological Essays*, "Proëmial Essay," in *Works* 1: 308; *Sceptical Chymist*, preface to the appendix, in *Works* 1: 591; *Chymical Paradox*, in *Works* 4: 498; and *Free Inquiry*, preface, in *Works* 5: 159.

17. Royal Society, MS 189, fol. 48r. This is a small notebook that begins with addresses of apothecaries and tradesmen and has a number of loose notes and observations along with some draft materials for prefaces to Boyle's works. Collins (*Changing Order*) discusses the experimenters' regress as a purely practical problem concerned with the determination of when an experiment can be considered a reliable check upon previous experimental work. Shapin and Schaffer (*Leviathan and the Air-Pump*, 226 ff.) use Collins's notion in their discussion of the problems surrounding the replication of Boyle's air-pump experiments and their analysis of the "negotiations between experimenters that were supposed to lead them out of this regress." They conclude that "knowledge production depends . . . on the practical social regulation of men and machines" (281). See Nickles, "Good Science as Bad History," 105–6, for a discussion of Collins's notion of the experimenters' regress and Pickering's newer, less positivist, version of the regress that looks at the "mutual dependence of experimental and theoretical practice." As we will see, Boyle's notion was most like Pickering's. More important, Boyle used this version of the regress in order to avoid the types of problems that Collins's version cannot resolve except by means of social negotiation.

18. Royal Society, *Boyle Papers* 9: fol. 56r.

19. Boyle, *Sceptical Chymist*, pt. 3, in *Works* 1: 510. See also *History of Cold*, preface, in *Works* 2: 471.

20. Boyle, *Experimenta et observationes physicae*, preamble, in *Works* 5: 564. He expressed here the hope that his readers would know his reputation from other works where hypotheses were prominent and thus excuse him for this deviation from his usual manner of writing.

21. Boyle, *Christian Virtuoso*, pt. 1, in *Works* 5: 538–39.

22. Royal Society, *Boyle Papers* 9: fol. 7r. The intended title for this work was "Use and Bounds of Experiment." It appeared as number 9 on the "list of Mr. Boyle's Philosophical Writings not yet Printed, set down July the 3rd 1691" (35: fol. 187r) as "Observations about the Uses and Bounds of Experience in Natural Philosophy"; and as number 8 on Birch's list in Boyle, *Works* 1: ccxxxvii. Boyle's notes, that cover *Boyle Papers* 9: fols. 1r–129v, are about the relations between experience, reason, and authority and include many repetitions, drafts, and pasted-up scraps. Notes for his "Discernment of Suppositions" are included here, as are his notes on probatory and exploratory experiments.

23. Royal Society, *Boyle Papers* 9: fol. 30v.

24. Ibid., 9: fol. 1r. Boyle listed eight types of "probable reasons" for the failure of others to perform experiments: "Arguments a majori"; "Arguments a pari"; "The not knowing or not considering how many causes one effect or phaenomena may

have"; "the not knowing or not considering how many effects or phaenomena one Body may produce upon several accounts"; "the arguing from false and doubtful Hypotheses"; "the supposing of rules either General or Absolute that admit of exceptions or are true but secundum quid"; "Erroneous inferences drawn from mistakes about Analogies and Proportions"; and "from undiscerned suppositions" (fol. 2r). He also noted that many would report experiments on the authority of tradition. When Boyle tested these experiments, he "found perhaps three parts of four to be false or at least uncertain" (fols. 38v–39r). This criticism was similar to that which he made of Pascal's work (see chapter 3 above). In correspondence with Hartlib, Oldenburg criticized the French for not being experimenters, and he wrote to Boyle in 1659 saying that the French were more discursive than active or experimental (in Boyle, *Works* 1: 286).

25. Boyle, *Certain Physiological Essays,* "Of Unsucceeding Experiments," in *Works* 1: 335–36.

26. Royal Society, *Boyle Papers* 9: fol. 13r. He added: "I take it to be a more considerable than observed cause of error, that even knowing men look upon many things too much in the bulk (if I may so speak) and do not sufficiently perceive and consider the particulars that are implicitly contained in that which they assent to."

27. Ibid., 10: fol. 7r. These notes are headed "Uses of Experiments."

28. Boyle, *Certain Physiological Essays,* "Of Unsucceeding Experiments," in *Works* 1: 341–42.

29. Boyle learned about the philosophical significance of experiment from Bacon, Galileo, and Harvey, but he learned about the practical problems from his own practice and the practices of tradesmen. In his preface to *Sceptical Chymist* (*Works* 1: 463), he noted that he had "had the good fortune to learn" chemical "operations from illiterate persons." This was appropriate, since as we saw in chapter 3, those who wrote books in chemistry did so in such an obscure fashion that it was not possible to follow their directions. While Boyle learned how to perform anatomical dissections from doctors such as Highmore and Petty, he also learned from butchers; see his *Usefulness of Natural Philosophy,* pt. 2, essay 2, in *Works* 2: 85.

30. Boyle, *Certain Physiological Essays,* "Of Unsucceeding Experiments," in *Works* 1: 339.

31. Ibid., 340, 341. Boyle frequently went out to these places with the explicit intention of learning about the tradesmen's practices. On another occasion, he recounted how on a "visit to the chief coperas work we have in England, one of the overseers of it, who went along with me to shew me the contrivance of it, assured me, that divers times, by the mistake or neglect of a circumstance in point of time, they had lost, and are yet subject to lose, some thousands of pounds of vitriol at a time" (340).

32. Ibid., "Concerning the Unsuccessfulness of Experiments," 318–33. Shapin and Schaffer (*Leviathan and the Air-Pump*) mention these essays in a dismissive tone, presumably because they have dismissed the relevance of epistemological reasons in their analysis of the performative nature of Boyle's work. They write that in these essays Boyle presented "two points vital to the performance, interpretation and reporting of experiments generally. First he offered a repertoire of excuses for the failure of particular experiments" as "a set of factors that could be brought into play so that a given expectation or theory need not be invalidated by experimen-

tal failure." Second, "Boyle stressed the importance of candid and total reporting of actual experimental outcomes" whether they were successful or not because this was basic to a "consensual natural philosophy" (185). Clearly such an analysis misses the epistemological complexity of Boyle's discussion, but it is also questionable whether it captures the performative nature of Boyle's work. It is misleading to see Boyle's justificatory procedures as based upon the simple achievement of consensus, and Boyle did not discuss the contingencies of experiments in order to save theories from experimental refutation. The contingencies that Boyle discussed were mostly those concerned with the practical problems that arise from performance.

33. Boyle, *Certain Physiological Essays*, "Concerning the Unsuccessfulness of Experiments," in *Works* 1: 319. Boyle said that he would fully trust only those chemicals that he manufactured himself.
34. Ibid.
35. Ibid., 329, 328. He also noted that seeds kept too long will lose "their germinating power, without losing any of their obvious qualities" (328).
36. Ibid., 327, 321, 322.
37. Ibid., 327.
38. Boyle, "Discernment of Suppositions," in Royal Society, *Boyle Papers* 9: fol. 15r.
39. Boyle, *Certain Physiological Essays*, "Concerning the Unsuccessfulness of Experiments," in *Works* 1: 330. He also noted that many medicines are "wont to be sophisticated with arsenick" (320).
40. Ibid., "Of Unsucceeding Experiments," 334.
41. See the discussion in chapter 3 above, at note 33, from Boyle's *Hydrostatical Paradoxes*, in *Works* 2: 743–44.
42. Boyle, *Certain Physiological Essays*, "Of Unsucceeding Experiments," in *Works* 1: 335.
43. Ibid., 340–46. A number of these contingencies, such as the effect that a patient's age, sex, and mental attitude could have on the administration of drugs, were discussed in chapter 3. Boyle also discussed such problems in *Usefulness of Natural Philosophy*, pt. 2, essay 5, in *Works* 2: 174, and appendix to pt. 2, in *Works* 2: 239. In addition to such "general uncertainty, to which most remedies are subject," there were other types of remedies "that seem obnoxious to contingencies of a particular nature," such as the "sympathetic powder" and "weapon-salve" that were not applied directly to a patient and "amulets" that were used for the prevention of illness, that clearly did not work on all occasions. Boyle's diffident attitude did not allow him to proclaim that these remedies "are never of any efficacy at all," yet he also refused to affirm "that they constantly perform what is promised of them" (*Works* 1: 346).
44. Boyle, *Certain Physiological Essays*, "Of Unsucceeding Experiments," in *Works* 1: 347.
45. Ibid., 347–48. Boyle noted that he was not the first to have noticed such contingencies concerning differences between instruments or between the places where the instruments were used. The "noble *Tycho*, who having laid out, besides his time and industry, much greater sums of money on instruments than any man we have heard of in later times," had written about the problems that resulted from the imperfections of instruments "quod hominum manibus paratur." So also had "the diligent *Fournier*," who found "upon trial of many instruments both at sea and

ashore . . . that no astronomer in the world can be sure to make his observation at sea within ten minutes of the precise truth, no not (says he) upon the sand itself, within one minute of it" (348).

46. Ibid., 348. He also discussed a "modern and famous writer or two, who have been so mistaken as to think that the weight of the water in comparison of the air is I know not how much under-reckoned, even by this last (overbold) estimate."

47. Ibid.

48. Ibid., 352. As we saw in chapter 2, a moral demonstration could be achieved in the absence of a specific reason to doubt. A general skepticism arising from considerations of the fallibility of a practice does not provide a specific reason to doubt.

49. Ibid., 352–53.

50. Royal Society, *Boyle Papers* 9: fol. 25r.

51. Boyle, *Certain Physiological Essays*, "Concerning the Unsuccessfulness of Experiments," in *Works* 1: 318–19.

52. Royal Society, *Boyle Papers* 9: fol. 52r. He also discussed this distinction in two notebooks: Royal Society, MSS 187, 189.

53. Royal Society, MS 187, fol. 147r. This notebook was written from both ends, working toward the middle. Shortly after this entry (fol. 142r), there is material for inclusion in Boyle's *Experimenta et observationes physicae*. The notion of proof is that which is involved in the proof of a coin or in legal proof. In *Sceptical Chymist* Boyle discussed the various types of "probation" used in experiment. In *Usefulness of Experimental Philosophy* ("Of Men's Great Ignorance," in *Works* 3: 471), he noted that there were many different ways in which one could investigate "the attributes of bodies, as chymical, optical, statical, etc., which being artificial and requiring skill, industry, and instruments, there are very few men that have had the curiosity and ability to examine them."

54. Royal Society, *Boyle Papers* 5: fol. 18v, repeated with slight changes at fol. 19v.

55. Boyle, *Certain Physiological Essays*, "Concerning the Unsuccessfulness of Experiments," in *Works* 1: 331. Boyle's development of indicator tests for acids and bases has been discussed in detail in M. B. Hall, *Boyle and Seventeenth-Century Chemistry*. In Boyle's *Usefulness of Experimental Philosophy* (*Works* 3: 420–21), he discussed a test that he had developed to detect the presence of copper in a sample of gold; in *Chymical Paradox* (*Works* 4: 498), he discussed the identification of substances by the ways they reacted in combination with other bodies.

56. Boyle, *Certain Physiological Essays*, "Concerning the Unsuccessfulness of Experiments," in *Works* 1: 320–21. A number of other examples are given immediately following this, up to 326.

57. Ibid., 321.

58. Royal Society, *Boyle Papers* 36: fols. 57v–58r.

59. Royal Society, MS 187, fol. 146v.

60. Royal Society, *Boyle Papers* 9: fol. 52r.

61. Ibid., 36: fol. 21v.

62. Throughout Boyle's published works, the hypothetical is mixed with the factual, and probatory experiments are mixed with exploratory. See *Origin and Forms of Qualities*, sec. 2, experiment 10, in *Works* 3: 112. Boyle discussed the mechanical

hypothesis that the texture of the parts of bodies is responsible for the qualities that they possess. As he said, he designed his experiments to show the likelihood of this hypothesis. In his notes (Royal Society, *Boyle Papers* 25: fols. 249r–50r), he designed experiments to show that "agitation changes the texture of bodies," in order to use agitation as a way in which to attribute a change of qualities in a body to the change made in its texture.

63. In Boyle's *New Experiments* (experiment 16, in *Works* 1: esp. 32), he noted that he had designed many more exploratory experiments, but "for want of leisure and conveniency to prosecute such trials, we were induced to reserve the rest for another time, and to content ourselves with making that which follows." Frederic L. Holmes, in "Do We Understand Historically How Experimental Knowledge Is Acquired?" *History of Science* 30 (1992): 119–36, provides a fascinating reconstruction of Boyle's experimental procedures from a sensitive reading of the text of *New Experiments* and concludes with a portrayal of "a playful, exploratory Boyle" (133), even though Holmes was apparently unaware of the unpublished material relating to this aspect of Boyle's practice. We will see further evidence of Boyle's "playfulness" in an analysis of his work on phosphorus discussed below in chapter 8.

64. Boyle, "Essay on the Aerial Noctiluca," "Additional Observations," in *Works* 4: 393–94. Many of the exploratory trials recorded in his notes are quite haphazard. See, for example, Royal Society, *Boyle Papers* 21: 213. In 1689 he put a number of different things, including a lump of poor gold ore, lapis lazuli, agate, rock crystal, the head of a tobacco pipe, chalk, amber, and a brick, into a "perforated copper box" that was then weighed in the air and in water in an attempt to find the "proportionate differences" between them. He repeated these experiments, replacing the copper box with a "hydrostatical jar" (214). He performed many exploratory experiments for the second edition of his *History of Blood* (which was never published); records of these experiments are preserved in his papers, 18: fols. 11r–59r.

65. Boyle, *New Experiments*, "To the Lord of Dungarvan," in *Works* 1: 10. He called them "fiat experimentum" in his preface to the *History of Cold* (*Works* 2: 472) and attributed this phrase to "the *Lord Verulam* himself."

66. Royal Society, *Boyle Papers* 25: fol. 11r. This letter has been published in Oldenburg, *Correspondence* 3: 160.

67. Royal Society, *Boyle Papers* 25: fol. 11r.

68. Royal Society, MS 187, fols. 147r–146v (from a section of the notebook that started from the last page).

69. Royal Society, *Boyle Papers* 28: 107r–108r.

70. Ibid., 18: fols. 129r–129v. He described thirteen experiments in all that were to be done on animals. He also had a list of "Enquiries and experiments about electrical bodies" (22: fols. 197r–200r). He had another list headed "Experiments proposed to be made in Mr. Boyle's Pneumatic engine" (26: fol. 217r). See also Royal Society, MS 43 (a small notebook identified as belonging to the Boyle Papers by Keith Moore, archivist of the Royal Society), for similar lists of things to be done, such as an experiment to discover the sphere of magnetic attraction that was to be made by freezing a lodestone to see if and how it would operate through ice.

71. Royal Society, MS 190. The notebook begins (fol. 1r) with the addresses of in-

strument makers; fol. 13r is headed "Experiments to be Done" and is followed by extensive dated lists of chemical, pneumatic, botanical, and hydrostatic experiments interspersed with lists of instruments to be made by tradesmen (e.g., at fol. 56r). This list concludes at fol. 142v. The book then reverses. Starting on the last page (fol. 176r), there are more addresses of tradesmen, and fol. 170v begins the "Experiments Done" section.

72. Boyle, *Origin of Forms and Qualities,* advertisements to sec. 2, in *Works* 3: 75.

73. As we saw in chapter 5, at note 118, Boyle maintained that a variety of experiments provided a naturalist with different "ways of *discovering* and *judging.*"

74. Boyle, *Certain Physiological Essays,* "Of Unsucceeding Experiments," in *Works* 1: 349.

75. Ibid., 349–50, 349. In his *Usefulness of Experimental Philosophy* ("Of Men's Great Ignorance," secs. 1 and 5, in *Works* 3: 472, 489), Boyle discussed all of the effects that an experiment may "concur to produce." In his *Experimental History of Colours* he performed a variety of tests on color change, and in his discussion of a liquid made according to Helmont's directions from a "preparation of steel" (pt. 3, experiment 45, in *Works* 1: 772–73), he noted that one must "in experiments about the changes of colors heedfully mind the circumstances of them."

76. Boyle, *History of Air,* title 13, in *Works* 5: 643–44. This passage, addressed to "Mr. Hartlib," would have been written many years prior to the publication of the history. See also Boyle's essay in *History of Cold* entitled "Thermometrical Experiments and Thoughts," in *Works* 2: 481–507.

77. Boyle, *Certain Physiological Essays,* "Of Unsucceeding Experiments," in *Works* 1: 348–49.

78. Royal Society, *Boyle Papers* 9: fol. 12r.

79. Boyle, *Certain Physiological Essays,* "Of Unsucceeding Experiments," in *Works* 1: 350–51. He also discussed an experiment in which he used spirit of salt to dissolve gold but noted that it only worked with one sample of the substance (350). Shapin and Schaffer, *Leviathan and the Air-Pump;* Shapin, "House of Experiment"; and Golinski, "Noble Spectacle," all treat replication as a social problem.

80. Royal Society, *Boyle Papers* 9: fol. 12r.

81. Boyle, *Usefulness of Experimental Philosophy,* "The Goods of Mankind," in *Works* 3: 445, 474. In "Discernment of Suppositions," in Royal Society, *Boyle Papers* 9: fol. 15r, Boyle had noted that time may be a "material condition."

82. Royal Society, *Boyle Papers* 9: fol. 12r.

83. Ibid., fol. 25r.

84. Boyle, *Certain Physiological Essays,* "Of Unsucceeding Experiments," in *Works* 1: 351. He noted that further investigation might show either that the original observation was faulty or that the "received hypothesis will not hold so universally as men presume." Bacon also spoke of how failed experiments could be instructive in *De augmentis* (bk. 5, in *Works of Bacon* 9: 83).

85. Boyle, *Certain Physiological Essays,* "Of Unsucceeding Experiments," in *Works* 1: 353. As Boyle said, because the "general scope of the naturalist" is "to discover truth" and because nature acts "conformably to her own lawes" and does "not miss her Aimes, whensoever we do ours," nature "therefore may instruct us. . . . Nay it may often happen that a good naturalist may learn more by a disappointment than by a prosperous event, since the truth he stumbles upon may be more considerable

than the thing he pursued" (Royal Society, *Boyle Papers* 9: fol. 38r). Shapin and Schaffer (*Leviathan and the Air-Pump*, 64–65) discuss how Boyle reported failures to show that he was reliable. This would certainly be a function of such a practice, but we can see here that the practice also had an important methodological function.

86. Royal Society, *Boyle Papers* 9: fol. 113r. In *Excellency of Theology* (sec. 3, in *Works* 4: 45), he discussed how "the soul being confined to the dark prison of the body, is capable . . . but of a dim knowledge." He discussed how future discoveries would lead to better methods in his advertisements to *Experimenta et observationes physicae*, in *Works* 5: 569; and his *Chymical Paradox*, in *Works* 4: 497–500.

87. Boyle, *Sceptical Chymist*, pt. 2, in *Works* 1: 499. He began reporting his failures in his first major work, *New Experiments Physico-Mechanical*. See, for example, experiment 15 (*Works* 1: 32), on his failed attempt to kindle fire in the evacuated receiver by a lens held outside. He reported the attempt although "it succeeded not" because "it is useful to recite what experiments miscarry as well as succeed" and also because "it is very possible, that what we endeavoured in vain may be performed by your lordship, or some other virtuoso, that shall have stauncher vessels than we had and more sunny days than the present winter allows us."

Chapter Eight

1. The first type of study is exemplified by Thomas M. Carr, Jr., *Descartes and the Resilience of Rhetoric* (Carbondale: Southern Illinois University Press, 1990); Harwood, "Science Writing and Writing Science"; and B. J. Shapiro, "Early Modern Intellectual Life." Boyle had some knowledge of traditional rhetorical theory, but as we will see, his writing style owes more to his negative reaction to such types of strategies and to the positive influence of Bacon's writing in such works as *Parasceve* and *Sylva sylvarum*. The other types of studies are represented by two recent anthologies: Peter Dear, ed., *The Literary Structure of Scientific Argument* (Philadelphia: University of Pennsylvania Press, 1991); and Marcello Pera and William R. Shea, eds., *Persuading Science: The Art of Scientific Rhetoric* (Canton, Mass.: Science History Publications, 1991). Often in the latter works the attempt is made to uncover aspects of the argument structure that are hidden by rhetorical devices. Boyle, unlike some of his contemporaries, wrote at length about the appropriate form of argument to be used in natural philosophy. His written reports were not meant so much as a way to establish the content of science as a way to exhibit this argument structure in its various practical applications.

2. The earliest works in this genre were Dear, "Totius in verba"; and Shapin, "Pump and Circumstance." Many other works have built upon these early analyses. For Boyle studies, see Golinski, "Noble Spectacle" and "Chemistry in the Scientific Revolution"; Shapin and Schaffer, *Leviathan and the Air-Pump*; and Shapin, "Boyle and Mathematics." For works on other scientists' writing styles, see Geoffrey Cantor, "The Rhetoric of Experiment," in *The Uses of Experiment*, ed. David Gooding, Trevor Pinch, and Simon Schaffer (Cambridge: Cambridge University Press, 1989), 159–80; Lissa Roberts, "A Word and the World: The Significance of Nam-

ing the Calorimeter," *Isis* 82 (1991): 198–222; and Frank J. Sulloway, "Reassessing Freud's Case Histories," *Isis* 82 (1991): 245–75.

It is certainly true that Boyle's habit of providing full accounts of all the circumstances surrounding his trials lent credibility to his reports, but the rhetorical studies cited above are problematic because they tend to portray his essays in an overly simplistic and calculated manner that is at odds with the chaotic nature of most of his works. Boyle did attempt to make the experimental process vivid for his readers, but he did not refer to them as additional witnesses to attest to the truth of the facts that he delivered. The closest Boyle came to a discussion of "virtual witnessing" was his account of "vicarious experience" in *Christian Virtuoso*, where he used that phrase to refer to historical and theological evidence (*Works* 5: 525). He never mentioned this notion in connection with his experimental essays. In *Parasceve* Bacon advised naturalists to provide complete accounts of all of the circumstances surrounding an experiment so that readers could *judge* "whether the information obtained from that experiment be trustworthy or fallacious" (*Works of Bacon* 8: 368). Boyle's friend Thomas Sydenham was also well known and respected for the vivid accounts that he provided of his clinical observations. See Meynell, *Bibliography of Sydenham*.

3. Boyle, *Sceptical Chymist*, conclusion, in *Works* 1: 585. In his preface to the appendix, Boyle repeated this point when he maintained that chemical experiments "ordered by a skilful naturalist . . . may far more conduce . . . to the speculative part of physics" (*Works* 1: 591).

4. Boyle, *History of Blood*, pt. 1, in *Works* 4: 598–99.

5. Boyle, *Certain Physiological Essays*, "Proëmial Essay," in *Works* 1: 313.

6. Ibid., 299, 317. Boyle's publishers often had to beg him for more copy. See the letters between Oldenburg and Boyle concerning the publication of *History of Cold*, in Boyle, *Works* 6: 65, 153, 155, 157, 160, 163, 172, 178; and the letters from Robert Sharrock concerning the publication of *The Usefulness of Natural Philosophy*, in *Works* 6: 321, and Royal Society, *Boyle Letters* 5: 83–102.

7. Boyle, *Certain Physiological Essays*, "Proëmial Essay," in *Works* 1: 299. This is the essay that Boyle asked Spinoza to read (see chapter 3 above). Harwood, in "Science Writing and Writing Science," discusses how Boyle's self-consciousness about writing emerged from his anxieties about the dangers of misreading and the importance of correctly reading a text.

8. Boyle, *Certain Physiological Essays*, "Proëmial Essay," in *Works* 1: 303. Paradis, in "Montaigne, Boyle, and the Essay of Experience," maintains that Boyle sought to base his style upon the passive reports of personal experience exemplified in the works of Montaigne. As we will see below, there was not much passivity in the way that Boyle reported his work. In addition, the influence of Montaigne is not clear. Although Boyle did say that he called his works essays "in imitation of the French," he seemed to be referring to the use of the word *essay* and not to any particular writing style. Boyle's style seems much closer to Bacon's and to that which Galileo promulgated, not only in his essays but also in his dialogues, which were still essays in the sense that they presented catalogs of experimental data together with speculations about their theoretical significance. See Westfall, "Galileo and Newton," for a discussion of Galileo's rhetorical style. Boyle made

his debt to Bacon explicit in *Experimenta et observationes physicae*, when he wrote that he would use the "authority and examples" of Bacon but "without confining myself to either," because Bacon was the "first, if not the only author I know of, that gave us a set of precepts of well writing natural histories" (advertisements, in *Works* 5: 567).

9. Boyle, *Certain Physiological Essays*, "Proëmial Essay," in *Works* 1: 303.

10. Boyle, *New Experiments*, conclusion, in *Works* 1: 117.

11. Boyle, *Certain Physiological Essays*, "Proëmial Essay," in *Works* 1: 304. He was opposed to the "dull and insipid way of writing, which is practiced by many chymists." He maintained that he would not use "exotic words and terms" except "for some peculiar significance of some such word, whose energy cannot be well expressed in our language," or if custom had made the terms "familiar and esteemed" (305). He also explained that his sentences might often be "over-long" when he wished to mention all the things "pertinent to my subject" in one place (305). See Charles Webster, *Samuel Hartlib and the Advancement of Learning* (Cambridge: Cambridge University Press, 1970), 68, on Hartlib's injunctions against ornate writing styles and the use of foreign expressions.

12. Boyle, *Certain Physiological Essays*, "Proëmial Essay," in *Works* 1: 303. Boyle added that it would be the "reader's own fault, if he be not a learner by them."

13. Boyle, *Sceptical Chymist*, "Physiological Considerations," in *Works* 1: 464. Cf. *Certain Physiological Essays*, "Proëmial Essay," in *Works* 1: 302–4.

14. See I. Bernard Cohen, *The Newtonian Revolution* (Cambridge: Cambridge University Press, 1983); Dear, "Totius in verba"; and Paradis, "Montaigne, Boyle, and the Essay of Experience." Dear, for example, wrote that Newton's 1672 letter to *Philosophical Transactions* was modeled on Boyle's records of experience and then itself became a model, but if Newton had actually intended to follow Boyle's style, he did not do a very good job of it. Newton sought to convey certainty in his writing, just as he sought to find certainty in his studies of Scripture. See Maurizio Mamiani, "The Rhetoric of Certainty in Newton's Science," in *Persuading Science: The Art of Scientific Rhetoric*, ed. Marcello Pera and William R. Shea (Canton, Mass.: Science History Publications, 1991), 157–72. Boyle, as we will see, was as tentative in his writing style as he was in his interpretations of Scripture.

15. *Philosophical Transactions* 80 (19 Feb. 1671/72): 3075–87. The *experimentum crucis* is discussed on 3078; the conclusion is on 3079. A number of the experiments reported by Newton are those that were first reported in Boyle's *Experimental History of Colours*. For a discussion of the problems associated with determining the accuracy of the "historical" account given by Newton, see John Hendry, "Newton's Theory of Colour," *Centaurus* 23 (1980): 230–51; Henry Guerlac, "Can We Date Newton's Early Optical Experiments?" *Isis* 74 (1983): 74–80; J. A. Lohne, "Isaac Newton: The Rise of a Scientist, 1661–1671," *Notes and Records of the Royal Society* 20 (1965): 125–39; and Richard S. Westfall, "The Development of Newton's Theory of Color," *Isis* 53 (1962): 339–58. For the most up-to-date analysis of Newton's optics, see Alan E. Shapiro, *Fits, Passions, and Paroxysms: Physics, Method, and Chemistry and Newton's Theories of Colored Bodies and Fits of Easy Reflection* (Cambridge: Cambridge University Press, 1993).

16. Boyle, *New Experiments*, experiment 37, in *Works* 1: 89–90. The full account of ex-

periment 37 covers 89–94. See experiment 36 (81–89), for an example of experiments performed to prove the weight of the air.

17. Ibid., 90. He went on to say that he had tried to discover the cause of this effect by additional "phaenomena laid together," but he noted that "he that would render a reason of the phaenomenon, . . . must do two things; whereof the one is difficult, and the other little less than impossible: for he must give an account not only whence the appearing whiteness proceeds, but whereof that whiteness doth sometimes appear, and sometimes not" (90–91). He admitted that he could do neither.

18. Boyle, *Experimental History of Colours*, preface, in *Works* 1: 662–63. There is no evidence that Boyle was ever persuaded by Newton's rhetoric. A copy of Hooke's original objections to Newton's letter is preserved in Royal Society, *Boyle Papers* 20: 1–17. Although Boyle had spoken of a particulate theory of light in his first work on colors, in subsequent works he refused to decide between Newton, Hooke, and Huygens. In his "Essay on the Porousness of Solid Bodies" (*Works* 4: 790), for example, he noted that light is "either a subtile and rapidly moving body, or at least requires such an one for its vehicle." In his essay "Efficacy of Effluviums" (*Works* 3: 685), he noted that the corporeal nature of light "is much disputed."

19. Boyle, *Experimental History of Colours*, pt. 1, chap. 1, in *Works* 1: 669. He discussed various modification theories of light (671–74), followed by an examination of Aristotelian and atomistic theories of colors. He listed six hypotheses in all (693–94) and concluded (695) that a better theory of light would be required before one could reach a final decision about the production of colors.

20. Ibid., pt. 2, "Experiment in Consort," in *Works* 1: 708–24. Part 3 (*Works* 1: 724–88) contains fifty numbered experiments along with numerous annotations concerning the production of all other colors. Here Boyle was even more tentative in his conclusions. See Marie Boas Hall, introduction to *Experiments and Considerations Touching Colours*, by Robert Boyle (New York: Johnson Reprint Corp., 1964), xiv, for a discussion of how Boyle put experiments together to support each other and present an experimental demonstration. Bacon had presented numerous experiments in consort in the "centuries" of his *Sylva sylvarum*, in *Works of Bacon* 4: 159–477, and 5: 7–164.

21. Boyle, *Experimenta et observationes physicae*, advertisements, in *Works* 5: 567. Boyle noted here that he called them experiments in consort "in imitation of" Bacon.

22. M. B. Hall, *Boyle on Natural Philosophy*, 68.

23. Boyle, *Certain Physiological Essays*, "Proëmial Essay," in *Works* 1: 315. Earlier he noted that he had "on some occasions, adventured to deliver my opinion," because it is "no disgrace in difficult matters rather to hazard the being sometimes mistaken, than not to afford inquisitive persons their best assistance towards the discovery of truth" (308).

24. Ibid., 303.

25. Boyle, *Usefulness of Experimental Philosophy*, pt. 2, preamble, in *Works* 3: 401.

26. Boyle, *Certain Physiological Essays*, "Proëmial Essay," in *Works* 1: 305–6. See Harwood's introduction to Boyle's *Early Essays*, for a discussion of Boyle's similar concern with persuading the reader about proper behavior in his ethical writings.

27. Boyle, *Sceptical Chymist,* preface, in *Works* 1: 459. As we saw in chapter 3, chemical experiments were not normally written in a clear style.

28. Boyle, *Experimental History of Colours,* preface, in *Works* 1: 663.

29. Boyle, *Experimenta et observationes physicae,* chap. 6, advertisement, in *Works* 5: 597–98. He added, however, that in this work he was more interested in teaching the novice and thus he would report "experiments, rather useful than specious," that is, "those that are proper to increase the reader's skill, to those that make an ostentation of the writer's."

30. Lawrence M. Principe, "Robert Boyle's Alchemical Secrecy: Codes, Ciphers, and Concealments," *Ambix* 39 (1992): 63–74, provides a good discussion of the secretive nature of alchemical treatises and how Boyle wrote in that style on some occasions. See also Clericuzio, "Carneades and the Chemists"; and Golinski, "Chemistry in the Scientific Revolution." Boyle was not breaking his rhetorical precepts by such practices but rather tailoring his style to a particular audience in order to fulfill the need for communication and collaboration with alchemists. He believed that obscurity was permissible in alchemical works, but not in works devoted to natural philosophy, even though he would often include alchemical hints in his experimental works by means of such strategies as "dispersion." See Webster, *Samuel Hartlib,* 39–40, on how Boyle's call for the effective communication of discoveries, especially in medicine, was motivated by altruism, not private gain.

31. Boyle, *Certain Physiological Essays,* "Proëmial Essay," in *Works* 1: 316. He added that even "trifles" are difficult and time-consuming to make.

32. In *New Experiments Physico-Mechanical* Boyle noted that the report of his failed experiment could lead to its successful performance by "some other virtuoso" (*Works* 1: 32).

33. Boyle, *Certain Physiological Essays,* "Proëmial Essay," in *Works* 1: 303–4. See M. B. Hall, *Boyle on Natural Philosophy,* 114, for a discussion of how the general trustworthiness of Boyle's reports led Lavoisier, a century later, to write that "such important experiments made by a physicist like Boyle naturally tended to put me on my guard against my own belief, however well demonstrated that was in my own eyes" (*Oeuvres de Lavoisier,* 6 vols. [Paris, 1862–93], 2: 105, quoted in Hall).

34. Boyle, *Certain Physiological Essays,* "Proëmial Essay," in *Works* 1: 304.

35. Ibid., 315. In a letter to Oldenburg dated 13 June 1666 (Royal Society, *Boyle Papers* 25: 3–5, printed in Oldenburg, *Correspondence* 3: 160), Boyle said that the inclusion of hypotheses was useful, "*partly,* because the knowledge of differing theories may admonish a man to observe divers such circumstances in an Experiment as otherwise 'tis like he would not heed" and partly because they would "conduce to make the History *both* more exact and compleat in itself, and more ready for use, and more acceptable to those that love to discourse upon *Hypotheses.*"

36. Boyle, *Experimenta et observationes physicae,* preamble, in *Works* 5: 565–66.

37. Ibid., chap. 6, advertisement, in *Works* 5: 598.

38. Boyle, *Sceptical Chymist,* preface to the appendix, in *Works* 1: 591. *Curious* can convey the meaning of "careful" when it is used in such phrases as "curious workmanship." Boyle used the term with this connotation in a number of places. In Royal

Society, *Boyle Papers* 27: 45, he spoke of a virtuosa who was "very curious both in the breeding and ordering of silkworms." Another instance of this usage of the term *curious* occurs below in the passage quoted at note 90.

39. Boyle, *Sceptical Chymist*, preface to the appendix, in *Works* 1: 591.

40. Boyle, *Certain Physiological Essays*, "Proëmial Essay," in *Works* 1: 308. He added that to "adhere" to a hypothesis when new information shows "it to be erroneous, is but a proud obstinacy, very injurious to truth, and very ill becoming the sense we ought to have of human frailities" (311).

41. Ibid., 311–12.

42. Boyle, *Certain Physiological Essays*, "Of Unsucceeding Experiments," in *Works* 1: 349.

43. Boyle, *Certain Physiological Essays*, "Proëmial Essay," in *Works* 1: 312. Golinski, "Chemistry in the Scientific Revolution"; and Shapin and Schaffer, *Leviathan and the Air-Pump*, include discussions of the social aspects surrounding the issue of civility in Boyle's day.

44. Quoted in R. E. W. Maddison, "Studies in the Life of Robert Boyle, F.R.S., Part IV: Robert Boyle and Some of His Foreign Visitors," *Notes and Records of the Royal Society* 11 (1954): 38, from BL, Add. MS 4229, fols. 57r–58r.

45. Birch relates this account from Bishop Burnet's funeral sermon for Boyle, in Boyle, *Works* 1: cxliv. The visits had begun while Boyle was at Oxford and increased as his fame did. See his letters to Oldenburg from 1664–65 in *Works* 6: 63–64, 66–67. He complained that "I have not had any opportunity to prosecute my studies vigorously, the perpetual resort of *English* strangers, and the want of glasses, and other mechanical employments, leaving me neither leisure nor accommodations." Two years before his death, Boyle did find it necessary to limit these visits somewhat on the advice of his doctor, Edmund King, and his "best friends." The notice that he had posted at that time (printed in *Works* 1: cxxix) asked that he be "excused from receiving visits" on Tuesday and Friday mornings (foreign post days) and Wednesday and Saturday afternoons.

46. Quoted in Maddison, "Part IV: Boyle and Foreign Visitors," 38, from BL, Add. MS 4229, fol. 51v. Dr. Dent, prebendary of Westminster, was described by Birch as a "particular friend" of Boyle's (Boyle, *Works* 1: cxli).

47. Quoted in R. E. W. Maddison, "Studies in the Life of Robert Boyle, F.R.S., Part I: Robert Boyle and Some of His Foreign Visitors," *Notes and Records of the Royal Society* 9 (1951): 2. In both this work and part IV, Maddison gives a detailed account of visits made to Boyle by Huygens, Leibniz, Cosmo III, and lesser-known figures such as Balthasar de Monconys, who wrote in his diary that Boyle "est autant civil et doux que scavant" (pt. I, 18). In the letters and diaries of these visitors, one can see evidence of the time that Boyle spent with them. Count Lorenzo Magalotti, secretary of the Academia del Cimento, met Boyle at Oxford in 1668. When Magalotti accompanied Cosmo III to London the next year, he visited Boyle twice at his home. Boyle returned these visits, and when Magalotti fell ill during his second stay in England, Boyle visited his "bedside for two or three hours daily" (pt. I, 22–29).

48. Royal Society, MS 188, fols. 112r–111v (from a section of a notebook that started from the last page). Among the "things to be sealed up" were bread, raw and roasted flesh, and milk.

49. Papin in Boyle, *The Second Continuation of Physico-Mechanical Experiments*, advertisement, in *Works* 4: 509. Boyle explained that he had met Papin through his acquaintance with Huygens, and in the trials that Papin performed for him, he "gave him the freedom to use his own [air-pump], because he best knew how to ply it." Boyle noted that he oversaw the method employed in a number of the experiments performed by Papin but that he was "purposely somewhat more incurious and remiss" about the experiments on the preservation of foods, whereas Papin had a "particular end of his own, somewhat different from mine in the other experiments; so I was very willing, that he should use his own method about them, not doubting that he would use his greatest industry therein, as I found, by the event that he had done" (506).

50. Boyle, *New Experiments*, in *Works* 1: 7. Hooke apparently had a troubled career and did not often receive the credit that he felt he deserved, except from Boyle. See the essays in *Robert Hooke: New Studies*, ed. Michael Hunter and Simon Schaffer (Woodbridge, Eng.: Boydell Press, 1989), particularly Steven Shapin, "Who Was Robert Hooke?" (253–85). In his will, Boyle bequeathed "my best microscope and my best loadstone" to Hooke (*Works* 1: clx).

51. See R. E. W. Maddison, "Studies in the Life of Robert Boyle, F.R.S., Part V: Boyle's Operator: Ambrose Godfrey Hanckwitz, F.R.S.," *Notes and Records of the Royal Society* 11 (1955): 159; and Robert Sharrock's letter to Boyle, in *Works* 6: 321, concerning the course on chemistry that Staehl taught at Oxford.

52. Maddison, "Part I: Boyle and Foreign Visitors," 32. Slare was described in Boyle's will as "late my servant" and was bequeathed "a ring of 8*l.*" (*Works* 1: clxi).

53. Maddison, "Boyle's Operator." Among Boyle's other assistants were Lawrence Rooke, Gresham Professor of Astronomy, John Milne, Hugh Greg, and Thomas Smith, an apothecary who lived with Boyle for seventeen years. Milne and Greg, described by Boyle as "once my servants," were both left rings in his will; Smith was left "30 pounds, in case he be with me, or in my service at the time of my death" (*Works* 1: clx).

   John Warre, Boyle's amanuensis, received forty pounds and half of Boyle's apparel and linen outright, in addition to a yearly sum of forty pounds for his "care and pains" with "the troublesome part of the execution of my said last will and testament" (*Works* 1: clxx). Boyle apparently trusted this servant more than he deserved. See R. E. W. Maddison, "Studies in the Life of Robert Boyle, F.R.S., Part III: The Charitable Disposal of Robert Boyle's Residuary Estate," *Notes and Records of the Royal Society* 10 (1952): 15–27.

54. Maddison, "Boyle's Operator," 170, from Ambrose Godfrey, *An Account of the New Method of Extinguishing Fires by Explosion and Suffocation* (London, 1724), x.

55. Boyle, "Aerial Noctiluca," in *Works* 4: 379–403. Boyle acknowledged his debt to Kraft (381, 402). See Maddison, "Part I: Boyle and Foreign Visitors," for a discussion of Boyle's acquaintance with Kraft.

56. Boyle, "Aerial Noctiluca," in *Works* 4: 382–84. Boyle said that he received additional help from "A.G.M.D.," a countryman of Kraft's (383). This could be a reference to Godfrey, who held a medical degree. See Maddison, "Boyle's Operator," 168–73, for Godfrey's account of how the phosphorus first came to be made.

57. Boyle, "Aerial Noctiluca," "Additional Observations," observation 6, in *Works* 4: 396–97. Boyle was here the corroborating witness for his assistant's observation.
58. Boyle, "Aerial Noctiluca," preface, in *Works* 4: 384.
59. Boyle, "New Experiments and Observations Made upon the Icy Noctiluca," in *Works* 4: 469–95.
60. Ibid., sec. 12, in *Works* 4: 488, 490, 489. A "worse mis-adventure" befell the hapless "laborant" not long after, when, having again prepared "some newly distilled grains of our noctiluca," he placed it in a vial, which he carried in his pocket on the way to Boyle's home. Unfortunately, the vial broke along the way, and the "heat of his body, increased by the motion his long walk had put it into, did so excite the matter, that was falled out of the broken phial, that it burned two or three great holes in his breeches." Except for the loss of his breeches, the assistant was not harmed, and Boyle noted that they looked upon this effect not "without some wonder as well as smiles."
61. Boyle, "Aerial Noctiluca," observation 13, in *Works* 4: 390. See also observation 19, in *Works* 4: 392, for another instance when Boyle took the substance to bed with him. These were apparently not the only times that Boyle performed experiments in his bedchamber. According to John Evelyn, "Glasses, pots, chemical and mathematical instruments, books and bundles of papers, did so fill and crowd his bed-chamber, that there was but just room for a few chairs; so as his whole equipage was very philosophical without formality" (quoted in Maddison, *Life*, 187).
62. Boyle, "Aerial Noctiluca," observations 16 and 17, in *Works* 4: 391–92.
63. Boyle, "Icy Noctiluca," sec. 12, in *Works* 4: 489.
64. Ibid., sec. 7, in *Works* 4: 480–81. In his preface to the *Second Continuation* Boyle gave Papin credit for making inferences as well as experiments (*Works* 4: 507). In his second edition of *New Experiments* Boyle also credited Richard Townley, who, upon a "perusal of my physico-mechanical experiments," discovered what has come to be known as Boyle's Law. Boyle wrote of Townley: "I wish in such attempts other ingenuous men would follow his example" (*Works* 1: 160).
65. Boyle, "Icy Noctiluca," sec. 7, in *Works* 4: 481. These experiments are recorded as observations 1–4, in *Works* 4: 481–82.
66. Boyle, "Aerial Noctiluca," in *Works* 4: 393–94. In "Icy Noctiluca" he began with an account of nine observations that he had made concerning the properties of the substance (*Works* 4: 475–76). He said that his assistant had been "very helpful to me in varying the preparation of the phosphorus" (sec. 12, in *Works* 4: 489).
67. His assistants were indicated by name, initials, title, or the use of *we* in places where Boyle clearly meant to emphasize the collaborative nature of the experiments performed. In his postscript to *Chymical Paradox* he also discussed the work of a "watchful laborant" who performed a number of exploratory experiments on oils and "kept a kind of journal of the number of rectifications, the quantities of pitchy matter from time to time afforded by them, and divers other phaenomena, or circumstances, that occurred in so tedious a prosecution, as I thought experiments of such moment deserved" (*Works* 4: 502).
68. In his preamble to *Experimenta et observationes physicae* (*Works* 5: 566), Boyle said that his *New Experiments* and his *History of Cold* were works in which he had presented his reports in accordance with his precepts on style. His *History of Cold* is similar,

of Chymistry, as he has thrown down the old; he has left us plentiful Matter, from whence we may draw out a true Explication of things, but the explication it self he has but very sparingly touch'd upon." John Freind, *Chymical Lectures* (London, 1712), 4, quoted in Maddison, *Life*, 193.

3. William Brande, "The Progress of Chemical Philosophy," 3d dissertation, *Encyclopaedia Britannica* (London, 1824), 15, quoted in Maddison, *Life*, 195. Marie Boas Hall reached a similar conclusion in this century concerning Boyle's eclecticism when she spoke of his "curious inability" to present a systematic treatment of any one subject (*Boyle on Natural Philosophy*, 68).

4. See, for example, Lavoisier's positive evaluations of Boyle's work, quoted in M. B. Hall, *Boyle on Natural Philosophy*, 43, 114.

5. Quoted in Maddison, *Life*, 186, from a letter dated 1696 to William Wotton.

6. Eustace Budgell, *Memoirs of the Lives and Characters of the Illustrious Family of the Boyles*, 3d ed. (London, 1737), 119–20, quoted in Maddison, *Life*, 195. Budgell also maintained that there is "one Particular, for which he can never be too much admired or commended; it is evident, that he made all his Experiments without any Design to confirm or establish any particular System." Peter Shaw's assessment, that Boyle "shewed philosophy in action," is discussed above in chapter 7.

7. Boyle's eclecticism, as well as his caution, have often been attributed to his personal character, especially by those who find these traits to be failings. His confessor, Gilbert Burnet, noted that Boyle was "humble and modest, almost to a fault" (Maddison, *Life*, 185). Hunter ("Conscience of Robert Boyle") has recently shown how Boyle's religious beliefs helped to shape his personal character.

8. Pérez-Ramos, *Bacon's Idea of Science*, 167–79.

9. Boyle, *Reconcileableness of Reason and Religion*, sec. 7, in *Works* 4: 181.

10. Ibid., sec. 8, in *Works* 4: 182.

11. Boyle, *Usefulness of Experimental Philosophy*, pt. 2, preamble, in *Works* 3: 401.

12. The idea that any method of interpretation will ultimately involve circularity has been described in this century as the "hermeneutic circle." See Gadamer, *Truth and Method*.

13. Boyle, *Things Said to Transcend Reason*, 1st advice, in *Works* 4: 450.

14. Boyle, *The Mechanical Origin of Qualities*, advertisement, in *Works* 4: 236.

15. Boyle was also sensitive to the fact that our customs often influence our evaluations of other people. See note 104 in chapter 6 above.

16. Boyle, *Christian Virtuoso*, pt. 1, in *Works* 5: 523; McGuire, "Boyle's Conception of Nature"; Osler, "Intellectual Sources of Boyle's Philosophy of Nature."

17. Wiener, "Experimental Philosophy of Boyle." See chapter 4 above, for Boyle's contention that corpuscular principles were "likely to be true."

18. Conant, "Boyle's Experiments"; Laudan, "Clock Metaphor and Probabilism"; and Mandelbaum, *Philosophy, Science, and Sense Perception*.

19. Shapin and Schaffer, *Leviathan and the Air-Pump*, 150. See also Dear, "Narratives, Anecdotes, and Experiments"; and Golinski, "Noble Spectacle," both of which follow Shapin and Schaffer's empiricist reading of Boyle.

20. Shapin and Schaffer, *Leviathan and the Air-Pump*, 225.

21. Ibid., 79, 150. See also Collins, *Changing Order*; Gooding, *Experiment and the Making of Meaning*; and Knorr-Cetina, *Manufacture of Knowledge*.

22. Shapin and Schaffer, *Leviathan and the Air-Pump*, 25.
23. Ernan McMullin, "Logicality and Rationality," in *Philosophical Foundations of Science*, ed. R. J. Seeger and R. S. Cohen (Dordrecht: D. Reidel, 1974), 415–30; Dudley Shapere, *Reason and the Search for Knowledge* (Dordrecht: D. Reidel, 1984); and Stephen Toulmin, "Scientific Strategies and Historical Change," in *Philosophical Foundations of Science*, 401–14. See also Arthur Fine, "Science Made Up: Constructivist Sociology of Scientific Knowledge," in *Disunity and Contextualism in the Philosophy of Science*, ed. Peter Galison and David Stump (Stanford, Calif.: Stanford University Press, 1992), 31–52; and Allan Franklin, review of *Experiment and the Making of Meaning*, by David Gooding, *Isis* 83 (1992): 177–78, for more recent discussions.
24. William Whewell, *Novum organon renovatum*, 3d ed. (London: John W. Parker and Son, 1858).
25. Thomas Nickles, "'Twixt Method and Madness," in *The Process of Science*, ed. Nancy Nersessian (Dordrecht: Martinus Nijhoff, 1987), 60–62. Earlier, Nickles noted that this type of process was also a component of Bacon's method of proof (43). See also Robert E. Butts, "Consilience of Inductions and the Problem of Conceptual Change in Science," in *Logic, Laws, and Life: Some Philosophical Complications*, ed. R. G. Colodny (Pittsburgh: University of Pittsburgh Press, 1977), 71–88; and Menachem Fisch, "Whewell's Consilience of Inductions—An Evaluation," *Philosophy of Science* 52 (1985): 239–55. While much less formal, there is some similarity between this type of confirmation procedure and the process of "bootstrapping" as discussed in Clark Glymour, *Theory and Evidence* (Princeton, N.J.: Princeton University Press, 1980).
26. Collins, *Changing Order*, esp. chap. 4. Collins's discussion of the experimenters' regress is primarily concerned with the problems involved in the calibration of instruments. But calibration is clearly more than a practical problem; it also has a theoretical dimension. See Shapin and Schaffer, *Leviathan and the Air-Pump*, 226, for one use of Collins's analysis.
27. Norwood Russell Hanson, *Patterns of Discovery* (Cambridge: Cambridge University Press, 1958); Mary Hesse, *Models and Analogies in Science* (London: Sheed and Ward, 1963); Thomas S. Kuhn, *The Structure of Scientific Revolutions*, 2d ed. (Chicago: University of Chicago Press, 1970); and Michael Polyani, *Personal Knowledge* (New York: Harper and Row, 1964). See also Popper, *Logic of Scientific Discovery*, 106–11; Dudley Shapere, "The Concept of Observation in Science and Philosophy," *Philosophy of Science* 49 (1982): 485–525; and Stephen Toulmin, *Human Understanding* (Princeton, N.J.: Princeton University Press, 1972).
28. Pierre Duhem, *The Aim and Structure of Physical Theory*, trans. Philip P. Wiener (New York: Atheneum, 1981), 200, 218.
29. Shapin and Schaffer, *Leviathan and the Air-Pump*, 23, 344.
30. Hacking, *Representing and Intervening*, 274.
31. Giere, *Explaining Science*, 125. See also 139, for Giere's "technological solution" to the Duhem-Quine thesis. Other works in this genre include Cartwright, *How the Laws of Physics Lie*; Franklin, *Neglect of Experiment*; Galison, *How Experiments End*; and Kockelmans, "Problem of Truth." There are significant differences between the approaches and conclusions of these works, but all emphasize the role of activity, and all harken back to the pragmatic aspects of scientific discovery and justifica-

tion developed by Charles Sanders Peirce in "The Fixation of Belief" and "How to Make Our Ideas Clear," both reprinted in *Charles S. Peirce: Selected Writings*, ed. Philip P. Wiener (New York: Dover Publications, 1966).

32. Gooding, "Putting Agency Back into Experiment," 65. He adds that philosophers then "order these narratives further as logically structured verbal activity. Manipulative practices, barely glimpsed through such texts, don't appear at all in philosophical discussions of science."

33. For various accounts of scientific explanation, see Wesley Salmon, *Scientific Explanation and the Causal Structure of the World* (Princeton, N.J.: Princeton University Press, 1984); Ernan McMullin, "Two Ideals of Explanation in Natural Science," in *Causation and Causal Theories*, ed. Peter A. French, Theodore E. Uehling Jr., and Howard K. Wettstein (Minneapolis: University of Minnesota Press, 1984), 205–20; and Brian Ellis, "What Science Aims to Do," in *Images of Science*, ed. Paul M. Churchland and Clifford A. Hooker (Chicago: University of Chicago Press, 1985), 48–74. Recently, explanation has been described as an "epistemic activity" in Barbara Tuchanska, "What Is Explained in Science?" *Philosophy of Science* 59 (1992): 102–19. The types of interests involved in the assessment of explanation may, of course, be epistemic. See Ernan McMullin, "Values in Science," in *PSA 1982*, ed. Peter D. Asquith and Thomas Nickles (East Lansing, Mich.: Philosophy of Science Association, 1983), 2: 3–27; and Larry Laudan, *Science and Values* (Berkeley and Los Angeles: University of California Press, 1984).

34. Boyle, "Excellency and Grounds of the Mechanical Hypothesis," recapitulation, in *Works* 4: 78; italics mine.

35. Boyle, *Final Causes*, sec. 3, in *Works* 5: 416.

36. Boyle, *Christian Virtuoso*, pt. 1, in *Works* 5: 536. He noted that this type of modesty was an "intellectual (as well as moral) virtue."

37. Boyle, *Certain Physiological Essays*, "Proëmial Essay," in *Works* 1: 311. See also *Experimenta et observationes physicae*, in *Works* 5: 569, for how the details of the experimental method itself may have to be altered; and *Excellency of Theology*, in *Works* 4: 56–60, on the "inconstancy" of doctrines and methods. Boyle wrote that "he that shall pretend to be infallible will make himself Ridiculous" (Royal Society, *Boyle Papers* 4: fol. 5r).

38. See, for example, Toulmin, *Cosmopolis*, 204–5. Some of these criticisms clearly motivate recent work in the sociology of scientific knowledge and in feminist philosophy of science.

39. Royal Society, MS 195, fol. 191v. This passage is from Boyle's "Aretology," printed in *Early Essays*.

40. Boyle, *Things Said to Transcend Reason*, 5th advice, in *Works* 4: 466. Boyle wrote that "there is such a relation betwixt the natural bodies . . . that he that makes a new experiment or discovers a new phaenomenon must not presently think that he has discovered a new truth" (Royal Society, *Boyle Papers* 8: fol. 155r).

41. Boyle, *Of the High Veneration Man's Intellect Owes to God*, par. 32, in *Works* 5: 150.

42. Boyle, *History of Blood*, postscript to the appendix, in *Works* 4: 758. This quotation is given in full in the introduction. See also Royal Society, *Boyle Papers* 10: fols. 17r–18r.

43. Boyle, *Excellency of Theology*, in *Works* 4: 36. Earlier in this same passage, Boyle used

the text analogy once again to illustrate his meaning. He said that "in the book of nature, as in a well-contrived romance, the parts have such a connection and relation to one another, and the things we would discover are so darkly or incompletely knowable by those that precede them, that the mind is never satisfied till it comes to the end of the book."

44. Ibid. Boyle labored over the exact wording of this metaphor, as can be seen from the way in which the draft for this passage is crossed out and rewritten. Royal Society, *Boyle Papers* 8: fols. 156r–57r.

BIBLIOGRAPHY

Aaron, Richard I. *John Locke.* Oxford: Clarendon Press, 1971.

Alexander, Peter. "Boyle and Locke on Primary and Secondary Qualities." *Ratio* 16 (1974): 51–67.

————. *Ideas, Qualities, and Corpuscles: Locke and Boyle on the External World.* Cambridge: Cambridge University Press, 1985.

Ariew, Roger. "The Duhem Thesis." *British Journal for the Philosophy of Science* 35 (1984): 313–25.

Ashworth, William B. "Catholicism and Early Modern Science." In *God and Nature,* edited by David C. Lindberg and Ronald L. Numbers, 136–66. Berkeley and Los Angeles: University of California Press, 1986.

Bacon, Francis. *The Works of Francis Bacon.* 15 vols. Edited by James Spedding, Robert Leslie Ellis, and Douglas Denon Heath. Boston: Taggard and Thompson, 1860–64.

Badcock, A. W. "Physics at the Royal Society, 1660–1800." *Annals of Science* 16 (1960): 95–115.

Baker, J. H. *An Introduction to English Legal History.* London: Butterworths, 1971.

————, ed. *The Reports of John Spelman.* Vol. 2. London: Selden Society, 1978.

Barker, Peter. "Jean Pena (1528–58) and Stoic Physics in the Sixteenth Century." *Southern Journal of Philosophy* 23 (1985): 93–107.

————. "Stoic Contributions to Early Modern Science." In *Atoms, "Pneuma," and Tranquility,* edited by Margaret J. Osler, 135–54. Cambridge: Cambridge University Press, 1991.

Barker, Peter, and Bernard R. Goldstein. "Is Seventeenth-Century Physics Indebted to the Stoics?" *Centaurus* 27 (1984): 148–64.

Barnes, Barry. *Scientific Knowledge and Social Imagery.* London: Routledge and Kegan Paul, 1974.

Barnes, Jonathan. "Aristotle's Theory of Demonstration." *Phronesis* 14 (1969): 123–52.

Baumgartner, Frederic J. "Galileo's French Correspondents." *Annals of Science* 45 (1988): 169–82.

Berman, Harold J. *Law and Revolution: The Formation of the Western Legal Tradition.* Cambridge: Cambridge University Press, 1983.

Birch, Thomas. *History of the Royal Society of London.* 1756–57. Reprint, New York: Johnson Reprint Corp., 1968.

Blackstone, William. *Commentaries on the Laws of England.* Vol. 1. 1765. Reprint, Chicago: University of Chicago Press, 1979.

Blackwell, Richard J. "Descartes' Concept of Matter." In *The Concept of Matter in Modern Philosophy,* edited by Ernan McMullin, 59–75. Notre Dame, Ind.: University of Notre Dame Press, 1978.

Blake, Ralph M. "The Role of Experience in Descartes' Theory of Method." In *Theories of Scientific Method,* edited by Ralph M. Blake, Curt J. Ducasse, and Edward H. Madden, 75–103. Seattle: University of Washington Press, 1960.

Bloor, David. "Durkheim and Mauss Revisited: Classification and the Sociology of Knowledge." *Studies in History and Philosophy of Science* 13 (1982): 267–97.

———. *Knowledge and Social Imagery.* London: Routledge and Kegan Paul. 1976.

———. "Left and Right Wittgensteinians." In *Science as Practice and Culture,* edited by Andrew Pickering, 266–82. Chicago: University of Chicago Press, 1992.

———. "Reply to J. W. Smith." *Studies in History and Philosophy of Science* 15 (1984): 245–49.

———. "The Sociology of Reasons; or Why 'Epistemic Factors' Are Really 'Social Factors.'" In *Scientific Rationality: The Sociological Turn,* edited by J. R. Brown, 295–324. Dordrecht: D. Reidel, 1984.

———. *Wittgenstein: A Social Theory of Knowledge.* New York: Columbia University Press, 1983.

Boyd, Richard N. "Lex orandi est lex credendi." In *Images of Science,* edited by Paul M. Churchland and Clifford A. Hooker, 3–34. Chicago: University of Chicago Press, 1985.

Boyle, Robert. *The Early Essays and Ethics of Robert Boyle.* Edited by John Harwood. Carbondale: Southern Illinois University Press, 1991.

———. *Letters and Papers of Robert Boyle.* Bethesda, Md.: University Publications of America, 1991. Microfilm.

———. *The Works of the Honourable Robert Boyle.* 6 vols. Edited by Thomas Birch. 1772. Reprint, Hildesheim: Georg Olms, 1965–66.

Brett, G. S. *The Philosophy of Gassendi.* London: Macmillan, 1908.

Briggs, John C. *Francis Bacon and the Rhetoric of Nature.* Cambridge: Harvard University Press, 1989.

Brown, Theodore. "The College of Physicians and the Acceptance of Iatro-Mechanism in England, 1665–1695." *Bulletin of the History of Medicine* 44 (1970): 12–30.

Butterfield, Herbert. *The Origins of Modern Science.* New York: Free Press, 1957.

Butts, Robert E. "Consilience of Inductions and the Problem of Conceptual Change in Science." In *Logic, Laws, and Life: Some Philosophical Complications,* edited by R. G. Colodny, 71–88. Pittsburgh: University of Pittsburgh Press, 1977.

———. "Whewell's Logic of Induction." In *Foundations of Scientific Method: The Nineteenth Century,* edited by Ronald N. Giere and Richard S. Westfall, 53–85. Bloomington: Indiana University Press, 1973.

Callon, Michel, and Bruno Latour. "Don't Throw the Baby Out with the Bath School! A Reply to Collins and Yearley." In *Science as Practice and Culture,* edited by Andrew Pickering, 343–68. Chicago: University of Chicago Press, 1992.

Campbell, Mary B. *The Witness and the Other World,* 400–1600. Ithaca, N.Y.: Cornell University Press, 1989.

Cantor, Geoffrey. "The Rhetoric of Experiment," in *The Uses of Experiment*, edited by David Gooding, Trevor Pinch, and Simon Schaffer, 159–80. Cambridge: Cambridge University Press, 1989.

Cardwell, Kenneth William. "Francis Bacon, Inquisitor." In *Francis Bacon's Legacy of Texts*, edited by William A. Sessions, 268–89. New York: AMS Press, 1990.

Carr, Thomas M., Jr. *Descartes and the Resilience of Rhetoric*. Carbondale: Southern Illinois University Press, 1990.

Cartwright, Nancy. *How the Laws of Physics Lie*. Oxford: Clarendon Press, 1983.

Charleton, Walter. *Physiologia Epicuro-Gassendo-Charltoniana*. 1654. Reprint, New York: Johnson Reprint Corp., 1966.

Churchland, Paul M., and Clifford A. Hooker, eds. *Images of Science*. Chicago: University of Chicago Press, 1985.

Clark, Sir George. *A History of the Royal College of Physicians of London*. Vol. 1. Oxford: Clarendon Press, 1964.

Clarke, Desmond. *Descartes' Philosophy of Science*. University Park: Pennsylvania State University Press, 1982.

———. "Descartes' Philosophy of Science and the Scientific Revolution." In *The Cambridge Companion to Descartes*, edited by John Cottingham, 258–85. Cambridge: Cambridge University Press, 1992.

Clericuzio, Antonio. "Carneades and the Chemists: A Study of the *Sceptical Chymist* and Its Impact on Seventeenth-Century Chemistry." In *Robert Boyle Reconsidered*, edited by Michael Hunter, 79–90. Cambridge: Cambridge University Press, 1994.

———. "A Redefinition of Boyle's Chemistry and Corpuscular Philosophy." *Annals of Science* 47 (1990): 561–89.

Cohen, I. Bernard. *The Newtonian Revolution*. Cambridge: Cambridge University Press, 1983.

———, ed. *Puritanism and the Rise of Modern Science*. New Brunswick, N.J.: Rutgers University Press, 1990.

Collingwood, Robin George. *The Idea of History*. Oxford: Clarendon Press, 1946.

Collins, H. M. *Changing Order: Replication and Induction in Scientific Practice*. Beverly Hills, Calif.: Sage, 1985.

———. "Knowledge, Norms, and Rules in the Sociology of Science." *Social Studies of Science* 12 (1982): 299–309.

Collins, H. M., and Steven Yearley. "Epistemological Chicken." In *Science as Practice and Culture*, edited by Andrew Pickering, 301–26. Chicago: University of Chicago Press, 1992.

———. "Journey into Space." In *Science as Practice and Culture*, edited by Andrew Pickering, 369–89. Chicago: University of Chicago Press, 1992.

Conant, James B. "Robert Boyle's Experiments in Pneumatics." In *Harvard Case Histories in Experimental Science*, vol. 1, edited by James B. Conant, 1–63. Cambridge: Harvard University Press, 1970.

Cook, Harold J. "The New Philosophy and Medicine in Seventeenth-Century England." In *Reappraisals of the Scientific Revolution*, edited by David C. Lindberg and Robert S. Westman, 397–436. Cambridge: Cambridge University Press, 1990.

Cottingham, John. Introduction to *The Cambridge Companion to Descartes*, edited by John Cottingham, 1–20. Cambridge: Cambridge University Press, 1992.

Daston, Lorraine. *Classical Probability in the Enlightenment.* Princeton, N.J.: Princeton University Press, 1988.

———. "The Factual Sensibility." *Isis* 79 (1988): 452–67.

Dear, Peter. "Jesuit Mathematical Science and the Reconstitution of Experience in the Early Seventeenth Century." *Studies in History and Philosophy of Science* 18 (1987): 133–75.

———. *Mersenne and the Learning of the Schools.* Ithaca, N.Y.: Cornell University Press, 1988.

———. "Miracles, Experiments, and the Ordinary Course of Nature." *Isis* 81 (1990): 663–83.

———. "Narratives, Anecdotes, and Experiments: Turning Experience into Science in the Seventeenth Century." In *The Literary Structure of Scientific Argument,* edited by Peter Dear, 135–63. Philadelphia: University of Pennsylvania Press, 1991.

———. "*Totius in verba:* Rhetoric and Authority in the Early Royal Society." *Isis* 76 (1985): 145–61.

———, ed. *The Literary Structure of Scientific Argument.* Philadelphia: University of Pennsylvania Press, 1991.

Deason, Gary B. "Reformation Theology and the Mechanistic Conception of Nature." In *God and Nature,* edited by David C. Lindberg and Ronald L. Numbers, 167–91. Berkeley and Los Angeles: University of California Press, 1986.

Debus, Allen G. *The English Paracelsians.* London: Oldbourne Book Co., 1965.

Delamont, Sara. "Three Blind Spots? A Comment on the Sociology of Science by a Puzzled Outsider." *Social Studies of Science* 17 (1987): 163–70.

Descartes, René. *Descartes: Philosophical Letters.* Edited and translated by Anthony Kenny. Oxford: Clarendon Press, 1970.

———. *The Philosophical Writings of Descartes.* 3 vols. Translated by John Cottingham, Robert Stoothoff, and Dugald Murdoch (vol. 3 only: also Anthony Kenny). Cambridge: Cambridge University Press, 1985–91.

Dewhurst, Kenneth. *Dr. Thomas Sydenham (1624–1689): His Life and Original Writings.* Berkeley and Los Angeles: University of California Press, 1966.

———. *John Locke (1632–1704): Physician and Philosopher.* London: Wellcome Historical Medical Library, 1963.

Dijksterhuis, E. J. *The Mechanization of the World Picture.* Oxford: Clarendon Press, 1961.

Dobbs, B. J. T. *The Foundations of Newton's Alchemy.* Cambridge: Cambridge University Press, 1975.

Doty, Ralph. "Carneades: A Forerunner of William James's Pragmatism." *Journal of the History of Ideas* 47 (1986): 133–38.

Drake, Stillman. *Cause, Experiment, and Science.* Chicago: University of Chicago Press, 1981.

Droysen, Johann Gustav. *Outline of the Principles of History.* Translation of *Grundriss der Historik.* Translated by E. B. Andrews. 1893. Reprint, New York: Fertig, 1967.

Ducasse, Curt J. "Francis Bacon's Philosophy of Science." In *Theories of Scientific Method,* edited by Ralph M. Blake, Curt J. Ducasse, and Edward H. Madden, 50–74. Seattle: University of Washington Press, 1960.

Duhem, Pierre. *The Aim and Structure of Physical Theory.* Translated by Philip P. Wiener. New York: Atheneum, 1981.

Ellis, Brian. "What Science Aims to Do." In *Images of Science*, edited by Paul M. Churchland and Clifford A. Hooker, 48–74. Chicago: University of Chicago Press, 1985.

Emerton, Norma E. *The Scientific Reinterpretation of Form.* Ithaca, N.Y.: Cornell University Press, 1984.

Farr, James. "The Way of Hypotheses: Locke on Method." *Journal of the History of Ideas* 48 (1987): 51–67.

Federman, Reinhard. *The Royal Art of Alchemy.* Translated by Richard H. Weber. Philadelphia: Chilton Book Co., 1969.

Feingold, Mordechai. "The Universities and the Scientific Revolution: The Case of England." In *New Trends in the History of Science*, edited by R. P. W. Visser, H. J. M. Bos, L. C. Palm, and H. A. M. Snelders, 29–52. Amsterdam: Rodopi, 1989.

Ferrerira, James. "Locke's 'Constructive Skepticism'—A Reappraisal." *Journal of the History of Philosophy* 24 (1986): 211–29.

Feyerabend, Paul. *Against Method.* London: New Left Books, 1975.

Fine, Arthur. "And Not Anti-realism Either." *Nous* 18 (1984): 51–65.

———. "Science Made Up: Constructivist Sociology of Scientific Knowledge." In *Disunity and Contextualism in the Philosophy of Science*, edited by Peter Galison and David Stump, 31–52. Stanford, Calif.: Stanford University Press, 1992.

———. *The Shaky Game: Einstein, Realism, and the Quantum Theory.* Chicago: University of Chicago Press, 1986.

Fisch, Menachem. "Necessity and Contingent Truth in William Whewell's Antithetical Theory of Knowledge." *Studies in History and Philosophy of Science* 16 (1985): 275–314.

———. "Whewell's Consilience of Inductions—An Evaluation." *Philosophy of Science* 52 (1985): 239–55.

Frank, Robert G., Jr. *Harvey and the Oxford Physiologists.* Berkeley and Los Angeles: University of California Press, 1980.

Franklin, Allan. "The Epistemology of Experiment." *British Journal for the Philosophy of Science* 35 (1984): 381–90.

———. *Experiment, Right or Wrong.* Cambridge: Cambridge University Press, 1990.

———. "Experimental Questions." *Perspectives on Science* 1 (1993): 127–46.

———. *The Neglect of Experiment.* Cambridge: Cambridge University Press, 1986.

———. Review of *Experiment and the Making of Meaning*, by David Gooding. *Isis* 83 (1992): 177–78.

Fraser, Antonia. *The Weaker Vessel.* New York: Knopf, 1984.

Fuller, Steve. "David Bloor's *Knowledge and Social Imagery* (2nd Edition)." *Philosophy of Science* 60 (1993): 158–70.

———. Review of *Science as Social Knowledge: Values and Objectivity in Scientific Inquiry*, by Helen E. Longino. *Philosophy of Science* 60 (1993): 361–62.

Fulton, John. *A Bibliography of Robert Boyle.* Oxford: Oxford University Press, 1961.

———. "Robert Boyle and His Influence on Thought in the Seventeenth Century." *Isis* 18 (1932): 77–102.

Gadamer, Hans-Georg. *Truth and Method.* Translation of *Wahrheit und Methode.* New York: Crossroads Publishing, 1975.

Galen. *On Anatomical Procedures.* Translated by Charles Singer. London: Oxford University Press, 1956.

———. *On the Usefulness of the Parts of the Body.* Translated by Margaret Tallmadge May. Ithaca, N.Y.: Cornell University Press, 1968.

Galilei, Galileo. *Dialogues Concerning Two New Sciences.* Translated by Henry Crew and Alfonso de Salvio. New York: Dover Publications, 1954.

Galison, Peter. "Experimental Probability." *Isis* 79 (1988): 467–70.

———. *How Experiments End.* Chicago: University of Chicago Press, 1987.

Garber, Daniel. *Descartes' Metaphysical Physics.* Chicago: University of Chicago Press, 1992.

———. "Descartes' Physics." In *The Cambridge Companion to Descartes,* edited by John Cottingham, 286–334. Cambridge: Cambridge University Press, 1992.

———. "Science and Certainty in Descartes." In *Descartes: Critical and Interpretive Essays,* edited by Michael Hooker, 114–51. Baltimore: Johns Hopkins University Press, 1978.

Gassendi, Pierre. *The Selected Works of Pierre Gassendi.* Edited by Craig Brush. New York: Johnson Reprint Corp., 1972.

Giere, Ronald N. "Constructive Realism." In *Images of Science,* edited by Paul M. Churchland and Clifford A. Hooker, 75–98. Chicago: University of Chicago Press, 1985.

———. *Explaining Science: A Cognitive Approach.* Chicago: University of Chicago Press, 1988.

———. "Foundations of Probability and Statistical Inference." In *Current Research in Philosophy of Science,* edited by Peter D. Asquith and Henry E. Kyburg Jr., 503–33. East Lansing, Mich.: Philosophy of Science Association, 1979.

Gieryn, Thomas F. "Relativist/Constructivist Programmes in the Sociology of Science: Redundance and Retreat." *Social Studies of Science* 12 (1982): 279–97.

Gilbert, Neal. *Renaissance Concepts of Method.* New York: Columbia University Press, 1960.

Glymour, Clark. "Explanation and Realism." In *Scientific Realism,* edited by Jarrett Leplin, 173–92. Berkeley and Los Angeles: University of California Press, 1984.

———. *Theory and Evidence.* Princeton, N.J.: Princeton University Press, 1980.

Golinski, Jan V. "Chemistry in the Scientific Revolution: Problems of Language and Communication." In *Reappraisals of the Scientific Revolution,* edited by David C. Lindberg and Robert S. Westman, 367–96. Cambridge: Cambridge University Press, 1990.

———. "A Noble Spectacle: Phosphorus and the Public Culture of Science in the Early Royal Society." *Isis* 80 (1989): 11–39.

Gooding, David. *Experiment and the Making of Meaning.* Boston: Kluwer, 1990.

———. "Putting Agency Back into Experiment." In *Science as Practice and Culture,* edited by Andrew Pickering, 65–112. Chicago: University of Chicago Press, 1992.

Grant, Edward. "Science and Theology in the Middle Ages." In *God and Nature,* edited by David C. Lindberg and Ronald L. Numbers, 49–75. Berkeley and Los Angeles: University of California Press, 1986.

Green, Thomas Andrew. *Verdict according to Conscience.* Chicago: University of Chicago Press, 1985.

Guerlac, Henry. "Can We Date Newton's Early Optical Experiments?" *Isis* 74 (1983): 74–80.

Guerrini, Anita. "The Ethics of Animal Experimentation in Seventeenth-Century England." *Journal of the History of Ideas* 50 (1989): 391–407.

Gutting, Gary. "The Logic of Invention." In *Scientific Discovery, Logic, and Rationality*, edited by Thomas Nickles, 221–34. Dordrecht: D. Reidel, 1980.

Hacking, Ian. "Do We See through a Microscope?" In *Images of Science*, edited by Paul M. Churchland and Clifford A. Hooker, 132–52. Chicago: University of Chicago Press, 1985.

———. *The Emergence of Probability*. Cambridge: Cambridge University Press, 1975.

———. *Representing and Intervening*. Cambridge: Cambridge University Press, 1983.

———. Review of *The Neglect of Experiment*, by Allan Franklin. *Philosophy of Science* 55 (1988): 306–8.

———. "The Self-Vindication of the Laboratory Sciences." In *Science as Practice and Culture*, edited by Andrew Pickering, 29–64. Chicago: University of Chicago Press, 1992.

———. "Statistical Language, Statistical Truth, and Statistical Reason." In *The Social Dimensions of Science*, edited by Ernan McMullin, 130–57. Notre Dame, Ind.: University of Notre Dame Press, 1992.

Hadden, Richard W. "Social Relations and the Content of Early Modern Science." *British Journal of Sociology* 39 (1988): 255–80.

Haldane, Elizabeth S., and G. R. T. Ross, trans. *The Philosophical Works of Descartes*. Vol. 1. Cambridge: Cambridge University Press, 1931.

Hall, A. Rupert. "Galileo's Thought and the History of Science." In *Galileo, Man of Science*, edited by Ernan McMullin, 67–81. New York: Basic Books, 1967.

———. *Henry More: Magic, Religion, and Experiment*. Oxford: Blackwell, 1990.

———. *The Scientific Revolution*. 2d ed. Boston: Beacon Press, 1983.

Hall, A. Rupert, and Marie Boas Hall. "Philosophy and Natural Philosophy: Boyle and Spinoza." In *Mélanges Alexandre Koyré* 2: 240–56. Paris: Harman, 1964.

Hall, Marie Boas. "Boyle as a Theoretical Scientist." *Isis* 46 (1950): 261–68.

———. "Galileo's Influence on Seventeenth-Century English Scientists." In *Galileo, Man of Science*, edited by Ernan McMullin, 405–14. New York: Basic Books, 1967.

———. Introduction to *Experiments and Considerations Touching Colours*, by Robert Boyle, vii–xxvi. New York: Johnson Reprint Corp., 1964.

———. "Matter in Seventeenth-Century Science." In *The Concept of Matter in Modern Philosophy*, edited by Ernan McMullin, 76–99. Notre Dame, Ind.: University of Notre Dame Press, 1978.

———. *Robert Boyle and Seventeenth-Century Chemistry*. Cambridge: Cambridge University Press, 1958.

———. *Robert Boyle on Natural Philosophy*. Bloomington: Indiana University Press, 1965.

Hall, Thomas S. *History of General Physiology*. Vol. 1. Chicago: University of Chicago Press, 1975.

Hamlyn, D. W. "Aristotelian Epagoge." *Phronesis* 21 (1976): 167–84.

Hanson, Norwood Russell. *Patterns of Discovery*. Cambridge: Cambridge University Press, 1958.

Harding, Sandra. *The Science Question in Feminism*. Ithaca, N.Y.: Cornell University Press, 1986.

Harding, Sandra, and Jean F. O'Barr, eds. *Sex and Scientific Inquiry*. Chicago: University of Chicago Press, 1987.

Harvey, William. *The Works of William Harvey*. Translated by Robert Willis. 1847. Reprint, Annapolis, Md.: St. John's College Press, 1949.

Harwood, John. Introduction to *The Early Essays and Ethics of Robert Boyle*, edited by John Harwood, xv–lxix. Carbondale: Southern Illinois University Press, 1991.

———. "Science Writing and Writing Science: Boyle and Rhetorical Theory." In *Robert Boyle Reconsidered*, edited by Michael Hunter, 37–56. Cambridge: Cambridge University Press, 1994.

Hatfield, Gary. "Metaphysics and the New Science." In *Reappraisals of the Scientific Revolution*, edited by David C. Lindberg and Robert S. Westman, 93–162. Cambridge: Cambridge University Press, 1990.

Heelan, Patrick. "Natural Science as a Hermeneutic of Instrumentation." *Philosophy of Science* 50 (1983): 181–204.

Heimann, P. M., and J. E. McGuire. "Newtonian Forces and Lockean Powers: Concepts of Matter in Eighteenth-Century Thought." In *Historical Studies in the Physical Sciences* 3: 233–306. Philadelphia: University of Pennsylvania Press, 1971.

Hempel, Carl G. *Aspects of Scientific Explanation*. New York: Free Press, 1965.

Hendry, John. "Newton's Theory of Colour." *Centaurus* 23 (1980): 230–51.

Henry, John. "Boyle and Cosmical Qualities." In *Robert Boyle Reconsidered*, edited by Michael Hunter, 119–38. Cambridge: Cambridge University Press, 1994.

———. "Medicine and Pneumatology: Henry More, Richard Baxter, and Francis Glisson's *Treatise on the Energetic Nature of Substance*." *Medical History* 31 (1987): 15–40.

———. "Occult Qualities and the Experimental Philosophy: Active Principles in Pre-Newtonian Matter Theory." *History of Science* 24 (1986): 335–81.

Hesse, Mary. "Francis Bacon." In *A Critical History of Western Philosophy*, edited by D. J. O'Connor, 141–52. New York: Free Press, 1964.

———. *Models and Analogies in Science*. London: Sheed and Ward, 1963.

———. *The Structure of Scientific Inference*. Berkeley and Los Angeles: University of California Press, 1974.

Heward, Edmund. *Matthew Hale*. London: Robert Hale and Co., 1972.

Hill, Christopher. *Change and Continuity in Seventeenth-Century England*. Cambridge: Harvard University Press, 1975.

———. "'A New Kind of Clergy': Ideology and the Experimental Method." *Social Studies of Science* 16 (1986): 726–35.

———. *The World Turned Upside Down: Radical Ideas during the Scientific Revolution*. Harmondsworth, Eng.: Penguin, 1975.

Hobbes, Thomas. *The English Works of Thomas Hobbes*. 11 vols. Edited by William Molesworth. London: John Bohn, 1839–45.

———. *Physical Dialogue*. Translation of *Dialogus physicus de natura aeris*. Translated by Simon Schaffer. In *Leviathan and the Air-Pump*, by Steven Shapin and Simon Schaffer, 345–91. Princeton, N.J.: Princeton University Press, 1985.

Holdsworth, William. *A History of English Law*. 4th ed. 12 vols. London: Methuen, 1936.

Holmes, Frederic L. "Do We Understand Historically How Experimental Knowledge Is Acquired?" *History of Science* 30 (1992): 119–36.

Hooke, Robert. *Micrographia*. London: J. Martyn and J. Allestry, 1665.

————. *The Posthumous Works.* 3 vols. Edited by Richard Waller. 1705. Reprint, Hildesheim: Georg Olms, 1970.

Horton, Mary. "In Defence of Francis Bacon." *Studies in History and Philosophy of Science* 4 (1973): 241–78.

Howell, T. B., comp. *State Trials.* Vol. 4. London: Longman, 1816.

Hull, David L. "In Defense of Presentism." *History and Theory* 18 (1979): 1–15.

————. *Science as Process: An Evolutionary Account of the Social and Conceptual Development of Science.* Chicago: University of Chicago Press, 1988.

Hunter, Michael. "Alchemy, Magic, and Moralism in the Thought of Robert Boyle." *British Journal for the History of Science* 23 (1990): 387–410.

————. "The Cabinet Institutionalized: The Royal Society's 'Repository' and Its Background." In *The Origins of Museums,* edited by Oliver Impey and Arthur MacGregor, 159–67. Oxford: Clarendon Press, 1985.

————. "Casuistry in Action: Robert Boyle's Confessional Interviews with Gilbert Burnet and Edward Stillingfleet, 1691." *Journal of Ecclesiastical History* 44 (1993): 80–98.

————. "The Conscience of Robert Boyle: Functionalism, 'Dysfunctionalism,' and the Task of Historical Understanding." In *Renaissance and Revolution: Humanists, Scholars, Craftsmen, and Natural Philosophers in Early Modern Europe,* edited by J. V. Field and Frank A. J. L. James, 147–59. Cambridge: Cambridge University Press, 1993.

————. Introduction to *Robert Boyle Reconsidered,* edited by Michael Hunter, 1–18. Cambridge: Cambridge University Press, 1994.

————. *Letters and Papers of Robert Boyle: A Guide to the Manuscripts and Microfilm.* Bethesda, Md.: University Publications of America, 1991.

————. "Promoting the New Science: Henry Oldenburg and the Early Royal Society." *History of Science* 26 (1988): 165–81.

————. "Science and Heterodoxy: An Early Modern Problem Reconsidered." In *Reappraisals of the Scientific Revolution,* edited by David C. Lindberg and Robert S. Westman, 437–68. Cambridge: Cambridge University Press, 1990.

————. *Science and Society in Restoration England.* Cambridge: Cambridge University Press, 1981.

Hunter, Michael, and Simon Schaffer, eds. *Robert Hooke: New Studies.* Woodbridge, Eng.: Boydell Press, 1989.

Hunter, Michael, and Paul B. Wood. "Towards Solomon's House: Rival Strategies for Reforming the Early Royal Society." *History of Science* 24 (1986): 49–108.

Hunter, Richard, and Ida Macalpine. "William Harvey and Robert Boyle." *Notes and Records of the Royal Society of London* 13 (1958): 115–27.

Hutchison, Keith. "Supernaturalism and the Mechanical Philosophy." *History of Science* 21 (1983): 297–333.

Impey, Oliver, and Arthur MacGregor. *The Origins of Museums.* Oxford: Clarendon Press, 1985.

Ishiguro, Hidé. "Pre-established Harmony versus Constant Conjunction: A Reconsideration of the Distinction between Rationalism and Empiricism." In *Rationalism, Empiricism, and Idealism,* edited by Anthony Kenny, 61–85. Oxford: Clarendon Press, 1986.

Ives, E. W. *The Common Lawyers of Pre-Reformation England.* Cambridge: Cambridge University Press, 1983.

Jacob, James R. "Boyle's Atomism and the Restoration Assault on Pagan Naturalism." *Social Studies of Science* 8 (1978): 211–33.

—————. "Boyle's Circle in the Protectorate: Revelation, Politics, and the Millennium." *Journal of the History of Ideas* 38 (1977): 131–40.

—————. "The Ideological Origins of Robert Boyle's Natural Philosophy." *Journal of European Studies* 2 (1972): 1–21.

—————. *Robert Boyle and the English Revolution.* New York: Burt Franklin, 1977.

Jacob, James R., and Margaret C. Jacob. "The Anglican Origins of Modern Science: The Metaphysical Foundations of the Whig Constitution." *Isis* 71 (1980): 251–67.

Jacob, Margaret C. *The Newtonians and the English Revolution.* Ithaca, N.Y.: Cornell University Press, 1976.

Jardine, Lisa. "*Experientia literata* or *Novum organum?* The Dilemma of Bacon's Scientific Method." In *Francis Bacon's Legacy of Texts,* edited by William A. Sessions, 47–67. New York: AMS Press, 1990.

—————. *Francis Bacon: Discovery and the Art of Discourse.* Cambridge: Cambridge University Press, 1974.

Jones, Roger. "Realism about What?" *Philosophy of Science* 58 (1991): 185–202.

Joy, Lynn. *Gassendi the Atomist.* Cambridge: Cambridge University Press, 1987.

Kargon, Robert Hugh. *Atomism in England from Hariot to Newton.* Oxford: Clarendon Press, 1966.

—————. "Walter Charleton, Robert Boyle, and the Acceptance of Epicurean Atomism in England." *Isis* 55 (1964): 184–92.

Keller, Evelyn Fox. *Reflections on Gender and Science.* New Haven, Conn.: Yale University Press, 1985.

Kenny, Anthony. Introduction to *Rationalism, Empiricism, and Idealism,* edited by Anthony Kenny, 1–5. Oxford: Clarendon Press, 1986.

Kitcher, Philip. "Persuasion." In *Persuading Science: The Art of Scientific Rhetoric,* edited by Marcello Pera and William R. Shea, 3–28. Canton, Mass.: Science History Publications, 1991.

Klaaren, Eugene M. *Religious Origins of Modern Science.* Grand Rapids, Mich.: Eerdmans, 1977.

Knorr-Cetina, Karin D. "The Constructivist Programme in the Sociology of Science: Retreats or Advances?" *Social Studies of Science* 12 (1982): 320–24.

—————. *The Manufacture of Knowledge.* Oxford: Pergamon Press, 1981.

Kocher, Paul H. "Francis Bacon on the Science of Jurisprudence." *Journal of the History of Ideas* 18 (1957): 3–26.

Kockelmans, Joseph J. "On the Problem of Truth in the Sciences." *Proceedings and Addresses of the American Philosophical Association* 61 (1987): 5–26.

Kroll, Richard W. F. "The Question of Locke's Relation to Gassendi." *Journal of the History of Ideas* 45 (1984): 339–60.

Krook, Dorothea. "Two Baconians: Robert Boyle and Joseph Glanvil." *Huntington Library Quarterly* 18 (1950): 261–78.

Kuhn, Thomas S. "Robert Boyle and Structural Chemistry in the Seventeenth Century." *Isis* 43 (1952): 12–36.

—————. *The Structure of Scientific Revolutions.* 2d ed. Chicago: University of Chicago Press, 1970.

Lakatos, Imre. "Falsification and the Methodology of Scientific Research Programmes." In *Criticism and the Growth of Knowledge,* edited by Imre Lakatos and Alan Musgrave, 91–196. Cambridge: Cambridge University Press, 1974.

Langbein, John H. *Prosecuting Crime in the Renaissance.* Cambridge: Harvard University Press, 1974.

Latour, Bruno. "One More Turn after the Social Turn . . ." In *The Social Dimensions of Science,* edited by Ernan McMullin, 272–94. Notre Dame, Ind.: University of Notre Dame Press, 1992.

———. "Postmodern? No, Simply Amodern! Steps Towards an Anthropology of Science." *Studies in History and Philosophy of Science* 21 (1990): 145–71.

Latour, Bruno, and Steve Woolgar. *Laboratory Life: The Social Construction of Scientific Facts.* Princeton, N.J.: Princeton University Press, 1979.

Laudan, Larry. "The Clock Metaphor and Hypotheses: The Impact of Descartes on English Methodological Thought." In *Science and Hypothesis: Historical Essays on Scientific Methodology,* by Larry Laudan, 27–58. Dordrecht: D. Reidel, 1981.

———. "The Clock Metaphor and Probabilism: The Impact of Descartes on English Methodological Thought, 1650–65." *Annals of Science* 22 (1966): 73–104.

———. "A Confutation of Convergent Realism." *Philosophy of Science* 48 (1981): 19–49.

———. "Explaining the Success of Science: Beyond Realism and Relativism." In *Science and Reality,* edited by J. Cushing, C. Delaney, and Gary Gutting, 83–105. Notre Dame, Ind.: University of Notre Dame Press, 1984.

———. "Problems, Truth, and Consistency." *Studies in History and Philosophy of Science* 13 (1982): 73–80.

———. *Progress and Its Problems.* Berkeley and Los Angeles: University of California Press, 1977.

———. "Progress or Rationality? The Prospects for Normative Naturalism." *American Philosophical Quarterly* 24 (1987): 19–31.

———. *Science and Values.* Berkeley and Los Angeles: University of California Press, 1984.

Leibniz, Gottfried Wilhelm. *Philosophical Essays.* Edited and translated by Roger Ariew and Daniel Garber. Indianapolis: Hackett Publishing, 1989.

Lennox, James G. "Boyle's Defense of Teleological Inference." *Isis* 74 (1983): 38–52.

Lenoir, Timothy. "Practical Reason and the Construction of Knowledge." In *The Social Dimensions of Science,* edited by Ernan McMullin, 158–97. Notre Dame, Ind.: University of Notre Dame Press, 1992.

Leplin, Jarrett. "Methodological Realism and Scientific Rationality." *Philosophy of Science* 53 (1986): 31–51.

Levack, Brian P. *The Civil Lawyers in England, 1603–1641.* Oxford: Clarendon Press, 1973.

Lindberg, David C. "Science and the Early Church." In *God and Nature,* edited by David C. Lindberg and Ronald L. Numbers, 19–48. Berkeley and Los Angeles: University of California Press, 1986.

Lloyd, G. E. R. *Magic, Reason, and Experience.* Cambridge: Cambridge University Press, 1979.

———. *Science, Folklore, and Ideology.* Cambridge: Cambridge University Press, 1983.

Locke, John. *The Correspondence of John Locke.* 8 vols. Edited by E. S. DeBeer. Oxford: Clarendon Press, 1976.

———. *An Essay Concerning Human Understanding.* Edited by Peter H. Nidditch. Oxford: Clarendon Press, 1979.

Loemker, Leroy E. "Boyle and Leibniz." *Journal of the History of Ideas* 16 (1955): 22–43.

Lohne, J. A. "Isaac Newton: The Rise of a Scientist, 1661–1671." *Notes and Records of the Royal Society* 20 (1965): 125–39.

Lower, Richard. *Tractatus de corde* (1669). Translated by K. J. Franklin. *Early Science in Oxford,* edited by R. T. Gunther, vol. 9. London: Dawson's, 1968.

Lynch, Michael. *Art and Artifact in Laboratory Science: Shop Work and Shop Talk in a Research Laboratory.* London: Routledge and Kegan Paul, 1985.

———. "Extending Wittgenstein: The Pivotal Move from Epistemology to the Sociology of Science." In *Science as Practice and Culture,* edited by Andrew Pickering, 215–65. Chicago: University of Chicago Press, 1992.

———. "From the 'Will to Theory' to the Discursive Collage: A Reply to Bloor's 'Left and Right Wittgensteinians.'" In *Science as Practice and Culture,* edited by Andrew Pickering, 283–300. Chicago: University of Chicago Press, 1992.

MacGregor, Arthur. "The Cabinet of Curiosities in Seventeenth-Century Britain." In *The Origins of Museums,* edited by Oliver Impey and Arthur MacGregor, 147–58. Oxford: Clarendon Press, 1985.

Machamer, Peter. "The Person-Centered Rhetoric of Seventeenth-Century Science." In *Persuading Science: The Art of Scientific Rhetoric,* edited by Marcello Pera and William R. Shea, 143–56. Canton, Mass.: Science History Publications, 1991.

Macintosh, J. J. "Robert Boyle on Epicurean Atheism and Atomism." In *Atoms, "Pneuma," and Tranquility,* edited by Margaret J. Osler, 197–219. Cambridge: Cambridge University Press, 1991.

Madden, Edward H. "Thomas Hobbes and the Rationalist Ideal." In *Theories of Scientific Method,* edited by Ralph M. Blake, Curt J. Ducasse, and Edward H. Madden, 104–18. Seattle: University of Washington Press, 1960.

Maddison, R. E. W. "The Earliest Writing of Robert Boyle." *Annals of Science* 17 (1961): 165–73.

———. "The Life of Robert Boyle: Addenda." *Annals of Science* 45 (1988): 193–95.

———. *The Life of the Honourable Robert Boyle, F.R.S.* New York: Barnes and Noble, 1969.

———. "The Portraiture of the Honourable Robert Boyle." *Annals of Science* 15 (1959): 141–214.

———. "Robert Boyle's Library." *Nature* 165 (1950): 981–82.

———. "Studies in the Life of Robert Boyle, F.R.S. Part I: Robert Boyle and Some of His Foreign Visitors." *Notes and Records of the Royal Society* 9 (1951): 1–35.

———. "Studies in the Life of Robert Boyle, F.R.S. Part II: Salt Water Freshened." *Notes and Records of the Royal Society* 9 (1952): 196–216.

———. "Studies in the Life of Robert Boyle, F.R.S. Part III: The Charitable Disposal of Robert Boyle's Residuary Estate." *Notes and Records of the Royal Society* 10 (1952): 15–27.

———. "Studies in the Life of Robert Boyle, F.R.S. Part IV: Robert Boyle and Some of His Foreign Visitors." *Notes and Records of the Royal Society* 11 (1954): 38–53.

———. "Studies in the Life of Robert Boyle, F.R.S. Part V: Boyle's Operator: Ambrose Godfrey Hanckwitz, F.R.S." *Notes and Records of the Royal Society* 11 (1955): 159–88.

———. "Studies in the Life of Robert Boyle, F.R.S. Part VI: The Stalbridge Period,

1645–55, and the Invisible College." *Notes and Records of the Royal Society* 18 (1963): 104–24.

———. "Studies in the Life of Robert Boyle, F.R.S. Part VII: The Grand Tour." *Notes and Records of the Royal Society* 20 (1965): 51–77.

Malcolm, Noel. "Hobbes and the Royal Society." In *Perspectives on Thomas Hobbes*, edited by G. A. J. Rogers and Alan Ryan, 43–66. Oxford: Clarendon Press, 1988.

Mamiani, Maurizio. "The Rhetoric of Certainty in Newton's Science." In *Persuading Science: The Art of Scientific Rhetoric*, edited by Marcello Pera and William R. Shea, 157–72. Canton, Mass.: Science History Publications, 1991.

Mancosu, Paolo, and Ezio Vailati. "Torricelli's Infinitely Long Solid and Its Philosophical Reception in the Seventeenth Century." *Isis* 82 (1991): 50–70.

Mandelbaum, Maurice. *Philosophy, Science, and Sense Perception: Historical and Critical Studies*. Baltimore: Johns Hopkins University Press, 1964.

Marek, Jiri. "Newton's Report and Its Relation to Results of His Predecessors." *Physis* 11 (1969): 390–407.

Martin, Gottfried. *Leibniz: Logic and Metaphysics*. Translated by K. J. Northcott and P. G. Lucas. Manchester: Manchester University Press, 1964.

Martin, Julian. *Francis Bacon, the State, and the Reform of Natural Philosophy*. Cambridge: Cambridge University Press, 1992.

Martin, R. Niall D. "Saving Duhem and Galileo: Duhemian Methodology and the Saving of the Phenomena." *History of Science* 25 (1987): 307–19.

Masson, Flora. *Robert Boyle: A Biography*. London: Constable and Co., 1914.

Mates, Benson. *The Philosophy of Leibniz: Metaphysics and Language*. Oxford: Oxford University Press, 1986.

McCabe, Bernard. "Francis Bacon and the Natural Law Tradition." *Natural Law Forum* 9 (1964): 111–21.

McGuire, J. E. "Boyle's Conception of Nature." *Journal of the History of Ideas* 33 (1972): 523–42.

———. "Neoplatonism and Active Principles: Newton and the *Corpus hermeticum.*" In *Hermeticism and the Scientific Revolution*, edited by Robert S. Westman and J. E. McGuire, 95–142. Berkeley and Los Angeles: University of California Press, 1977.

McGuire, J. E., and P. M. Rattansi. "Newton and the 'Pipes of Pan.'" *Notes and Records of the Royal Society* 21 (1966): 108–43.

McKie, Douglas. Introduction to *The Works of the Honourable Robert Boyle*, 6 vols., edited by Thomas Birch, 1: v*–xx*. 1772. Reprint, Hildesheim: Georg Olms, 1965–66.

McKirahan, Richard D. "Aristotelian Epagoge in *Prior Analytics* 2.21 and *Posterior Analytics* 1.1." *Journal of the History of Philosophy* 21 (1983): 1–13.

McMullin, Ernan. "The Conception of Science in Galileo's Work." In *New Perspectives on Galileo*, edited by Robert E. Butts and J. C. Pitt, 209–57. Dordrecht: D. Reidel, 1978.

———. "Conceptions of Science in the Scientific Revolution." In *Reappraisals of the Scientific Revolution*, edited by David C. Lindberg and Robert S. Westman, 27–92. Cambridge: Cambridge University Press, 1990.

———. "Empiricism and the Scientific Revolution." In *Art, Science, and History in the Renaissance*, edited by Charles S. Singleton, 331–69. Baltimore: Johns Hopkins University Press, 1968.

———. "Galilean Idealization." *Studies in History and Philosophy of Science* 16 (1985): 247–73.

———. "The Goals of Natural Science." *Proceedings and Addresses of the American Philosophical Association* 58 (1984): 37–64.

———. Introduction to *Galileo, Man of Science*, edited by Ernan McMullin, 3–51. New York: Basic Books, 1967.

———. Introduction to *The Social Dimensions of Science*, edited by Ernan McMullin, 1–26. Notre Dame, Ind.: University of Notre Dame Press, 1992.

———. "Logicality and Rationality." In *Philosophical Foundations of Science*, edited by R. J. Seeger and R. S. Cohen, 415–30. Dordrecht: D. Reidel, 1974.

———. *Newton on Matter and Activity*. Notre Dame, Ind.: University of Notre Dame Press, 1978.

———. "The Rational and the Social in the History of Science." In *Scientific Rationality: The Sociological Turn*, edited by J. R. Brown, 127–63. Dordrecht: D. Reidel, 1984.

———. "Rhetoric and Theory Choice in Science." In *Persuading Science: The Art of Scientific Rhetoric*, edited by Marcello Pera and William R. Shea, 55–76. Canton, Mass.: Science History Publications, 1991.

———. "Structural Explanation." *American Philosophical Quarterly* 15 (1978): 139–47.

———. "Two Ideals of Explanation in Natural Science." In *Causation and Causal Theories*, edited by Peter A. French, Theodore E. Uehling Jr., and Howard K. Wettstein, 205–20. Minneapolis: University of Minnesota Press, 1984.

———. "Values in Science." In *PSA 1982*, edited by Peter D. Asquith and Thomas Nickles, 2: 3–27. East Lansing, Mich.: Philosophy of Science Association, 1983.

Merchant, Carolyn. *The Death of Nature*. New York: Harper and Row, 1980.

Mersenne, Marin. *Les nouvelles pensées de Galilée*. Edited by Pierre Costabel and Michel-Pierre Lerner. Paris: J. Vrin, 1973.

Meynell, G. G. *A Bibliography of Dr. Thomas Sydenham (1624–1689)*. Folkestone, Eng.: Winterdown Books, 1990.

More, Louis Trenchard. *The Life and Works of the Honourable Robert Boyle*. London: Oxford University Press, 1944.

Morgan, John. "Puritanism and Science: A Reinterpretation." *Historical Journal* 22 (1979): 535–60.

Muir, M. M. Pattison. Introduction to *The Skeptical Chemist*, by Robert Boyle, ix–xxii. London: J. M. Dent, 1949.

Mulkay, Michael, and G. Nigel Gilbert. "What Is the Ultimate Question?" *Social Studies of Science* 12 (1982): 309–19.

Nakajima, Hideto. "Two Kinds of Modification Theories of Light: Some New Observations on the Newton-Hooke Controversy of 1672 concerning the Nature of Light." *Annals of Science* 41 (1984): 261–78.

Newman, William. "Boyle's Debt to Corpuscular Alchemy." In *Robert Boyle Reconsidered*, edited by Michael Hunter, 107–18. Cambridge: Cambridge University Press 1994.

———. *Gehennical Fire: The Lives of George Starkey, an Alchemist of Harvard in the Scientific Revolution*. Cambridge: Harvard University Press, forthcoming.

———. "Newton's *Clavis* as Starkey's *Key*." *Isis* 78 (1987): 564–74.

Newton, Isaac. *Certain Philosophical Questions: Newton's Trinity Notebook.* Edited by J. E. McGuire and Martin Tamny. Cambridge: Cambridge University Press, 1983.

———. *The Correspondence of Isaac Newton.* 7 vols. Edited by H. W. Turnbull. Cambridge: Cambridge University Press, 1959–77.

———. "New Theory about *Light* and *Colours.*" *Philosophical Transactions of the Royal Society* 6 (1672): 3075–87.

———. *The Optical Papers of Isaac Newton.* Vol. 1, *The Optical Lectures, 1670–1672.* Edited by Alan E. Shapiro. Cambridge: Cambridge University Press, 1984.

———. *Opticks.* New York: Dover Publications, 1952.

Nickles, Thomas. "Beyond Divorce: Current Status of the Discovery Debate." *Philosophy of Science* 52 (1985): 177–206.

———. "Good Science as Bad History: From Order of Knowing to Order of Being." In *The Social Dimensions of Science,* edited by Ernan McMullin, 85–129. Notre Dame, Ind.: University of Notre Dame Press, 1992.

———. "'Twixt Method and Madness." In *The Process of Science,* edited by Nancy Nersessian, 41–67. Dordrecht: Martinus Nijhoff, 1987.

O'Brien, John J. "Samuel Hartlib's Influence on Robert Boyle's Scientific Development." *Annals of Science* 21 (1965): 1–14, 257–76.

Ogg, David. *England in the Reign of Charles II.* Vol. 1. Oxford: Clarendon Press, 1955.

Oldenburg, Henry. *The Correspondence of Henry Oldenburg.* 11 vols. Edited by A. Rupert Hall and Marie Boas Hall. Madison: University of Wisconsin Press, 1965.

Oldroyd, David. "Why Not a Whiggish Social Studies of Science?" *Social Epistemology* 3 (1989): 355–72.

Olson, Richard. "Historical Reflections on Feminist Critiques of Science: The Scientific Background to Modern Feminism." *History of Science* 28 (1990): 125–47.

———. "On the Nature of God's Existence, Wisdom, and Power: The Interplay between Organic and Mechanistic Imagery in Anglican Natural Theology, 1640–1740." In *Approaches to Organic Form,* edited by Frederick Burwick, 1–48. Dordrecht: D. Reidel, 1987.

Osler, Margaret J. "The Intellectual Sources of Robert Boyle's Philosophy of Nature: Gassendi's Voluntarism and Boyle's Physico-Theological Project." In *Philosophy, Science, and Religion in England, 1640–1700,* edited by Richard W. F. Kroll, Richard Ashcraft, and Perez Zagorin, 178–98. Cambridge: Cambridge University Press, 1992.

———. "John Locke and the Changing Ideal of Scientific Knowledge." *Journal of the History of Ideas* 31 (1970): 3–16.

———. "Providence and Divine Will in Gassendi's Views on Scientific Knowledge." *Journal of the History of Ideas* 44 (1983): 549–60.

Oster, Malcolm. "The 'Beame of Diuinity': Animal Suffering in the Early Thought of Robert Boyle." *British Journal for the History of Science* 22 (1989): 151–80.

———. "Biography, Culture, and Science: The Formative Years of Robert Boyle." *History of Science* 31 (1993): 177–226.

———. "Virtue, Providence, and Political Neutralism: Boyle and Interregnum Politics." In *Robert Boyle Reconsidered,* edited by Michael Hunter, 19–36. Cambridge: Cambridge University Press, 1994.

O'Toole, F. J. "Qualities and Powers in the Corpuscular Philosophy of Robert Boyle." *Journal of the History of Philosophy* 12 (1974): 295–315.

Pacchi, Arrigo. "Hobbes and the Problem of God." In *Perspectives on Thomas Hobbes*, edited by G. A. J. Rogers and Alan Ryan, 171–87. Oxford: Clarendon Press, 1988.

Pagel, Walter. *New Light on William Harvey*. New York: S. Karger, 1976.

———. *Paracelsus*. Basel: S. Karger, 1958.

———. *William Harvey's Biological Ideas*. New York: S. Karger, 1967.

Paradis, James. "Montaigne, Boyle, and the Essay of Experience." In *One Culture*, edited by George Levine, 59–91. Madison: University of Wisconsin Press, 1987.

Park, Katherine, and Lorraine J. Daston. "Unnatural Conceptions: The Study of Monsters in Sixteenth- and Seventeenth-Century France and England." *Past and Present* 92 (1981): 20–54.

Pascal, Blaise. *The Physical Treatises of Pascal*. Translated by I. H. B. Spiers and A. G. H. Spiers. New York: Columbia University Press, 1937.

Patey, Douglas Lane. *Probability and Literary Form*. Cambridge: Cambridge University Press, 1984.

Peirce, Charles Sanders. *Charles S. Peirce: Selected Writings*. Edited by Philip P. Wiener. New York: Dover Publications, 1966.

Peltonen, Markku. "Politics and Science: Francis Bacon and the True Greatness of States." *Historical Journal* 35 (1992): 279–305.

Pera, Marcello. "The Role and Value of Rhetoric in Science." In *Persuading Science: The Art of Scientific Rhetoric*, edited by Marcello Pera and William R. Shea, 29–54. Canton, Mass.: Science History Publications, 1991.

Pera, Marcello, and William R. Shea, eds. *Persuading Science: The Art of Scientific Rhetoric*. Canton, Mass.: Science History Publications, 1991.

Pérez-Ramos, Antonio. "Francis Bacon and the Disputations of the Learned." *British Journal for the Philosophy of Science* 42 (1991): 577–88.

———. *Francis Bacon's Idea of Science and the Makers' Knowledge Tradition*. Oxford: Clarendon Press, 1988.

Pickering, Andrew. "Against Putting the Phenomena First: The Discovery of the Weak Neutral Current." *Studies in History and Philosophy of Science* 15 (1984): 85–117.

———. *Constructing Quarks: A Sociological History of Particle Physics*. Chicago: University of Chicago Press, 1984.

———. "From Science as Knowledge to Science as Practice." In *Science as Practice and Culture*, edited by Andrew Pickering, 1–26. Chicago: University of Chicago Press, 1992.

———. Review of *How Experiments End*, by Peter Galison. *Isis* 79 (1988): 472–73.

Pickering, Andrew, and Adam Stephanides. "Constructing Quaternions: On the Analysis of Conceptual Practice." In *Science as Practice and Culture*, edited by Andrew Pickering, 139–67. Chicago: University of Chicago Press, 1992.

Polyani, Michael. *Personal Knowledge*. New York: Harper and Row, 1964.

Popkin, Richard H. *The History of Skepticism from Erasmus to Spinoza*. Berkeley and Los Angeles: University of California Press, 1979.

———. "Hobbes and Skepticism." In *History of Philosophy in the Making*, edited by Linus J. Thro, 133–48. Washington, D.C.: University Press of America, 1982.

Popper, Karl R. *The Logic of Scientific Discovery*. New York: Harper and Row, 1968.

Potter, Elizabeth. "Modeling the Gender Politics in Science." *Hypatia* 3 (1988): 19–33.

Power, Henry. *Experimental Philosophy.* 1664. Reprint, New York: Johnson Reprint Corp., 1966.

Prest, Wilfred R. *The Inns of Court under Elizabeth I and the Early Stuarts, 1590–1640.* Totowa, N.J.: Rowman and Littlefield, 1972.

Principe, Lawrence M. "Boyle's Alchemical Pursuits." In *Robert Boyle Reconsidered,* edited by Michael Hunter, 91–105. Cambridge: Cambridge University Press, 1994.

———. "Robert Boyle's Alchemical Secrecy: Codes, Ciphers, and Concealments." *Ambix* 39 (1992): 63–74.

Purver, Margery. *The Royal Society: Concept and Creation.* London: Routledge and Kegan Paul, 1967.

Quinn, Philip L. "What Duhem Really Meant." *Boston Studies in the Philosophy of Science* 14 (1974): 33–56.

Randall, J. H., Jr. *The School of Padua and the Emergence of Modern Science.* Padua: Editrice Antenore, 1961.

Rattansi, P. M. "The Intellectual Origins of the Royal Society." *Notes and Records of the Royal Society* 23 (1968): 129–43.

Ray, John. *The Wisdom of God Manifested in the Works of the Creation.* London, 1727.

Rees, Graham. "The Transmission of Bacon Texts: Some Unanswered Questions." In *Francis Bacon's Legacy of Texts,* edited by William A. Sessions, 311–23. New York: AMS Press, 1990.

Richardson, W. C. *A History of the Inns of Court.* Baton Rouge: Louisiana State University Press, 1975.

Roberts, Lissa. "A Word and the World: The Significance of Naming the Calorimeter." *Isis* 82 (1991): 198–222.

Rodis-Lewis, Geneviève. "Descartes' Life and the Development of His Philosophy." In *The Cambridge Companion to Descartes,* edited by John Cottingham, 21–57. Cambridge: Cambridge University Press, 1992.

Rogers, G. A. J. "Boyle, Locke, and Reason." *Journal of the History of Ideas* 27 (1966): 205–16.

———. "Descartes and the Method of English Science." *Annals of Science* 29 (1972): 237–55.

———. "Hobbes's Hidden Influence." In *Perspectives on Thomas Hobbes,* edited by G. A. J. Rogers and Alan Ryan, 189–205. Oxford: Clarendon Press, 1988.

Romanell, Patrick. *John Locke and Medicine.* Buffalo: Prometheus Books, 1984.

Rorty, Richard. *Philosophy and the Mirror of Nature.* Princeton, N.J.: Princeton University Press, 1979.

Rouse, Joseph. "The Politics of Postmodern Philosophy of Science." *Philosophy of Science* 58 (1991): 607–27.

Rowbottam, Margaret E. "The Earliest Published Writing of Robert Boyle." *Annals of Science* 6 (1948–50): 376–89.

Ruby, Jane E. "The Origins of Scientific 'Law.'" *Journal of the History of Ideas* 47 (1986): 341–60.

Sabra, A. I. *Theories of Light from Descartes to Newton.* Cambridge: Cambridge University Press, 1981.

Salmon, Wesley. *Scientific Explanation and the Causal Structure of the World.* Princeton, N.J.: Princeton University Press, 1984.

Sanchez-Gonzalez, Miguel A. "Medicine in John Locke's Philosophy." *Journal of Medicine and Philosophy* 15 (1990): 675–95.

Sarasohn, Lisa T. "Motion and Morality: Pierre Gassendi, Thomas Hobbes, and the Mechanical World View." *Journal of the History of Ideas* 46 (1985): 363–79.

Sargent, Rose-Mary. "Explaining the Success of Science." In *PSA 1988*, edited by Arthur Fine and Jarrett Leplin, 1: 55–63. East Lansing, Mich.: Philosophy of Science Association, 1988.

———. "Learning from Experience: Boyle's Construction of an Experimental Philosophy." In *Robert Boyle Reconsidered*, edited by Michael Hunter, 57–78. Cambridge: Cambridge University Press, 1994.

———. "Robert Boyle's Baconian Inheritance: A Response to Laudan's Cartesian Thesis." *Studies in History and Philosophy of Science* 17 (1986): 469–86.

———. "Scientific Experiment and Legal Expertise: The Way of Experience in Seventeenth-Century England." *Studies in History and Philosophy of Science* 20 (1989): 19–45.

Schlagel, Richard H. "Fine's 'Shaky Game' (and Why NOA is No Ark for Science)." *Philosophy of Science* 58 (1991): 307–23.

Schoeman, Ferdinand. "Cohen on Inductive Probability and the Law of Evidence." *Philosophy of Science* 54 (1987): 76–91.

Shanahan, Timothy. "God and Nature in the Thought of Robert Boyle." *Journal of the History of Philosophy* 26 (1988): 547–69.

———. "Teleological Reasoning in Boyle's *Disquisition about Final Causes*." In *Robert Boyle Reconsidered*, edited by Michael Hunter, 177–92. Cambridge: Cambridge University Press, 1994.

Shapere, Dudley. "The Concept of Observation in Science and Philosophy." *Philosophy of Science* 49 (1982): 485–525.

———. "Discussion: Astronomy and Antirealism." *Philosophy of Science* 60 (1993): 134–50.

———. *Reason and the Search for Knowledge.* Dordrecht: D. Reidel, 1984.

Shapin, Steven. "Discipline and Bounding: The History and Sociology of Science as Seen through the Externalism-Internalism Debate." *History of Science* 30 (1992): 333–69.

———. "History of Science and Its Sociological Reconstructions." *History of Science* 20 (1982): 157–211.

———. "The House of Experiment in Seventeenth-Century England." *Isis* 79 (1988): 373–404.

———. "Pump and Circumstance: Robert Boyle's Literary Technology." *Social Studies of Science* 14 (1984): 481–520.

———. "Robert Boyle and Mathematics: Reality, Representation, and Experimental Practice." *Science in Context* 2 (1988): 23–58.

———. "Social Uses of Science." In *Ferment of Knowledge*, edited by G. S. Rousseau and R. Porter, 93–139. Cambridge: Cambridge University Press, 1980.

———. "Who Was Robert Hooke?" In *Robert Hooke: New Studies*, edited by Michael Hunter and Simon Schaffer, 253–85. Woodbridge, Eng.: Boydell Press, 1989.

Shapin, Steven, and Simon Schaffer. *Leviathan and the Air-Pump*. Princeton, N.J.: Princeton University Press, 1985.

Shapiro, Alan E. "The Evolving Structure of Newton's Theory of White Light and Color." *Isis* 71 (1980): 211–35.

———. *Fits, Passions, and Paroxysms: Physics, Method, and Chemistry and Newton's Theories of Colored Bodies and Fits of Easy Reflection*. Cambridge: Cambridge University Press, 1993.

Shapiro, Barbara J. "Early Modern Intellectual Life: Humanism, Religion, and Science in Seventeenth-Century England." *History of Science* 29 (1991): 45–71.

———. "Law and Science in Seventeenth-Century England." *Stanford Law Review* 21 (1969): 727–65.

———. "Law Reform in Seventeenth-Century England." *American Journal of Legal History* 19 (1975): 280–312.

———. *Probability and Certainty in Seventeenth-Century England*. Princeton, N.J.: Princeton University Press, 1983.

———. "Sir Francis Bacon and the Mid-Seventeenth-Century Movement for Law Reform." *American Journal of Legal History* 24 (1980): 331–62.

———. "'To a Moral Certainty': Theories of Knowledge and Anglo-American Juries, 1600–1850." *Hastings Law Journal* 38 (1986): 153–93.

Siegel, Harvey. "What Is the Question concerning the Rationality of Science?" *Philosophy of Science* 52 (1985): 517–37.

Sloan, Phillip R. "Descartes, the Skeptics, and the Rejection of Vitalism in Seventeenth-Century Physiology." *Studies in History and Philosophy of Science* 8 (1977): 1–28.

———. "John Locke, John Ray, and the Problem of the Natural System." *Journal of the History of Biology* 5 (1972): 1–53.

Smith, Joseph Wayne. "Primitive Classification and the Sociology of Knowledge: A Response to Bloor." *Studies in History and Philosophy of Science* 15 (1984): 237–43.

Söderqvist, Thomas. "Existential Projects and Existential Choice in Science: Scientific Biography as an Edifying Genre." In *Telling Lives: Studies of Scientific Biography*, edited by Richard Yeo and Michael Shortland. Cambridge: Cambridge University Press, forthcoming.

Spinoza, Baruch. *The Correspondence of Spinoza*. Edited by A. Wolf. New York: Lincoln MacVeagh, 1927.

———. *Principles of Cartesian Philosophy*. Translated by Harry E. Wedeck. New York: Philosophical Library, 1961.

Sprat, Thomas. *The History of the Royal Society of London, for the Improving of Natural Knowledge*. Edited by Jackson I. Cope and Harold Whitmore Jones. 1667. Reprint, St. Louis: Washington University Press, 1958.

Steneck, Nicholas H. "Greatrakes the Stroker: The Interpretation of Historians." *Isis* 73 (1982): 161–77.

Stewart, M. A. *Selected Philosophical Papers of Robert Boyle*. New York: Barnes and Noble, 1979.

Stubbe, Henry. *Censures upon Certaine Passages Contained in a History of the Royal Society as Being Destructive to the Established Religion and Church of England*. Oxford, 1670.

Sulloway, Frank J. "Reassessing Freud's Case Histories." *Isis* 82 (1991): 245–75.

Temkin, Owsei. *Galenism: Rise and Decline of a Medical Philosophy.* Ithaca, N.Y.: Cornell University Press, 1973.

Toulmin, Stephen. *Cosmopolis: The Hidden Agenda of Modernity.* Chicago: University of Chicago Press, 1990.

———. *Human Understanding.* Princeton, N.J.: Princeton University Press, 1972.

———. "Scientific Strategies and Historical Change." In *Philosophical Foundations of Science,* edited by R. J. Seeger and R. S. Cohen, 401–14. Dordrecht: D. Reidel, 1984.

Tuchanska, Barbara. "What Is Explained in Science?" *Philosophy of Science* 59 (1992): 102–19.

Tuck, Richard. "Hobbes and Descartes." In *Perspectives on Thomas Hobbes,* edited by G. A. J. Rogers and Alan Ryan, 11–41. Oxford: Clarendon Press, 1988.

Urbach, Peter. "Francis Bacon as a Precursor to Popper." *British Journal for the Philosophy of Science* 33 (1982): 113–32.

———. *Francis Bacon's Philosophy of Science.* La Salle, Ill.: Open Court, 1987.

van den Daele, Wolfgang. "The Social Construction of Science: Institutionalization and Definition of Positive Science in the Latter Half of the Seventeenth Century." In *The Social Production of Scientific Knowledge,* edited by Everett Mendelsohn, Peter Weingart, and Richard Whitley, 27–54. Dordrecht: D. Reidel, 1977.

van Fraassen, Bas C. "Empiricism in the Philosophy of Science." In *Images of Science,* edited by Paul M. Churchland and Clifford A. Hooker, 245–308. Chicago: University of Chicago Press, 1985.

———. *The Scientific Image.* Oxford: Clarendon Press, 1980.

van Leeuwen, Henry G. *The Problem of Certainty in English Thought, 1630–1690.* The Hague: Martinus Nijhoff, 1963.

Vickers, Brian, ed. *Essential Articles for the Study of Francis Bacon.* Hamden, Conn.: Archon Books, 1968.

Waite, Arthur Edward. *The Secret Tradition in Alchemy.* New York: Samuel Weiser, 1969.

Waldman, Theodore. "Origins of the Legal Doctrine of Reasonable Doubt." *Journal of the History of Ideas* 20 (1959): 299–316.

Wear, Andrew. "William Harvey and the 'Way of the Anatomists.'" *History of Science* 21 (1983): 223–49.

Webster, Charles. *The Great Instauration: Science, Medicine, and Reform, 1626–1660.* New York: Holmes and Meier, 1975.

———. *The Intellectual Revolution of the Seventeenth Century.* London: Routledge and Kegan Paul, 1974.

———. *Samuel Hartlib and the Advancement of Learning.* Cambridge: Cambridge University Press, 1970.

Westfall, Richard S. *The Construction of Modern Science.* Cambridge: Cambridge University Press, 1977.

———. "The Development of Newton's Theory of Color." *Isis* 53 (1962): 339–58.

———. "Galileo and Newton: Different Rhetorical Strategies." In *Persuading Science: The Art of Scientific Rhetoric,* edited by Marcello Pera and William R. Shea, 107–24. Canton, Mass.: Science History Publications, 1991.

———. Review of *Leviathan and the Air-Pump,* by Steven Shapin and Simon Schaffer. *Philosophy of Science* 54 (1987): 128–30.

————. *Science and Religion in Seventeenth-Century England.* Ann Arbor: University of Michigan Press, 1973.

————. "Unpublished Boyle Papers Relating to Scientific Method." *Annals of Science* 12 (1956): 63–73, 103–17.

Westman, Robert S. "The Copernicans and the Churches." In *God and Nature,* edited by David C. Lindberg and Ronald L. Numbers, 76–113. Berkeley and Los Angeles: University of California Press, 1986.

Weston, Thomas. "Approximate Truth and Scientific Realism." *Philosophy of Science* 59 (1992): 53–74.

Wheeler, Harvey. "Francis Bacon's *New Atlantis:* The 'Mould' of a Lawfinding Commonwealth." In *Francis Bacon's Legacy of Texts,* edited by William A. Sessions, 291–310. New York: AMS Press, 1990.

Whewell, William. *Novum organon renovatum.* 3d ed. London: John W. Parker and Son, 1858.

Whitney, Charles. *Francis Bacon and Modernity.* New Haven, Conn.: Yale University Press, 1986.

Wiener, Philip P. "The Experimental Philosophy of Robert Boyle." *Philosophical Review* 41 (1932): 594–609.

Willis, Robert. Introduction to *The Works of William Harvey,* xvii–xcvi. Annapolis, Md.: St. John's College Press, 1949.

Wojcik, Jan. "Robert Boyle and the Limits of Reason: A Study in the Relationship between Science and Religion in Seventeenth-Century England." Ph.D. diss., University of Kentucky, 1992.

————. "The Theological Context of Boyle's *Things above Reason.*" In *Robert Boyle Reconsidered,* edited by Michael Hunter, 139–55. Cambridge: Cambridge University Press, 1994.

Woolgar, Steve. "Some Remarks about Positionism: A Reply to Collins and Yearley." In *Science as Practice and Culture,* edited by Andrew Pickering, 327–42. Chicago: University of Chicago Press, 1992.

Woolhouse, R. S. *Locke's Philosophy of Science and Knowledge.* New York: Barnes and Noble, 1971.

Wright, Peter W. G. "On the Boundaries of Science in Seventeenth-Century England." In *Sciences and Cultures: Anthropological and Historical Studies of the Sciences,* edited by Everett Mendelsohn and Yehuda Elkana, 82–95. Dordrecht: D. Reidel, 1981.

Zagorin, Perez. "Hobbes on Our Mind." *Journal of the History of Ideas* 51 (1990): 317–35.

INDEX

Entries under Boyle are limited to citations for personal and biographical information. Extended discussions concerning his views on particular topics and persons are listed under those headings. Short titles of Boyle's works appear as main entries, and references are confined primarily to places where his works are discussed in the text. No citations have been given for references to his works in endnotes unless the note includes additional information concerning a particular work.

absolute propositions, 119, 275n70
  See also innate ideas
accommodation theory. *See* interpretation
"Account of Philaretus," 23
activity
  epistemic value of, 35, 40–41, 159–65, 211, 291n3
  as goal of experiment, 214
activity of matter
  Boyle's conception of, 100, 259n3, 266n85, 268n100, 271n20
  Leibniz's conception of, 100–101, 268nn99–100
"Aerial Noctiluca," 191–92, 206nn60–61
air-pump
  construction of, 63, 190, 282n42
  experiments with, 26, 59–61, 134, 174, 190, 279n15, 302nn16–17, 305n49
alchemy
  Boyle's criticism of, 71–72, 73–74, 111, 253n64, 273n31, 301n11
  Boyle's interest in, 8, 187, 222n41, 251n41, 252n50

medical opposition to, 75–76, 254n83
  rhetorical style, 252n51, 294n29, 301n11, 303nn27, 30
  See also chemistry
Alexander, Peter, 2, 217nn4, 5, 229n16, 234n98, 245n96, 268n94, 269n111
analogy, reasoning from, 135–36, 157, 173, 245n96, 280n22
Anglicanism, 88, 260nn8, 9, 263n47, 272n28, 275n68
antiperistasis, theory of, 196, 200
apothecaries. *See* artisans and tradesmen; alchemy
Aquinas, Thomas, 260n7
Archimedes, 67, 280n29
Aristotle
  Boyle's criticism of, 27, 149, 150
  Boyle's perception of, 23–27, 92, 113, 228–29nn16, 20, 29, 31, 285n66, 287n82
  cold, theory of, 196
  induction, theory of, 229n20
  Nicomachean Ethics, 239n26
  See also scholastic philosophy

artificiality
  Aristotelian criticism of, 160, 291n4
  Bacon's defense of, 292n8
  Boyle's defense of, 159–65, 291–
    92nn4–6, 8
  epistemic value of, 18, 70, 159–62,
    296n53
  of experimental materials, 167–68,
    295n43
  of instruments, 169, 295–96n45
  as notional attribute, 160
  opposed to natural, 160–61, 165
  See also experiment
artisans and tradesmen, 138, 149–52,
    160–61, 165–67, 170, 187, 252n54,
    281n34, 288n101, 293n17,
    294nn29, 31, 297n71, 309n115
Ashmole, Elias, 284–85n64
atomism
  ancient, 229n31
  Boyle's criticism of, 39–40, 69, 98,
    113
  See also corpuscularianism
Augustine, 91, 260n7, 262nn31, 35,
    266n83, 272n21, 273n41

Bacon, Francis, 2, 7–8, 24, 26
  Advancement of Learning, 230nn39, 41,
    233nn74, 76, 234nn84, 90,
    263n43, 292n8
  atomism, 229n31, 271n15
  Boyle's perception of, 15, 28–29, 35–
    41, 52–53, 61, 109, 124, 206–7,
    230n50, 234n93
  Commission for the Union, 45,
    239nn20–21
  and common law, 44–46, 48
  and credibility of witnesses, 287n90
  and Descartes, 232n54
  De augmentis, 31, 232nn60–61, 239n16,
    240n40, 242nn53, 54, 298n84
  and experience, 33–35, 50–52,
    231n47, 242n56
  and experiments, 30, 34–35, 51–52,

    165, 176, 231n50, 242nn53–54,
    298n84, 302nn20–21
  and facts, 132, 142, 224n53, 278n7
  heat, theory of, 307nn68, 71, 308n90,
    310n129
  and hypotheses, 31
  idols of the mind, 34, 233n76,
    243n72
  on knowledge, 31, 33–35, 40, 206,
    232n61, 233n82
  legal analogy, 43, 236n6, 237n9
  legal reform, 44–45
  Maxims of the Law, 44, 239nn15–16, 18
  natural histories, 51–52
  New Atlantis, 249n12
  New Organon, 30, 229n31, 230–35nn
    passim, 242nn53–63, 243n72,
    244n93, 251n43, 261n25, 276n89
  Parasceve, 233nn70, 82, 235n111,
    287n90, 300n2
  and religion, 93, 112, 261nn18, 25,
    263n43, 271n15, 275n68
  and scholastic philosophy, 28, 33
  Sylva Sylvarum, 278n7, 299n1, 302n20
  as underbuilder, 235n103
  writing style, 299n1, 300n8
Badcock, A. W., 308n105, 309nn109,
    115, 310n128
Baker, J. H., 238–39nn17, 19, 25, 27,
    240n35, 241nn45, 47, 242n52
Barker, Peter, 262n35
Barlow, Thomas, 112
Barnes, Jonathan, 229n20
Baxter, Richard, 260nn3, 7
Berman, Harold J., 238n13
Beverwijck, John, 258n117
Bible. *See* Scripture
Birch, Thomas, 19, 217n2
Black, Joseph, 308n105
Blackstone, William, 237n7, 238n12
Blake, Ralph M., 231n47, 233n64
Bloor, David, 4–6, 219nn15–17,
    220nn20, 22, 221n26, 225n65
Boas, Marie. *See* Hall, Marie Boas

bodies
   as corporeal agents, 102, 159, 162
   nature of, 95, 104
   See also qualities of bodies
book of nature, 17, 90, 93, 101, 110–13,
      116, 122–23, 126, 271nn10, 12,
      272n25, 314n43
Boyle, Francis, 227n2, 273n41
Boyle, Katherine. *See* Ranelagh, Kather-
      ine Jones, Viscountess
Boyle, Richard, first earl of Cork, 23–24,
      227n2
Boyle, Robert
   and assistants, 189–93, 305nn49–53,
      306nn57, 60, 64, 66–67, 308n99
   curiosity of, 20, 38, 138, 140, 189,
      193, 197
   diffidence of, 12, 19–20, 125, 145,
      170, 185, 188, 205–9, 215–16,
      227n75, 261n10, 295n43
   eclecticism of, 1, 10, 15, 61, 88, 115,
      123, 124, 205–9, 215–16, 260–
      61nn7, 9–10
   as empiricist, 2–3, 42, 210, 211,
      219nn12–13, 236nn2–3, 5,
      264n63, 277n118
   ethical essays, 62, 139, 223n43,
      228n14, 236n115, 248n4, 281n37,
      282n39
   at Eton, 23
   and foreign visitors, 189, 191,
      288n95, 304n45
   health of, 75, 254n82
   indexes of works, 140–43, 282n46,
      284n56
   in Ireland, 62–63, 139, 256n101
   linguistic abilities, 274n57
   mathematical abilities, 23, 37, 39, 66–
      68, 250n26
   and metaphysics, 69, 264n66, 274n54
   modesty of, 8, 14, 215–16, 313n36
   at Oxford, 63, 112, 139, 190, 255n86,
      282n41
   papers and notebooks, 19, 139–43,
      226n72, 238n12, 281–82nn38, 41–
      45, 283nn47, 51, 284nn58–59,
      293nn17, 22, 296n53
   as philosopher, 1, 11, 14, 19, 205
   and politics, 25, 228n15
   as pragmatist, 2, 210, 218n7
   as rationalist, 2, 218nn9–10
   religious affiliations, 88, 260n9,
      275n68
   at Stalbridge, 62, 138–39, 248n3
   as underbuilder, 20, 38, 52
   youth and education, 23–25, 62–63,
      75, 89, 112, 115–16, 138–39,
      227n2
Boyle, Roger, Lord Broghill, 248n6
Boyle lectures, 274n55
Boyle's law, 265n69
Brahe, Tycho, 295n45
Brande, William, 205, 311n3
Brazil, medicine in, 151
Brett, G. S., 257n113
Brown, Theodore, 258n126
Brush, Craig, 250n29
Budgell, Eustace, 205, 311n6
Burnet, Gilbert, 235n115, 240n33,
      260n9, 273n41, 304n45, 311n7
Butterfield, Herbert, 251n42
Butts, Robert E., 312n25

Callon, Michel, 220n22, 221nn28, 30
Calvin, John, 260n7, 273n41
Campbell, Mary B., 283n55
Cantor, Geoffrey, 299n2
Carew, John, 227n2
Carneades, 244n85
Carr, Thomas M., 299n1
Cartwright, Nancy, 12, 224n59, 312n31
causality
   and activity of matter, 100, 102,
      268nn94, 99–100
   as Boyle's goal, 67, 69, 97–98, 103–8
   chain of causes, 93, 111, 134–35
   and facts, 132–35, 136
   and experiments, 15, 17, 39–40, 42–

causality (*continued*)
43, 51, 54, 70–71, 74, 78, 80, 81–
83, 126, 159, 161–62, 165–66,
172–73, 178, 210, 214
as independent of human perception,
105, 107, 146
intermediate causes, 40, 52, 54, 58,
69, 133–35, 206, 279n13
and laws, 97, 100
as necessary, 99, 102, 110
and qualities of bodies, 100, 105–8
and relational properties of bodies,
106–8, 111, 126, 162
secondary causes, 104
Spinoza's conception of, 68
Cavendish, Margaret, 310n1
Certain Physiological Essays
"Proëmial Essay," 38–39, 69, 133, 137,
149, 183, 186, 188
"Redintegration of Nitre," 39
"Some Specimens," 69
"Unsucceeding Experiments," 168–70,
177, 287n85
"Unsuccessfulness of Experiments,"
167–68, 171
Charles II, 191
Charleton, Walter, 230n35, 247n1
chemistry, 9, 16
and corpuscular philsophy, 69–70, 73,
74, 251nn41, 45, 265n71
experiments in, 62, 70–75, 139, 160,
167–68, 171–72, 174, 177,
292n11
improvements required, 71–72, 74,
252n50, 253n68, 254n74
medicinal uses, 75–77, 79, 167–68,
171–72, 255nn90, 97, 258n126
principles too narrow, 73–74
and reasoning from signs, 70–71, 72,
74, 252n49, 253n63, 256n100
rhetorical style in, 71–72, 252n51,
294n29, 301n11, 303n27
and theology, 92, 262n28
See also alchemy

China, medicine in, 145, 151
Christian Virtuoso, 64, 118, 119, 162, 215,
263n45, 275n74
"Chromatic Examen," 175
Cicero, 262n35
circulation of the blood
Boyle's discussion of, 79–84, 207–8,
258nn118, 122
cause of, 81–83
Descartes's theory of, 81–82,
257n113, 258n120
experiments on, 82–83
Harvey's theory of, 79, 80, 82–84,
109, 207, 256n103, 258n122
civility, 189–90
Clarendon Act, 242n50
Clark, George, 254n83
Clarke, Desmond, 228n16, 231n47,
232nn52, 54, 59, 61–62, 235n99
Clarke, Samuel, 112
Clericuzio, Antonio, 223n41, 233n65,
251n41, 265n71, 266n85,
268n100, 303n30
Clerselier, Claude, 233n65
clock metaphor, 270n119
Clod, Frederick, 142, 248n7, 256n101
Cohen, I. Bernard, 259n2, 301n14
coin analogy, 162–63, 171, 188
Coke, Edward, 46, 56, 239nn23–24, 26,
240nn30–31, 37, 263n37
cold
artificial production of, 201–2,
310n135
Boyle's investigation of, 193–204
cause of, 202–4
degrees of, 198–99
effects of, 194, 198, 308n91
experiments on, 198, 200–204,
309n14
philosophically considered, 197–98
as a sensible quality, 197, 308n90
theories of, 194, 199, 200–202
collaboration, 180, 210
in Boyle's laboratory, 189, 191–93,

305nn49, 53, 56, 306nn60, 66–67, 308n99
civility needed for, 189–90
epistemic value of, 211–12
writing to achieve, 182, 185, 189, 193, 303n30
Collingwood, Robin George, 224n56
Collins, Harry M., 4–5, 211, 213, 220–22nn21–22, 28, 30–31, 293n17, 311n21, 312n26
common law, 15–16, 236n6
  Bacon's defense of, 44–45, 48
  certainty in, 46, 50
  Coke's defense of, 46
  experience as foundation, 43–50, 238n14, 240n41
  facts in, 49–50, 241n46
  flexibility of, 45
  Hale's defense of, 47–50
  Hobbes's criticism of, 46–48
  moral demonstration in, 48–50, 241n42
  as precedent law, 44–46, 239n27
  reason in, 46–48, 240n41
  rules of testimony, 241n45, 242n50
Conant, James B., 2, 210, 217n4, 218n9, 250n26, 311n18
concurrence of probabilities, 16, 74
  Boyle's conception of, 54–56, 58, 61, 182, 207–8, 211, 212–13, 243nn79–80, 247n116
  for circulation of the blood, 82–84, 207–8
  for facts, 171, 181, 199, 200–201
  for spring of the air, 83, 134–35, 207
  in theology, 89–92
  See also moral demonstration
conditional propositions, 119–20, 126
Confucius, 260n7
consensus, 153, 212, 295n32
consilience of inductions, 212–13, 312n25
contextualist history, 2–5, 9–10, 43, 181, 211

See also sociology of scientific knowledge
conventions, 212–13
Cook, Harold J., 254n83
Corporation for Propagating the Gospel, 142, 283n56
corpuscularianism, 2, 4, 16
  ancient, 27–28
  Boyle's conception of, 28–29, 39–40, 69–70, 98–103, 104, 106, 125, 215, 265n75
  and chemistry, 70, 73–74, 251nn41, 45, 265n71
corroboration
  and consensus, 153, 212
  for observation reports, 153–54, 209, 307n86
cosmical qualities, 105–8
"Cosmical Suspicions." See *History of Particular Qualities*
Cottingham, John, 235n99
credibility, 18
  of authors, 149
  of divers, 156–58, 290nn133–34
  of experimental reports, 177–78
  of illiterate observers, 149–52, 197, 208
  of literate observers, 147–49, 152, 197, 208, 288nn93, 101
  of observation reports, 147, 153–55, 197–98, 209, 287n87
  of legal testimony, 51–53
  of witnesses in court, 49–50, 242n50
  of women, 152, 289nn110–11
curiosity
  of Boyle, 20, 38, 138, 140, 189, 193, 197
  of Boyle's assistants, 192–93
  as diligence, 188–89, 197, 287n92, 303n38
  excited in readers, 182, 186–89

Daston, Lorraine, 10, 224n52, 243n80, 278nn4, 6, 285–86nn67, 70–71

Davis, Edward B., 226n72

Dear, Peter, 131, 219n12, 224n45,
    238n11, 245n93, 255n100, 260n6,
    262n31, 278n4, 290n123, 299nn1–
    2, 301n14, 311n19

Deason, Gary B., 2, 218n11

Debus, Allen G., 254n83, 262n37

Democritus, 27, 51

de Monconys, Balthasar, 304n47

demonstration
    inverted order of, 15–16, 35–36, 38,
        39, 84, 133, 252n49
    mathematical, 32–33, 37–39, 184
    See also moral demonstration

Dent, Thomas, 190, 304n46

Descartes, René, 2, 7–8, 14, 24, 26, 104,
    184
    and analogical reasoning, 280n22
    and Aristotle, 228n16
    and Bacon, 30, 232n54
    Boyle's perception of, 28–29, 35–37,
        69–70, 235n99, 257n117,
        269n116, 286n76
    circulation of the blood, theory of,
        81–82, 257n113, 258n120
    cold, theory of, 310n129
    Discourse on the Method, 30–32,
        230nn37, 40, 231nn51–52,
        232nn58–62, 257nn115–16,
        292n14
    and empiricism, 231n47
    and experiments, 30, 32–33, 231n51,
        258n120
    and hypotheses, 31, 232n59
    immortality of the soul, 114, 272n26
    on knowledge, 32–33, 232n61
    Mersenne correspondence, 232n54,
        233n65, 234n84, 235n99
    Plempius correspondence, 258n120
    Principles of Philosophy, 30, 32,
        232nn53, 61, 63
    qualities of bodies, 269n116
    Regulae, 31, 33, 232n59, 233n66
    and scholastic philosophy, 28–29, 32
    and voluntarism, 88, 260n5

Dewhurst, Kenneth, 254n83, 255n93

Digby, Kenelm, 62, 142, 248n7

diffidence. *See under* Boyle

"Diffident Naturalist, The," 19, 127,
    143, 146, 147, 227n74

Dijksterhuis, E. G., 261n14, 273n35,
    276n97

Diodati, Jean, 112, 260n7, 271n14,
    273n41

"Discernment of Suppositions," 146,
    298n81

divers, Boyle's assessment of, 155–58,
    209, 290nn133–34

Doty, Ralph, 244n85

Drake, Stillman, 250n24

drugs. *See under* medicine

Ducasse, Curt J., 231n47

Duhem, Pierre, 213, 312n28

Duhem-Quine thesis, 246n100,
    312n31

Dury, John, 116, 238n12, 260n9,
    273nn39–40

East India Trading Company, 142, 143

East Indies, observations from, 143, 148,
    284n57, 291n139

Eliot, John, 283n56, 284n58

Ellis, Brian, 313n33

emergent axioms, 52, 54, 251n39
    See also intermediate causes, *under* cau-
        sality

Emerton, Norma E., 2, 217n4, 229n16,
    247n1, 253n69

empiricism, 2, 7, 42, 51, 88, 108, 161–
    62, 210, 217n6, 236n1

Empiricus, Sextus, 37

Epicurus, 23, 113, 227n7

Erasmus, 260n7

"Essay of the Holy Scripture," 112,
    271n15, 273n41

essays. *See* writing

eternal truths, 119

ethnomethodology, 6

Eton, 23

Euclid, 37

Evelyn, John, 189, 205, 306n61
"Excellency and Grounds of the Mechanical Hypothesis," 98, 271n10
Excellency of Theology, 112, 113–14, 117, 229n31, 249n17, 268n94, 299n86
experience
  Bacon's conception of, 50–52, 239n16
  Boyle's conception of, 15–16, 42, 44, 61, 52–54, 245n93
  in common law, 15–16, 43–44, 46–50, 238n14
  twentieth-century conception of, 42, 50
experiment
  as active knowledge, 35, 40–41, 160, 206, 213–14, 311n6
  artificiality of, 159–63, 291n4, 292nn6, 8
  Bacon's conception of, 18, 30, 34–35
  in chemistry, 70–75, 167–68, 171–72, 174, 177
  on cold, 198–204
  and collaboration, 18–19, 189, 191–93, 305nn49, 53, 56, 306nn60, 66–67
  in consort, 185, 199, 200–201, 302nn20–21
  contingency of, 18, 165–71, 176–77, 179, 185, 192, 198, 209, 294–95nn32, 43, 45, 307n74
  to correct senses, 147, 161–63, 168, 179
  as demonstration, 39–41, 54, 57, 59, 296n53
  Descartes's conception of, 30, 32–33, 81–82
  to discover causes (*see* experiments, *under* causality)
  as dynamic practice, 10–13, 17, 20, 179
  expedience of, 159, 161, 178
  exploratory, 137, 170, 172–76, 184, 192, 193, 209, 211, 297nn63–64, 70–71, 306n67
  and facts, 174, 177, 184, 187

  failures of, 178–79, 294n32, 298–99nn84–85, 87, 302n32
  fiat experimentum, 174–75, 297n65, 307n74
  as form of life, 9
  Galileo's conception of, 64–65
  Hobbes's criticism of, 39, 56, 58–60, 109
  and hypotheses, 173, 176
  interpretation of, 165, 170, 177, 179–80, 187, 209
  materials used in, 167–68, 171–72, 295n43
  and mathematics, 169
  in mechanics, 64–69
  in medicine, 76–84, 168, 172
  and observation, 137–38, 158, 159, 161, 178
  opposed to empiricism, 161–63, 179
  Pascal's conception of, 66–67
  on phosphorus, 191–93
  probatory, 170–76, 192, 209, 296n53
  reason as warrant for, 123, 125–26, 170, 179–80, 242n63
  repetition of, 18, 176, 177–78, 186, 187, 200, 209
  replication of, 165, 174, 175–76
  Spinoza's criticism of, 68–69
  strategies, 18, 176–80
  and theory, 18, 163–65, 168, 171–72, 179, 184–85, 188, 213
  thought experiments, 164, 293n24
  variation of, 18, 176–77, 179, 203, 209
  and writing, 180–85, 186–88, 303n29, 309n108
Experimenta et Observationes Physicae, 38, 142, 143, 163, 185, 186, 187, 280n25, 283n47, 286n74, 313n37
experimental philosophy
  Boyle's goal, 54, 58, 69, 214
  difficulty of, 84, 170, 179
  dynamic nature of, 10–13, 17, 20, 115, 122–23, 127–29, 136, 179, 182, 183, 185, 206–8, 210–11, 213–16, 313n37

experimenters' regress, 163, 213,
  293n17, 312n26
experimentum crucis, 184, 310n15
expertise
  of Boyle's assistants, 192–93,
    305nn49, 56, 306nn57, 64, 66,
    308n99
  of experimenters, 166–67, 170, 178
  in law, 46–48, 50
  of observers, 146–47, 148–53,
    154–55
  of writers, 181
explication, 275n81
  See also interpretation

factual foundation
  construction of, 131–32, 136, 138,
    145, 151, 163, 185, 203
  flexibility of, 145, 154, 158
facts
  Bacon's conception of, 132, 224n53,
    278n7
  Boyle's conception of, 10, 17, 131–
    38, 214
  Boyle's use of, 125, 137
  collations of, 181, 285n70
  dynamic nature of, 136, 145, 158,
    163
  as effects, 131–35, 136, 145, 156,
    278n7
  as epistemic category, 132–36, 145,
    163, 182, 278nn6, 8
  and experiment, 174, 177–78, 245n93
  and human agency, 132–35, 214
  and hypotheses, 131–35, 154, 156,
    207
  in legal cases, 49–51, 132, 237n9,
    278n6
  as linguistic category, 132, 136, 278–
    79nn4, 8
  in natural histories, 51, 140, 142
  rational construction of, 17, 137, 146,
    154–55, 207, 280n25, 285n70
  as regularities, 132, 136, 278n5

social construction of, 155–58,
    213–14
  twentieth-century conception of, 10,
    131–32, 224n51
Feingold, Mordechai, 228n13
Ferguson, Robert, 260nn3, 7
Final Causes, 93, 257n117, 262n26,
    265n79, 267–68n94
Fine, Arthur, 12–13, 225nn62–63, 65,
    312n23
Fisch, Menachem, 312n25
Foreign Plantations, Council for, 142
forms. *See* qualities; Scholastic
    philosophy
Frank, Robert G., Jr., 254n83, 256n101,
    257n117
Franklin, Allan, 12–13, 224n59, 225n66,
    312n23
Fraser, Antonia, 228n13
Free Inquiry, 26, 93–98, 99, 101–3, 105,
    143, 144, 237n7, 263nn47–49,
    266n85
Fraunce, Abrahm, 241n42
Freind, John, 310n2
Fuller, Steve, 221n26, 222n32, 225n65
Fulton, John, 248n7

Gadamer, Hans-Georg, 224n56, 311n12
Galen, 16, 76, 79
  On Anatomical Procedures, 256n102
  and religion, 91, 262n28, 263n39
Galenic medicine, 256n103
  Boyle's criticisms of, 75–76, 151,
    255nn87, 90
  opposition to chemists, 75, 254n83
  uroscopy, 77
Galileo, 1, 16, 281n29
  biblical interpretation, 272n21,
    273n41
  Boyle's perception of, 24, 26, 64–68,
    248–49nn11, 17
  experiments of, 64–65, 109
  and facts, 285n70
  and hypotheses, 65

laws of motion, 65, 250n24, 265n69
 rhetorical style of, 186, 228n12,
  300n8
 Two New Sciences, 249nn12–17, 250n24
 as underbuilder, 65, 235n103
Galison, Peter, 12, 224–25nn59, 61,
  312n31
Garber, Daniel, 7–8, 222nn38–40,
  231nn47, 52, 232nn61–62, 235n99
Gassendi, Pierre, 1, 24, 36, 104, 261n14
 Boyle's perception of, 62, 230n35,
  247n1, 262n36, 287n82
 circulation of the blood, 257n113
 cold, theory of, 203
 and empiricism, 236nn1, 3, 250n29,
  259n1, 265n78
 as influence on Boyle, 219n13,
  230n35, 236n3, 247n1
 and voluntarism, 99, 259n1
Giere, Ronald N., 11–13, 214, 219n15,
  224n57, 225nn62, 64–65, 312n31
Gilbert, Neal, 237n10, 256n102
Gilbert, William, 230n35
Glanvil, Joseph, 9, 260n3
Glymour, Clark, 312n25
God
 as author of the Bible, 91, 120
 as author of nature, 89, 92, 101–2,
  110–11
 as divine architect, 90, 91–92, 101
 as first cause, 100, 104, 113, 268n100,
  272n30, 286n76
 as outside time, 99, 265n79
 power of, 99–100, 111, 116, 276n86
Godfrey, Ambrose, 190–91, 305nn51,
  54, 56
Goldstein, Bernard R., 262n35
Golinski, Jan V., 224n54, 278n4,
  298n79, 299n2, 303n30, 304n43,
  311n19
Gooding, David, 4, 211, 214, 220n22,
  222n32, 291n3, 311n21, 313n32
Green, Thomas Andrew, 239n23,
  241nn45, 47–48, 243n78

Greg, Hugh, 305n53
Grey, Elizabeth, Countess of Kent,
  289n111
Guerlac, Henry, 301n15

Hacking, Ian, 12, 214, 224–25nn59–61,
  65, 243n82, 312n30
Hale Commission, 242n50
Hale, Matthew, 16, 56, 78, 237n7,
  255n96
 credibility of testimony, 242nn49–50
 criticism of Hobbes, 47–50,
  240nn33–34, 36–39, 41
Hall, A. Rupert, 249n14, 250–51nn26,
  34, 267n87, 280n20
Hall, Marie Boas, 2, 8, 217nn3, 5,
  227n6, 249n14, 250n34, 253nn65,
  68, 254n76, 268n94, 296n55,
  302nn20, 22, 303n33, 311n4
Hall, Thomas S., 256n103
Hamlyn, D. W., 229n20
Hanckwitz, Ambrose Godfrey. *See* Godfrey, Ambrose
Hanson, N. R., 213, 312n27
Harding, Sandra, 289n110
Harrison, John, 23
Hartlib, Samuel, 142
 Boyle correspondence, 37, 63, 137,
  230n35, 235n100, 247–48nn1–2,
  6–7, 9, 280n23, 298n76
 Hartlib Circle, 24, 228n14, 248n7
 and Oldenburg, 194n24
 and religion, 89, 260n9
 writing style, 301n11
Harvey, William, 16, 92, 230n35
 Boyle's acquaintance with, 256n101,
  257n117
 and chemistry, 257n117
 circulation of the blood, 79, 80–83,
  109, 207–8
 experimental proof, 79–80
 De generatione, 79–80, 257n104
 De motu cordis, 257n115
Harwood, John, 8, 223n43, 228n14,

Harwood, John (*continued*)
    236n115, 277n109, 289n111,
    299n1, 300n7, 302n26
Helmont, J. B. van, 16
    acids and alkalies, theory of, 72–73,
    110
    Boyle's perception of, 72–73, 74, 76,
    110, 310n128
    laws of nature, 95, 264n62, 267n88
Hempel, Carl G., 224n51
Hendry, John, 301n15
Henry, John, 8, 223n42, 254n83, 259n3,
    260n5, 265n71, 266n85, 268n100,
    269n119
hermeneutics, 16–17, 271n15, 311n12
    See also interpretation
Hesse, Mary, 213, 231n47, 312n27
Heward, Edmund, 237n7
Highmore, Nathaniel, 62, 92, 248n7,
    255n86
High Veneration, 93
Hill, Christopher, 4, 219n18, 221n24,
    247n115, 260n8
historical presentism, 11–12, 224n56
History of Air, 135, 142, 177, 194,
    280n21, 284n64, 287n85, 298n76
History of Blood, 138, 182, 280n24
History of Cold, 19, 193–204, 205,
    282n46, 306–7nn68–75
History of Colours, 132, 152, 184–85,
    252n52, 289n75, 302nn19–20
History of Fluidity and Firmness
    "Intestine Motions of Particles," 100,
    152
History of Particular Qualities
    "Cosmical Qualities," 269n113
    "Cosmical Suspicions," 97–98, 105–8
Hobbes, Thomas, 9, 16, 104
    Boyle's criticisms of, 57–61, 161–62,
    202
    circulation of the blood, 270n4
    cold, theory of, 196, 202, 309n120
    criticism of Boyle, 39, 56, 58–60,
    109, 205

    criticism of common law, 46–48,
    240nn29–32
    Dialogus Physicus, 39, 56, 58–60, 161,
    235n105, 244nn88–90, 246–
    47nn99, 105–10, 112, 115
    as empiricist, 161–62, 292n13
    Leviathan, 240n31, 244n90
    Mathematicae Hordiernae, 246n98
Holdsworth, William, 238n13,
    239nn16–17, 23–24, 240n33,
    241nn43–45, 47–48
Holmes, Frederic L., 297n63
Hooke, Robert, 9, 246n104, 286n71,
    287n81, 291n5
    as Boyle's assistant, 63, 190, 282n42,
    305n50
    criticism of Newton, 302n18
Horton, Mary, 231n47, 232n55
Hudson's Bay Company, 142, 284n57
Hull, David L., 11–13, 224nn56–57,
    225nn62, 64
humanism, 13–14
Hume, David, 265n72, 267n87
Hunter, Michael, 7–8, 14, 43, 222–
    23nn33, 41, 43–44, 49, 226–
    27nn70, 72, 74–75, 237n10,
    259n3, 261nn10, 14, 268n94,
    271n15, 273n40, 284n57, 286n71,
    311n7
Hunter, Richard A., 256n101
Hutchison, Keith, 218n11, 268n94,
    272n28
Huygens, Christian, 58, 205, 246n97,
    304n47, 305n49, 310n2
Hyde, Thomas, 112
"Hydrostatical Discourse," 155–58
Hydrostatical Paradoxes, 67–68, 245n94,
    249n20
hydrostatics
    Boyle, 66–68, 155–58, 169
    Pascal, 67, 156
hypotheses
    in Bacon, 31, 36
    in Boyle, 36, 55–58, 126, 127,

250n30, 293n20, 296n62, 303n35, 304n40
  in Descartes, 31, 36
  and facts, 131–35, 154
  and experiment, 173
  in Galileo, 65
  in Hobbes, 56
  and mathematics, 132–33

"Icy Noctiluca," 191–92, 306nn65–66
imagination, 161–62, 276n100
immortality, 114, 272n26
indicator tests, 171–72, 175, 253nn64–65, 296n55
innate ideas, 119, 274n64, 275n69
intellect, 161–62
intellectualist history, 2–3, 5, 10, 210
interpretation
  accommodation theory, 113–14, 272n21
  in common law, 45–46
  of experiments, 165, 170, 177, 179–80
  of nature, 31, 111–12, 122–28, 144–45, 181, 207, 275n81
  of Scripture, 112, 115–22
Ishiguro, Hidé, 267n87, 276n86, 292n13
Israel, Menasseh ben, 90, 112, 261n21, 271n14, 273n41
Ives, E. W., 238nn13–14, 239nn17, 25

Jacob, James R., 4, 219–20nn18–20, 228n15, 259n3, 260n8, 263n47, 264n63, 270n124, 272n26, 273n35
Jacob, Margaret C., 4, 219–20nn18–20, 259n3
James I, 45
Jardine, Lisa, 231n47, 241n42, 256n102
Joy, Lynn, 230n35, 259n1

Kargon, Robert Hugh, 218n8, 230n35, 247n1
Keller, Evelyn Fox, 289n110

King, Edmund, 304n45
Klaaren, Eugene M., 2, 218n11, 259nn1, 3, 260n4
Knorr-Cetina, Karin, 211, 224n61, 311n21
knowledge
  Bacon's conception of, 31, 33–35, 40, 206, 232n61, 233n82, 291n2
  Boyle's conception of, 35–41, 163, 206, 210, 213–16
  Descartes's conception of, 32–33, 232n61
  usefulness of, 40–41, 77, 206, 215, 244n86, 291n2
Kocher, Paul H., 236n6
Kockelmans, Joseph J., 218n6, 236n1, 312n31
Kraft, Johann Daniel, 191, 305nn55–56
Kuhn, Thomas S., 2, 11, 213, 217n4, 251n42, 253n68, 312n27

laboratory assistants, 189–93, 305nn49–53, 306nn60, 64, 66–67
Lady Kent's Powder, 289n111
Laertius, Diogenes, 227n7
Langbein, John H., 238n13, 241nn43–46
Latitudinarianism, 88, 260nn8–9
Latour, Bruno, 4–6, 220nn21–22, 221nn24, 28–30
Laudan, Larry, 2, 210, 218nn9–10, 231n46, 234nn85–86, 235n102, 236n3, 281n31, 311n18, 313n33
Lavoisier, Antoine, 303n33, 311nn3–4
Laws of nature
  Bacon's conception of, 236n6
  Boyle's conception of, 95–98, 160, 237n7, 264n62
  as customs of nature, 97
  as figurative expression, 95–96
  Helmont's conception of, 264n62
  as instituted by God, 96–97, 98–100, 102, 103

Laws of nature (*continued*)
  as necessary, 99–100, 102, 110
  as universal, 97
legal analogy, 15–16, 110–11
  in Bacon, 43, 236n6, 237n9
  in Boyle, 43–44, 53, 61, 110–11, 127,
    237nn9–10, 238n12, 241n42,
    243n78
  See also common law
Leibniz, Gottfried Wilhelm
  activity of matter, 100–101, 266–
    68nn87, 93, 99–100
  and Boyle, 100–103, 266n87, 268n97,
    304n47
  criticism of Boyle, 205, 310n2
  criticism of Cartesians, 292n13
  criticism of Locke, 100, 266n87
  criticism of Spinoza, 260n5, 267n88
  eclecticism of, 267n87
  "On Nature Itself," 100–101, 267n88
  nature of bodies, 264n60
  Philosophical Essays, 260n5, 266–
    68nn87, 88, 93, 97, 100
  pre-established harmony, 101,
    276n86
  as rationalist, 88
  and teleological reasoning, 268n97
  and voluntarism, 88, 260n5
Lennox, James G., 2, 217n4, 257n107,
  110, 112
Lenoir, Timothy, 9, 223n47
Levack, Brian P., 238n13, 239nn23–24
Levellers, 46
Leviathan and the Air-Pump, issues discussed
  in
  artificiality, problem of, 211, 225n61,
    291n3, 312n29
  Boyle as empiricist, 211, 219nn12–13,
    236n2, 264n63, 292n13, 311nn19–
    20, 312n22
  Boyle/Hobbes debate, 58–59,
    244n88, 246n100, 247n115,
    310n1
  civility, purpose of, 304n43

consensus, 295n32
experimenters' regress, 293n17,
  312n26
explanation of Boyle's success, 4–6,
  9–10, 43, 220–21nn22–26,
  223nn47–48, 230n42
failed experiments, reporting of,
  294n32, 299n85
matters of fact, 132, 213–14, 278–
  79nn8–9
passivity of matter, 266n85, 270n124
politics and law, 237–38n11, 241n46,
  242n50
politics and religion, 259n3, 262n26
relativism thesis, 211–12, 311n21
replication, 298n79
rhetorical analysis, 224n54, 299n2
social status of witnesses, 155–58,
  288n87, 290n129
spring of the air, 279n9
virtual witnessing, 290n123, 300n2
Lilburne, John, 241n45
Lloyd, G. E. R., 218n6, 236n1, 237n10,
  252n49
local reason. *See* prejudice
lock and key analogy, 106–7
Locke, John, 238n12
  and Boyle, 63, 142
  credibility assessments, 287n87
  and innate ideas, 275n69
  Leibniz's criticism of, 100, 266n87
  and probability, 245n96, 274n64
  as underbuilder, 235n103
Loemker, Leroy E., 218n8
Lohne, J. A., 301n15
Lower, Richard, 63, 82, 255n86,
  258n121
Lucretius, 113
Luther, Martin, 260n7, 273n41
Lynch, Michael, 4–6, 220nn21–22,
  221n27, 222n35

Macalpine, Ida, 356n101
MacGregor, Arthur, 284–85n64

Maddison, R. E. W., 217n2, 227nn1–2, 7, 228nn12–13, 238n12, 248n5, 251n42, 261n21, 271n14, 204nn44, 46–47, 305nn51–56, 310–11nn1–7
Magalotti, Lorenzo, 304n47
Malcolm, Noel, 292n13
Mallet, John, 228n15, 273n38, 274n57
Mamiani, Maurizio, 301n14
Mancosu, Paolo, 292n13
Mandelbaum, Maurice, 210, 218n9, 311n18
Marcombes, Isaac, 23–24, 227n2, 228n14, 235n115, 260n9, 273n37, 281n35
Martin, Gottfried, 264n60
Martin, Julian, 43, 237n11
Masson, Flora, 228n13
Mates, Benson, 264n60, 269n106
mathematics
  and Bacon, 244n93
  Boyle's criticism of, 15–16, 23, 37–39, 57, 66–68, 98, 169, 184, 235n99, 245n94, 250n26
  and Descartes, 235n99
  and experiments, 169
  Hobbes's use of, 46–48, 56
  lawyers' criticism of, 47–48, 50
  rhetorical style of, 184
matter
  activity of, 100–101, 259n3, 266n85, 268n100, 271n20
  passivity of, 3–4, 10, 220n20, 259n3, 263n47, 266n85, 270n124
McCabe, Bernard, 236n6, 239n22
McGuire, J. E., 3, 94, 210, 219n13, 260n4, 264nn57, 62–63, 267n88, 270n124, 276n93, 311n16
McKie, Douglas, 253n67
McKirahan, Richard D., 229n20
McMullin, Ernan, 212, 233n65, 234n85, 249nn13, 15, 20, 250n25, 312n23, 313n33
Mechanical Origin of Qualities, 194, 253n64

"Origin of Electricity," 289n114
"Origin of Gems," 152
mechanical philosophy. *See* corpuscularianism
medicine, 9, 16
  Boyle's views on, 75–84
  Boyle's collection of recipes, 139, 254n82, 281n38, 289n111
  in Brazil, 151
  and chemistry, 75–77, 78, 254n83, 255n90
  in China, 145, 151
  diagnostics, 76–77
  dietetics, 151
  drugs, 75, 76, 78, 168, 172, 174, 255n97, 258n126, 295n43
  experience in, 76
  and experiments, 76–84, 168, 172
  and illiterate practitioners, 150–51
  legal analogy, 77–78
  reasoning from signs, 76–77, 79, 80, 252n49, 256n100
  therapeutics, 75–76, 77–78, 151, 255n87
Merchant, Carolyn, 289n110
Mercurius Trismegistus, 90
Merret, Christopher, 196, 200, 309n109
Mersenne, Marin, 1, 16, 26, 104
  Boyle's perception of, 62, 64, 248n2
  Descartes correspondence, 30, 32–33, 232n54, 233n65, 234n84, 235n99
  as empiricist, 88, 259n1, 265n78
  and Hartlib, 247–48n2
  on medicine, 79
  Les nouvelles pensées de Galilée, 248n11, 270n2
  on sound, 64, 70
  universal harmony, 88, 260n6
  voluntarism, 99, 259n1
Merton thesis, 3, 259nn2–3
Meynell, G. G., 255n93, 300n2
Milne, John, 305n53
Milton, John, 24, 228n13
miracles, 267–68n94, 272n28

modernism, 6, 8, 13–14
Monconys, Balthasar de, 304n47
moral demonstration
    Boyle's conception of, 54–55, 111,
        243n78
    in common law, 48–50, 241n42
    See also concurrence of probabilities
Moray, Robert, 246n97
More, Henry, 115
More, Louis Trenchard, 217n2, 228n13,
    240n33, 248n5, 270n1, 271n14,
    273n35

natural histories
    in Bacon, 51–52
    collections for, 53, 124–25, 140, 142–
        44, 149, 163, 208
    composition of, 174, 182, 194,
        284nn62, 64, 300n8, 303n35
    incompleteness of, 52, 56, 114–15,
        131, 144, 151
nature
    ambiguity of term, 94
    Boyle's notion of, 16–17, 26, 93–98,
        185, 194, 207, 216, 263–64nn47–
        49, 57, 60, 265n68, 267n88,
        313n40
    contingency of, 166, 168
    as created by God, 98–99, 144–45
    as essences of bodies, 94, 264n60
    Leibniz's notion of, 264n60, 267nn88,
        93
    as notional entity, 94, 96
    as semi-deity, 94
    as the universe, 94–95
necessarianism, 87–88, 259n1
New England Corporation, 142, 283n56
New Experiments Physico-Mechanical, 26,
    38–39, 57–61, 133–35, 174, 184,
    190, 194, 246nn101–4, 247n111,
    279n15, 297n63, 299n87, 301n16,
    306n64, 307n69
Newman, William, 8, 222n41, 248n7,
    251n41, 265n71

Newton, Isaac, 8, 14, 205, 248n7,
    265n69
    optics, 184, 301nn14–15, 302n18
Nickles, Thomas, 12, 212, 224nn56, 58,
    291n3, 293n17, 312n25
nominalism, 97, 102, 229n29, 264n57,
    269n106
nullius in verba, 270n1

O'Brien, John, 218n8
Observations
    assessing credibility of, 148–58, 197
    Boyle's collection of, 17–18, 138–45,
        143, 281nn30, 32, 282nn42–45,
        283nn47, 51, 284n58,
        290nn120–23
    and experiments, 137–38, 158, 159
    importance of, 137
    about oddities, 143–45, 148, 153,
        196–97, 204
    as rational activity, 136–38, 146–47,
        154–55, 208, 213, 287nn84–85
    and theory, 154, 207
    from travelers, 142–45, 148–49, 196–
        97, 307nn74, 82, 84
    use of, 137–38
    from women, 138, 152, 289nn110–
        12, 114, 304n38
Occam's razor, 310n129
occasionalism, 102
Occasional Reflections, 93, 289n104
occult qualities, 8, 265n71
oddities, 143–45, 148, 153, 204, 208,
    284n58, 285–86nn67, 70–71,
    307n86
    as tests, 144, 196–97, 280n26,
        285n66
Ogg, David, 239n23
Oldenburg, Henry, 254n80
    Boyle correspondence, 135, 139, 142,
        174, 280n21, 282n43, 283–
        84nn54–56, 294n24, 300n6,
        303n35, 304n45, 309n109
    and Hartlib, 294n24
    and Menasseh ben Israel, 271n14

Spinoza correspondence, 39, 68–70, 235n106, 250–51nn34–38, 270n5

Olson, Richard, 259n3

optics, 301n15, 302nn18–20

Origin of Forms and Qualities, 25–26, 103–8, 137, 159, 194, 268n98, 272n30, 296n62

"Relative Nature of Physical Qualities," 106–8, 269nn114–15

"Of Substantial Forms," 229n30

Osler, Margaret, 210, 219nn12–13, 230n35, 236n3, 259n1, 264n63, 276n93, 311n16

Oster, Malcolm, 8, 223n43, 228n15

O'Toole, F. J., 217n4

"Outlandish Book," 143, 184n58, 287n81, 290n133

Owen, John, 260nn3, 7

Oxford, 63, 82, 112, 190, 255n86, 282n41

Pacchi, Arrigo, 292n13

Padua, medical school, 75, 79, 80, 227n9, 256n102

Pagel, Walter, 255n91, 256n103, 257n117

Papin, Denis, 190, 305n49, 306n64

Paracelsus, 16, 76, 143, 258n117
  Boyle's criticism of, 71–72, 74, 83
  and theology, 91
  and tria prima, 71–72, 73

Paradis, James, 224n54, 300n8, 301n14

Park, Katherine, 285n67

Pascal, Blaise, 16, 246n101
  Boyle's perception of, 67, 109, 250n28, 270n3
  and experimental proof, 66, 67, 109, 156
  Physical Treatises, 66, 249–50nn21–22
  Puy-de-Dôme, 174, 249n21
  wager argument, 55, 243n80

passivity of matter, 3–4, 10, 220n20, 259n3, 263n47, 266n85, 270n124

Patey, Douglas Lane, 243nn82, 85,

252n49, 255nn93–95, 100, 257n107, 275n69

pearl divers. *See* divers

Peirce, Charles Sanders, 312n31

Peltonen, Markku, 237n11, 239n22

pendulum experiments, 60–61, 247n111

Pera, Marcello, 299n1

Perez-Ramos, Antonio, 7–8, 40, 206, 222nn36–37, 40, 224n53, 231n47, 233n82, 235n112, 278n7, 291n2, 311n8

perfection of the world
  Boyle's conception of, 101–3, 207, 216
  Leibniz's conception of, 101–2

Petty, William, 62, 248n6, 254n82, 255n86, 256n101

phenomena. *See* facts

philosophical systems
  Bacon's criticism of, 33, 35, 79, 206
  Boyle's criticism of, 14, 35–37, 124, 183–84, 185, 234n89, 235n99
  Boyle's use of, 35, 123

philosophical worship of nature, 90–92

phosphorus, 174, 190–93

Pickering, Andrew, 4–5, 13, 219n15, 220n21, 221nn25–26, 224n55, 225n66, 293n17

Piso, Gulielmus, 142, 289n108

Plato, 90

Pliny, 149, 150

Plutarch, 90

pneumatics
  Boyle, 59–61, 63, 66, 133–36, 174, 190, 206
  Pascal, 66
  See also air-pump; spring of the air

Pococke, Edward, 112

Polyani, Michael, 213, 312n27

Popkin, Richard H., 218n6, 236n1, 259n1, 265n78

Popper, Karl R., 224n51, 312n27

postmodernism, 13–14

post-nati, case of, 45, 239n22

Potter, Elizabeth, 220n20, 289n110

Power, Henry, 291n5
pragmatism, 2, 210
predestination, 265–66n79
pre-established harmony. *See* universal
  harmony
prejudice
  in interpretation of scripture, 117–18,
    120, 121, 275n70
  in legal decisions, 53
  in natural philosophy, 124–25, 137,
    145, 149, 155, 158, 208, 276nn89,
    91, 285–86nn71–77, 289n104
pressure of the air, 279–80n20
  See also spring of the air
Prest, Wilfred R., 238n13, 241n42
priests of nature, 90–92, 261n26
primum frigidum, 196, 203–4
Principe, Lawrence, 8, 222n41, 251n41,
    252n50, 265n71, 303n30
probability, 54–56, 111, 229n30,
    245n96, 277n118, 296n53
propositions
  absolute, 119
  conditional, 119–20, 126
Puy-de-Dôme experiment, 174, 222n39,
    249n21
Pythagoras, 122

qualities of bodies, 71, 100, 104–8
  as causal powers, 71, 105–8
  cosmical, 107–8, 269n113
  as essential, 104–5, 107–8
  as independent of human perception,
    105, 107
  learned from artisans, 150
  occult, 8, 265n71
  primary and secondary, 105, 269n111
  as relational properties, 105–8, 111,
    126, 162, 265n71, 269n113
Quine, W. V. O. *See* Duhem-Quine
  thesis

Randall, J. H., Jr., 256n102
Ranelagh, Katherine Jones, Viscountess

as Boyle's confidante, 226n71,
    228n13, 248n3, 289n110
  and Hartlib circle, 24, 228n13
  letter on nonconformists, 273n39
  medicinal recipes of, 289n111
  and Menasseh ben Israel, 271n14
  and Parliamentarian party, 228n15,
    248n5
rationalism, 2, 7, 88
Ray, John, 263n37
reasoning
  analogical, 135–36, 157, 245n96,
    280n22
  to correct senses, 127, 146, 154
  in experiment, 123, 125–26, 170,
    179–80, 242n63
  improved by experience, 53–54
  from signs, 70–72, 74, 76–77, 79–80,
    252n49, 253n63, 256n100
  as source of error, 146
  as warrant for facts, 127, 154
  as warrant for hypotheses, 126
Reconcileableness of Reason and Religion, 93,
    272n24
reconciliation
  in natural philosophy, 123–25, 127,
    145, 154–55, 156–58, 179–80, 201,
    290n122
  among religions, 116, 118, 119
  in scriptural interpretation, 121–22,
    275nn70, 74
Rees, Graham, 232n54
relativism, 211–14
replication, 165, 174, 175–76
  See also experiment, repetition of
rhetoric, 181, 183, 299nn1–2
  See also writing
Richardson, W. C., 238n13, 239nn23–
    24, 26, 241n42
Roberts, Lissa, 299n2
Rodis-Lewis, Geneviève, 232n54
Rogers, G. A. J., 2, 218n10, 234n85,
    292n13
Roman law, 44, 46, 49, 238n13
Rooke, Lawrence, 305n53

Rorty, Richard, 219n12
Rouse, Joseph, 13, 225n66
Royal Mines, Company of, 142, 284n56
Royal Society, 19, 109, 190, 270n1,
    286n71
Ruby, Jane E., 237n8

Salmon, Wesley, 313n33
Sanchez-Gonzalez, Miguel A., 254n83
Sanctorius, 154
Sarasohn, Lisa T., 259n1
Sceptical Chymist, 26, 29, 71, 73, 163,
    182, 186, 188, 253n68
"Sceptical Dialogue on Cold," 194, 202,
    204, 205, 310n129
Schaffer, Simon. See *Leviathan and the Air-
    Pump*
scholastic philosophy
    Bacon's criticism of, 28, 33
    Boyle's criticism of, 24–29, 65, 103–4
    Descartes's criticism of, 28–29
Scripture
    as a coherent system, 120–21
    contradictions in, 117, 118, 120
    critics of, 116
    interpretation of, 112, 115–22
    prejudiced readings of, 117, 118, 120,
        121
    style of, 116–19
    to teach divinity, 113–14
    translations of, 118–20
seminal principles, 98, 99, 103, 268n99
Semmedo, Alvaro, 142, 286n75
Seneca, 90, 262n35
sensation, Boyle's investigation of, 146–
    47, 154–55, 286–87nn81, 84
senses
    Bacon's assessment of, 30, 34, 50–51
    Boyle's assessment of, 52–53, 125–26,
        127, 145–47, 154, 197–98
    corrected by reason, 125–26, 145–47,
        154–55
    Descartes's assessment of, 32–33
    improved by experiments, 168,
        292n11

as source of error, 32–34, 125–26,
    145–47, 197, 198, 287n81,
    291n137
Shanahan, Timothy, 257n107, 259n1,
    260nn4, 7
Shapere, Dudley, 212, 312nn23, 27
Shapin, Steven, 6, 155, 219nn15, 18,
    222–24nn31, 47, 54, 250n26,
    278n4, 287n87, 290nn123, 129,
    299n2, 305n50
    See also *Leviathan and the Air-Pump*
Shapiro, Alan E., 301n15
Shapiro, Barbara J., 2, 43, 218n11,
    237n10, 240–41nn28, 41,
    242nn49–51, 259n1, 299n1,
    301n15
Sharrock, Robert, 300n6, 305n51
Shaw, Peter, 147, 153, 205, 287n86,
    290n117
Shea, William R., 299n1
skepticism, 28–29, 34, 213
Slare, Frederick, 190, 191, 305n52
Sloan, Phillip R., 232n63, 258n126
Smith, Thomas, 305n53
sociology of scientific knowledge, 2–7,
    11, 219n15, 220n20, 221nn28, 30,
    222n32
    See also contextualist history
Söderqvist, Thomas, 7, 222n34, 223n46,
    224nn56, 58,
    227n75
Solomon, 91, 262n30
Spinoza, Baruch
    Boyle's perception of, 69–70, 184
    criticism of Boyle, 39, 68–69, 109–10
    heat, theory of, 69, 250n37, 270n4
    as necessarian, 88, 259n1
    Oldenburg correspondence, 39, 68–
        70, 235n106, 250–51nn34–38,
        270n5
    Principles of Cartesian Philosophy, 251n34
Sprat, Thomas, 9
spring of the air, 59–60
    as causal, 133–35, 206, 279n9,
        280n20

spring of the air (*continued*)
  as factual, 133–35, 279n9
  as a model, 135–36
Staehl, Peter, 190, 305n51
Starkey, George, 62, 139, 142, 248n7
Steneck, Nicholas H., 272n28
Stillingfleet, Edward, 260n7
Stoicism, 262n35
Stubbe, Henry, 255nn86–87
Sturm, Christopher, 267n88
Style of the Holy Scriptures, 112, 117,
  273n41, 275n68
Sulloway, Frank J., 299n2
Sydenham, Thomas, 129–30, 255nn86,
  92–95, 300n2

Tallents, Francis, 273n38
teleological reasoning
  in natural philosophy, 80–81,
    257nn107–8, 117, 268n97
  in theology, 92–93, 263n39
Temkin, Owsei, 256n103
Tertullian, 260n7
testimony
  in law, 49–50
  in natural philosophy, 51–52, 53
  See also credibility; witnesses
theology
  Boyle's eclecticism, 88
  Boyle's essays, 62, 112, 139
  dynamic nature of, 115–17, 121–22
  natural, 16–17, 89–93, 102, 215
  relation to natural philosophy, 3, 12,
    93, 112–15, 268n94
  revealed, 16–17, 91, 272nn24–25 (*see
    also* scripture)
thermometers, 177, 198–99,
  308nn97–98
Things Said to Transcend Reason, 122, 162
Toulmin, Stephen, 13–14, 212, 225–
  26nn67–69, 312nn23, 27, 313n38
Townley, Richard, 306n64
tradesmen. *See* artisans and tradesmen
translation, principles of, 118–20
tria prima, 71–72, 73

Tuchanska, Barbara, 313n33
Tuck, Richard, 292n13
Tymme, Thomas, 92, 262n37, 267n89

underdetermination. *See* Duhem-Quine
  thesis
universal harmony
  in Boyle, 101–3, 108, 111, 113, 126,
    144–45, 207, 216, 276n86,
    277n109, 313n40
  in Leibniz, 101, 276n86
  in Mersenne, 88, 260n6
Urbach, Peter, 231n47, 232n55
Urban VIII, 273n41
Usefulness of Experimental Philosophy, 40,
  149, 160, 282n46, 286n77, 296n53
Usefulness of Natural Philosophy
  Part 1, 89–92, 93, 111, 112, 123,
    163–64, 177–78
  Part 2, 76, 82, 150, 154, 255n90
"Uses and Bounds of Experience," 163–
  64, 177–78, 293n22

vacuum, experimental, 59, 246n101
Vailati, Ezio, 292n13
van Fraassen, Bas, 218n6, 236nn1–2
van Helmont. *See* Helmont, J. B. van
van Leeuwen, Henry G., 3, 218n11,
  259n1
virtual witnesses, 181, 290n123, 300n2
voluntarism, 87–88, 102, 259nn1, 3,
  260n6

wager argument, 55, 243n80
Waite, Arthur Edward, 252n51
Warre, John, 260n7, 305n53
Wear, Andrew, 256n102
Webster, Charles, 4, 219n18, 259n2,
  301n11, 303n30
Westfall, Richard S., 2, 217nn4–5,
  218n11, 222n24, 228n12, 231n46,
  244n85, 260n8, 300n8, 301n15
Westman, Robert S., 271n15
Wheeler, Harvey, 225n67
Whewell, William, 212, 312n24

Whitney, Charles, 43, 237n11
Wiener, Philip P., 2, 210, 218n7, 250n26, 311n17
Wilkins, John, 63, 240n33, 248n8
Willis, Robert, 257n113
Willis, Thomas, 63
Winthrop, John, 283n56
witnesses, 18
  in common law, 49–50
  credibility of (*see under* credibility)
  to experiments, 177–78
  in Roman law, 49
  social status of, 155–58, 288n87, 290n129
  virtual, 181, 290n123, 300n2

Wojcik, Jan, 223n43, 259n3, 261n14, 266n79, 271n15, 273n40, 274n41
women
  credibility of, 152
  equality of, 289n110
  observations from, 138, 152, 289nn110–12, 114, 304n38
Wood, Paul B., 223n49
Woolgar, Steve, 4–6, 220n21, 221nn28, 30
Wotton, Henry, 23, 227n2, 311n5
Wren, Christopher, 63

writing
  Bacon's style, 299n1, 300n8
  Boyle's style, 182–89, 191–93, 195–96, 226nn71, 73, 299n1, 301n11
  civility in, 189–90
  for collaboration, 182, 185, 193, 200, 303n30
  for collation of data, 181–82, 185
  as communication, 182–83, 194, 303n30
  dynamic nature of, 183, 185
  essays, 18–19, 183–85, 300n8
  to excite curiosity, 182, 185–89
  as experimental activity, 181
  experimental reports, 182–87, 300n2, 303n29, 309n108
  Galileo's style, 186, 228n12, 300n8
  natural histories, 174, 182, 194, 284nn62, 64, 300n8, 303n35
  Newton's style, 184
  to report facts, 181, 194
  to teach, 183, 184, 187, 194, 303n29
  and theory, 186–88

Yearley, Steven, 220–21nn22, 28, 30

Zagorin, Perez, 292n13